Springer Tracts in Modern Physics
Volume 177

Springer
Berlin
Heidelberg
New York
Barcelona
Hong Kong
London
Milan
Paris
Tokyo

Springer Tracts in Modern Physics

Springer Tracts in Modern Physics provides comprehensive and critical reviews of topics of current interest in physics. The following fields are emphasized: elementary particle physics, solid-state physics, complex systems, and fundamental astrophysics.
Suitable reviews of other fields can also be accepted. The editors encourage prospective authors to correspond with them in advance of submitting an article. For reviews of topics belonging to the above mentioned fields, they should address the responsible editor, otherwise the managing editor.
See also http://www.springer.de/phys/books/stmp.html

Vladimir A. Smirnov

Applied Asymptotic Expansions in Momenta and Masses

With 52 Figures

Springer

Vladimir A. Smirnov
Lomonosov Moscow State University
Skobeltsyn Institute of Nuclear Physics
Moscow 119899, Russia
E-mail: smirnov@theory.sinp.msu.ru

Physics and Astronomy Classification Scheme (PACS): 11.25.Db, 11.10.Jj, 12.38.-t, 11.10.Gh

ISSN print edition: 0081-3869
ISSN electronic edition: 1615-0430
ISBN 3-540-42334-6 Springer-Verlag Berlin Heidelberg New York

Springer-Verlag Berlin Heidelberg New York
a member of BertelsmannSpringer Science+Business Media GmbH

http://www.springer.de

Typesetting: Camera-ready copy from the author using a Springer LaTeX macro package
Cover design: *design & production* GmbH, Heidelberg

Printed on acid-free paper SPIN: 10695661 56/3141/tr 5 4 3 2 1 0

Preface

'The sturgeon they sent was second-grade-fresh,' said the barman.
'Really, what nonsense!'
'Why nonsense?'
'"Second-grade-fresh" – that's what I call nonsense! There's only one degree of freshness – the first, and it's the last.[1]

(M.A. Bulgakov, *The Master and Margarita*)

The goal of this book is to describe in detail how Feynman integrals can be expanded in suitable parameters, when various momenta or masses are small or large. In a narrow sense, this problem is connected with practical calculations. In a situation where a given Feynman integral depends on parameters of very different scales, a natural idea is to replace it by a sufficiently large number of terms of an expansion of it in ratios of small and large scales. It will be explained how this problem of expansion can be systematically solved, by formulating universal prescriptions that express terms of the expansion by using the original Feynman integral with its integrand expanded into a Taylor series in appropriate momenta and masses.

It turns out that knowledge of the structure of the asymptotic expansion at the diagrammatic level is a key point in understanding how to perform expansions at the operator level. There are various examples of these expansions: the operator product expansion, the large-mass expansion, Heavy Quark Effective Theory, Non-Relativistic QCD, etc. Each of them serves as a realization of the factorization of contributions of different scales. For example, an effective Lagrangian in the large-mass limit describes fields with small masses, while the information about the large scale is encoded in its coefficients.

When expanding Feynman integrals in various limits of momenta and masses, we shall apply two strategies. One of them is based on a graph-theoretical language and renormalization techniques. Therefore the mathematical proof of the prescriptions for this strategy, which we shall call *expansion by subgraphs*, resembles very much the standard proofs of the Bogoliubov–Parasiuk theorem on renormalization. This strategy can be applied only to limits typical of Euclidean space, in particular the off-shell large-momentum limit and the large-mass limit.

For limits typical of Minkowski space, i.e. those which cannot be formulated for Euclidean Feynman integrals, we shall apply a physical, heuristic strategy [18], which we call the strategy of *expansion by regions*. It is quite general and can be applied to any limit. Here we do not have a mathematical justification similar to that for expansion by subgraphs. The method of

[1]Translated from the Russian by M. Glenny. Collins and Harvill Press, London, 1967.

expansion by regions has, however, been checked through numerous examples and successfully applied in practice to problems of phenomenological importance.

A word of caution is needed about the word 'asymptotic', which is also commonly applied to perturbative series with a zero radius of convergence. In our problems, this word means just that the remainder of an asymptotic expansion satisfies a desired estimate provided we pick up a sufficiently large number of leading terms of the expansion. I would like to stress that the radius of convergence of any series in any expansion in momenta and masses is always non-zero. This is not a rigorously proven mathematical theorem, but I do not know of any examples where such a radius of convergence is zero.

I promise, henceforward, not to use the word 'rigorous' in this book and hence will never say that some proof is more rigorous than another. Rather, I will try to distinguish between results which have been proven, have been confirmed by comparison with known analytic results, have been checked numerically, or merely seem plausible owing to some hand-waving argument that one may or may not believe.

Although all the necessary definitions concerning Feynman integrals and their expansions are provided, it would be helpful for the reader to know the basics of perturbative quantum field theory, e.g. by following the first chapters of [27] or [190].

I am grateful to M. Beneke for our fruitful collaboration during the years 1997–1999, starting with [18], which has formed the basis for this book. It is my pleasure to acknowledge the valuable assistance of K.G. Chetyrkin and A.A. Penin without which it would have been more difficult to write the book. I am very much indebted to P.A. Baikov, A. Czarnecki, A.G. Grozin and R. Harlander for numerous stimulating discussions of the preliminary versions of the book, and to T. Binoth, G. Heinrich and O.L. Veretin for their critical comments on the drafts. Many thanks to Z. Bern, D. Broadhurst, A.I. Davydychev, L. Dixon, M.A. Ivanov, K. Melnikov and M. Steinhauser for various pieces of advice, to J.H. Kühn for his encouragement and for the very idea of writing this book, and to my family for love, sympathy, patience and understanding.

Moscow, August 2001 *V.A. Smirnov*

Contents

1 Introduction

The problem of obtaining asymptotic expansions of Feynman integrals in various limits of momenta and masses is a typical mathematical problem in elementary-particle physics. In this book, it is explained how this problem is solved. To characterize, briefly, the main steps of the solution let us consider, for example, the process of e^+e^- annihilation, where an incoming electron and positron produce, according to the quantum-theoretical description, a virtual photon, $e^+e^- \to \gamma^*$, which in turn produces some particles, for example a quark–antiquark pair, $\gamma^* \to q\bar{q}$, which may then be transformed into mesons. The process of quark production is described, in perturbative quantum field theory, by Feynman integrals corresponding to various graphs generated by the Feynman rules. One of the three external legs of such a diagram corresponds to a triple vertex associated with the quark vector current, and the other two external legs correspond to the external quarks. If we are interested in the total cross-section for the production of the quarks the problem reduces to the evaluation of the imaginary part of diagrams contributing to the vacuum polarization and containing only two external vertices for the two vector currents.

Suppose that the Euclidean momentum squared of the intermediate photon, $Q^2 = -q^2$, is much larger than the other parameters involved in the problem. In operator language, the problem of the large-Q^2 behaviour reduces to the analysis of the product of the two currents when the momentum transfer between the currents is large. The corresponding diagrams are generated by the two vertices for the currents and a family of vertices associated with interactions via the QCD Lagrangian. At the diagrammatic level, we arrive at the problem of determining the asymptotic behaviour of such diagrams in the limit when Q^2 and also the scalar products $Q \cdot q_i$ with other external momenta are much larger than the remaining kinematical invariants and the squares of masses.

Let us take an example of such a diagram expanded in the large-momentum limit. To simplify the situation let us turn to the scalar massless ϕ^4 theory and replace the quark currents by two composite operators of the form ϕ^2. In this case the simplest one-loop diagram with two external vertices for the composite operators and one external vertex generated by the interaction Lagrangian is that of Fig. 1.1. According to the Feynman rules,

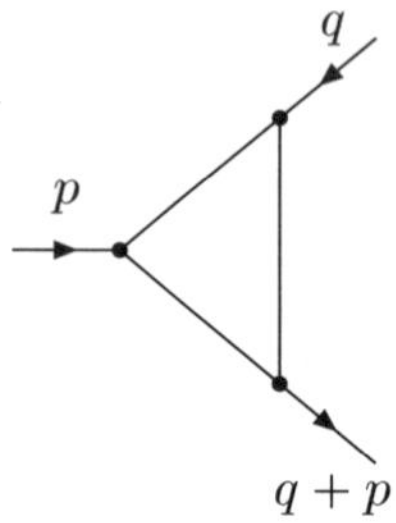

Fig. 1.1. One-loop triangle diagram with a large external momentum q and a small external momentum p

this diagram is given by the following integral:

$$F_\Gamma(q,p) = \int \frac{\mathrm{d}^4k}{k^2(q-k)^2(p+k)^2} \,. \tag{1.1}$$

The integration is over four components of the loop momentum $k = (k_0, \mathbf{k})$, where $\mathbf{k} = (k_1, k_2, k_3)$. The squares of four-vectors are implied in the pseudo-Euclidean metrics: $k^2 = k_0^2 - \mathbf{k}^2$, etc. This Feynman integral, depending on two four-vectors p and q, is a scalar quantity, so that it is a function of three kinematical invariants, q^2, p^2 and $p{\cdot}q = p_0q_0 - \mathbf{p}{\cdot}\mathbf{q}$.

Let us suppose that the external momentum q is large with respect to the external momentum p in the Euclidean sense, i.e. $|p| \ll |q|$ where $|p| = \sqrt{p_0^2+\mathbf{p}^2}$. Therefore at least one four-component of q is larger than any component of p. We can measure the order of magnitude in powers of the small momentum p so that q^2 is large, $p{\cdot}q$ is small and of the first order, and p^2 is of the second order.

Although Feynman integrals are generally divergent, this specific integral turns out to be convergent, which can be seen by power counting at small and large values (in the Euclidean sense) of the four-dimensional momentum k.

Let us now pose the problem of expanding integral (1.1) in the given limit without computing it for general values of the arguments. Moreover, we would like to write down a result for the expansion in terms of Feynman integrals that depend on a lower number of scales. To simplify the situation as much as possible, let us confine ourselves to the leading asymptotics in this limit. To further simplify the subsequent discussion let us consider the Euclidean variant[1] of the integral (1.1). The form of the integral stays the same but all the scalar products are now defined in accordance with Euclidean metrics, i.e. $k^2 = k_0^2 + \mathbf{k}^2$, $k{\cdot}q = k_0q_0 + \mathbf{k}{\cdot}\mathbf{q}$, etc.

Let us observe that if we naively take the limit $p \to 0$ *in the integrand* we obtain the integral

$$\int \frac{\mathrm{d}^4k}{(k^2)^2(q-k)^2} \,, \tag{1.2}$$

[1] We shall later see what will cause additional difficulties in the pseudo-Euclidean case.

which is divergent at small loop momenta k. This can be seen by introducing generalized spherical coordinates and counting powers of the corresponding radial variable, r. The divergence is of the logarithmic type, $\int_0^\Lambda \mathrm{d}r/r$. The non-existence of such a naive limit and the character of the divergence in (only) one of the local coordinates (i.e. r) provide a hint that the true asymptotic behaviour, when $p \to 0$, is of the logarithmic type.

When one is solving the problem of the asymptotic behaviour of Feynman diagrams, the goal can be quite different: it can range from the evaluation of the coefficients of the leading logarithms to the evaluation of the coefficients of *all* powers and logarithms. Suppose for a moment that we are interested in only the leading logarithmic behaviour of our triangle diagram. In this case the (standard) strategy based on regions can be used: one concentrates on the contributions of only those regions of the loop momenta which are responsible for the leading logarithmic behaviour. In our case, the relevant region is that of small k so that we can restrict ourselves to the integral

$$\int_{|k|\leq\Lambda} \frac{\mathrm{d}^4k}{k^2(q-k)^2(p+k)^2} ,$$

with a cut-off at some non-zero value of Λ. Since the logarithmic divergence, when $p \to 0$, arises only at the point $k = 0$, can we put $k = 0$ in the factor $1/(q-k)^2$, which is independent of p.

Let us choose, for convenience, $p = (p_0, \mathbf{0})$, where $p_0 = |p|$, and introduce the spherical coordinates

$$k_0 = r\cos\theta ,$$
$$k_1 = r\sin\theta\sin\psi\cos\phi ,\; k_2 = r\sin\theta\sin\psi\sin\phi ,\; k_3 = r\sin\theta\cos\psi ,$$

where $\mathrm{d}^4k = r^3\,\mathrm{d}r\,\sin^2\theta\,\mathrm{d}\theta\,\sin\psi \mathrm{d}\psi\,\mathrm{d}\phi$. Integrating over the first two angles ϕ and ψ, we obtain

$$\frac{4\pi}{q^2}\int_0^\pi \sin^2\theta\,\mathrm{d}\theta \int_0^\Lambda \frac{r\,\mathrm{d}r}{r^2+2|p|r\cos\theta+p^2} . \tag{1.3}$$

The internal integral in r can be evaluated in terms of elementary functions with a result which has the asymptotic behaviour $-\ln|p|$ for general θ and Λ. Multiplying this result by the integral $\int_0^\pi \sin^2\theta\,\mathrm{d}\theta = \pi/2$, we arrive at the following asymptotic behaviour of the given Euclidean Feynman integral when $p \to 0$:

$$F_\Gamma(q,p) \sim -\frac{\pi^2}{q^2}\ln\left(p^2/\Lambda^2\right) . \tag{1.4}$$

If we want to obtain not only the leading logarithm but also a constant term in the leading power, $1/q^2$, we have to 'honestly' take into account the contribution of the whole integration domain to the loop integral. Let us choose an intermediate scale Λ, where $|p| \ll \Lambda \ll |q|$, and divide[2] our integral into two parts corresponding to integrations over small and large loop momenta:

$$F_\Gamma = f_{\text{small}} + f_{\text{large}} , \tag{1.5}$$

where

$$f_{\text{small}} = \int_{|k|\leq\Lambda} \frac{\mathrm{d}^4 k}{k^2(q-k)^2(p+k)^2} , \tag{1.6a}$$

$$f_{\text{large}} = \int_{|k|\geq\Lambda} \frac{\mathrm{d}^4 k}{k^2(q-k)^2(p+k)^2} . \tag{1.6b}$$

The limit $p \to 0$ can be taken safely in the integrand of the second integral. On the other hand, the terms in the combination $(q-k)^2 = q^2 - 2q{\cdot}k + k^2$ in the first integral are of different orders so that is legitimate to take the limit $k \to 0$ there. Thus we have

$$f_{\text{small}} \sim f^{(0)}_{\text{small}} \equiv \frac{1}{q^2} \int_{|k|\leq\Lambda} \frac{\mathrm{d}^4 k}{k^2(p+k)^2} , \tag{1.7a}$$

$$f_{\text{large}} \sim f^{(0)}_{\text{large}} \equiv \int_{|k|\geq\Lambda} \frac{\mathrm{d}^4 k}{(k^2)^2(q-k)^2} , \tag{1.7b}$$

and

$$F_\Gamma \sim f^{(0)}_{\text{small}} + f^{(0)}_{\text{large}} \tag{1.8}$$

in the limit $p \to 0$.

Up to now we have dealt with convergent integrals. Now we want to arrive at a simple result for the behaviour of the initial integral in the given limit at the cost of running into divergences. To handle such divergent quantities we use a standard procedure called 'regularization' which means that, instead of a given divergent integral, we turn our attention to a quantity which depends on a (regularization) parameter, is well defined in some domain of the values of this parameter and formally coincides with the initial divergent integral at some limiting value of the regularization parameter. We shall use dimensional regularization [28, 144], which, formally, reduces to replacing the space dimension 4 by a complex number $d \equiv 4 - 2\varepsilon$. The integrals regularized in this manner are denoted by switching from d^4 to d^d. It will be explained in the next chapter how dimensional regularization is systematically introduced but here we make use of the following practical recipe: evaluate the integral

[2] If we were dealing with the Feynman integral in Minkowski space then it would be necessary to use the same decomposition with $|k|$ understood to be in the Euclidean metrics. (Still, this would look unnatural because the decomposition is not Lorentz invariant.) On the other hand it would be necessary to suppose that the external momentum is, for example, space-like. The statements that we have made after the decomposition are also true in the Minkowski case but not so obvious as in the Euclidean case.

in a general *integer* number of dimensions and then replace the number of dimensions by the complex number d in the corresponding explicit result.

Let us rewrite the result obtained by extending the integration, in both $f^{(0)}_{\text{small}}$ and $f^{(0)}_{\text{large}}$, to all values of the loop momentum k and subtracting the corresponding additional terms:

$$\begin{aligned} F_\Gamma \sim \frac{1}{q^2} \int \frac{\mathrm{d}^d k}{k^2(p+k)^2} + \int \frac{\mathrm{d}^d k}{(k^2)^2(q-k)^2} \\ - \frac{1}{q^2} \int_{|k| \geq \Lambda} \frac{\mathrm{d}^d k}{k^2(p+k)^2} - \int_{|k| \leq \Lambda} \frac{\mathrm{d}^d k}{(k^2)^2(q-k)^2} \, . \end{aligned} \tag{1.9}$$

Now we may take the limit $p \to 0$ in the first integral in the second line and neglect k with respect to q in the factor $1/(q-k)^2$ in the second integral. Thus the sum of the two integrals in the second line reduces, in the given limit, to the integral

$$-\frac{1}{q^2} \int \frac{\mathrm{d}^d k}{(k^2)^2} \, ,$$

which is an integral without scale. Such integrals are generally set to zero within dimensional regularization in accordance with a well-known convention. However, as we shall see later, such Feynman integrals are automatically set to zero in the case of limits typical of Euclidean space, without referring to additional prescriptions. On the other hand, for typical pseudo-Euclidean limits, this convention still looks like an ad hoc prescription; it will be implied for any integral without scale. So the second line is zero and we obtain the following leading-power behaviour:

$$F_\Gamma \sim \frac{1}{q^2} \int \frac{\mathrm{d}^d k}{k^2(p+k)^2} + \int \frac{\mathrm{d}^d k}{(k^2)^2(q-k)^2} \, . \tag{1.10}$$

Let us stop for a moment and observe that we have already solved the problem of the expansion (in the leading order) because the initial quantity depending on three kinematical invariants can be replaced, in the given limit and order, by two integrals that depend only on one scale each, so that both terms are homogeneous with respect to the expansion parameter. The homogeneity degree (e.g. with respect to p) is easily computed by power counting before integration: it equals -2ε and 0 for the first and second term, respectively. Moreover, both terms on the right-hand side are nothing but familiar Feynman integrals, rather than functions from another class, Feynman integrals with additional integrations, etc. However, we have paid a (negligible) price to obtain this result by introducing the regularization. Without it, i.e. for $d = 4$, both terms would be ill-defined because the first integral would be divergent at large k and the second integral at small k.

The two integrals in (1.10) are indeed much simpler quantities: in particular, they can be evaluated for general values of the regularization parameter d. Following the above recipe, we can compute them at general integer values of d. Results for these integrals (which are of the same type) can be obtained from the corresponding integral with general values of powers of propagators. Results of this kind and similar ones are presented in the next chapter.[3] In accordance with the recipe formulated above, when evaluating this general integral we can proceed in an integer number of dimensions, then arrive at an explicit result in terms of gamma functions, insert the relevant values of the powers of propagators and, finally, consider the number of dimensions, d, as a complex number. Thus we obtain

$$F_\Gamma \sim \pi^{d/2} \left(\frac{\Gamma(1-\varepsilon)^2 \Gamma(\varepsilon)}{\Gamma(2-2\varepsilon)} \frac{1}{q^2 (p^2)^\varepsilon} - \frac{\Gamma(1-\varepsilon)^2 \Gamma(\varepsilon)}{\Gamma(1-2\varepsilon)} \frac{1}{(q^2)^{1+\varepsilon}} \right) . \quad (1.11)$$

If we allow d to tend to 4 (i.e. $\varepsilon \to 0$), we see that the poles in the two contributions cancel and we obtain the following leading-power behaviour:

$$F_\Gamma \sim \pi^2 \frac{\ln(p^2/q^2) - 2}{q^2} . \quad (1.12)$$

Let us now observe that we could arrive at our result (1.10) and, further, at (1.11) and (1.12), by use of the following prescriptions:

(*i*) divide the integration domain into regions of small and large momenta and use leading approximations for the integrand (i.e. take the zero-order terms in Taylor series in the small parameters) (see (1.8));
(*ii*) extend the integration to *all* loop momenta in the expanded contribution of every region.

In this book, the strategy of expansion by regions will be formulated for

- arbitrary order of expansion,
- a general diagram,
- any limit.

The greatest part of the book is devoted to details and subtleties of the method, which are explained through numerous characteristic examples.

In our first example, we have justified this strategy. It will be justified for some other examples also. But mathematical proofs for general diagrams are unknown at present. It turns out, however, that for limits typical of Euclidean space, expansion by regions leads to the same prescriptions for expanding Feynman integrals as does another strategy [47, 126, 207, 208], which we shall call the 'strategy of expansion by subgraphs'. This second strategy has been mathematically justified (in this simpler case of limits) for general diagrams so that we obtain here an indirect confirmation of the strategy of expansion by regions.

[3] Another way to compute this general integral is to observe that it is of convolution type, and then apply a well-known theorem which says that the Fourier transform of a convolution equals the product of the Fourier transforms of the initial factors, and also formula (A.51) for the Fourier transform of the function $(q^2)^\lambda$ (taken, for example, from [124]).

Let us illustrate expansion by subgraphs, using the same diagram as in Fig. 1.1. The key point of this strategy is to reduce the problem of the asymptotic expansion to the problem of the R-operation, i.e. diagrammatic renormalization. The goal of the R-operation is to make a Feynman integral finite by removing its ultraviolet divergences. When a diagram is only overall divergent and does not involve subdivergences, the R-operation is of the form $R = 1 - M$, where M is a subtraction operator. When there is also a divergent one-particle-irreducible subgraph then R takes the form $R = (1-M)(1-M_1)$, where M_1 performs subtractions in the subgraph. In a general situation, the R-operation is given by the forest formula [245, 247]. It happens that, for typical Euclidean limits, the remainder of the asymptotic expansion can be also determined by the forest formula written in terms of appropriate subtraction operators which, in this case, perform subtractions in a certain family of subgraphs associated with the given limit and provide the desired asymptotic behaviour of the remainder.

In the case of our one-loop triangle diagram, in the leading order of expansion in the limit of large q, the remainder can be written as

$$\mathcal{R}F_\Gamma = (1 - \mathcal{M})(1 - \mathcal{M}_1)F_\Gamma \,, \tag{1.13}$$

where the subtraction operators $\mathcal{M}$ and $\mathcal{M}_1$ are defined by

$$\mathcal{M}F_\Gamma = \int \mathrm{d}^d k \frac{1}{k^2(q-k)^2} \mathcal{T}_p \frac{1}{(p+k)^2} \equiv \int \mathrm{d}^d k \frac{1}{(k^2)^2(q-k)^2} \,, \tag{1.14a}$$

$$\mathcal{M}_1 F_\Gamma = \int \frac{\mathrm{d}^d k}{k^2(p+k)^2} \mathcal{T}_k \frac{1}{(q-k)^2} \equiv \frac{1}{q^2} \int \frac{\mathrm{d}^d k}{k^2(p+k)^2} \,, \tag{1.14b}$$

and $\mathcal{T}_{\ldots}$ are operators that perform Taylor expansion in the corresponding variable. In this example, these are Taylor expansions of zero order, i.e. they just set the corresponding variable to zero.

The rule for choosing the operators $\mathcal{M}$ is very simple: they are operators that perform Taylor expansion of the integrand of the Feynman integrals in parameters that are considered *small* for a given subgraph. The operator $\mathcal{M}$ performs Taylor expansion with respect to the small parameter of the whole graph, i.e. the four-momentum p, and the operator $\mathcal{M}_1$ performs Taylor expansion with respect to the small parameter of the subgraph that consists of the line with momentum $q-k$. (By definition, all the loop momenta of the whole graph that are external for a given subgraph (e.g. k in our case) are considered small.) Note that we have introduced a regularization to make it possible to consider individual subtraction operators and their products, and we always choose this regularization to be dimensional.

The remainder defined in this way can be rewritten as

$$\begin{aligned} \mathcal{R}F_\Gamma &= \int \mathrm{d}^d k \frac{1}{k^2} \left(\frac{1}{(p+k)^2} - \frac{1}{k^2} \right) \left(\frac{1}{(q-k)^2} - \frac{1}{q^2} \right) \\ &\equiv \frac{1}{q^2} \int \mathrm{d}^d k \frac{(p^2 + 2p\cdot k)(k^2 - 2q\cdot k)}{(k^2)^2(q-k)^2(p+k)^2} \,. \end{aligned} \tag{1.15}$$

We observe, by power counting at small and large values of k, that the remainder is both infrared and ultraviolet finite. On the other hand, it manifestly involves the first power of the small expansion parameter p, so that it behaves, when $p \to 0$, as $|p|$ modulo logarithms (and, in any case, as $|p|^{1-\alpha}$ for arbitrarily small α).

Starting from the remainder (1.13) with these properties, we represent our Feynman integral as

$$\begin{aligned} F_\Gamma &= [1 - (1 - \mathcal{M})(1 - \mathcal{M}_1)]\, F_\Gamma + \mathcal{R} F_\Gamma \\ &= \mathcal{M} F_\Gamma + \mathcal{M}_1 F_\Gamma - \mathcal{M}\mathcal{M}_1 F_\Gamma + \mathcal{R} F_\Gamma \, . \end{aligned} \tag{1.16}$$

The third term,

$$- \mathcal{M}\mathcal{M}_1 F_\Gamma = -\frac{1}{q^2} \int \frac{\mathrm{d}^d k}{(k^2)^2} \, , \tag{1.17}$$

is a massless vacuum integral, which we set to zero.

Hence the asymptotic expansion is given by the first two terms in the second line of (1.16), and we recognize our previous result (1.10) obtained above by expansion by regions. Let us also stress that, this time, all the arguments used in the derivation of this result are equally applicable to a Feynman integral in Minkowski space.

In this book, it will be explained how expansion by subgraphs can be applied to

- arbitrary order of expansion,
- a general diagram,
- any limit typical of Euclidean space.

It is this strategy which we shall prefer to apply to limits of this kind, with the corresponding prescriptions written by means of a simple formula containing a summation over an appropriate family of subgraphs.

Let us remember that we have arrived at expansion of the triangle Feynman diagram from the problem of the expansion of two vector quark currents. Let us come back to the operator level. It turns out that the techniques for expanding individual Feynman diagrams enable us to perform operator expansions. In this particular case, the prescriptions for expanding Feynman integrals $F_\Gamma(q, p_1, \ldots, p_n)$ in the limit where one of the external momenta, q, is large and the other external momenta, p_i, and all the masses are small can be naturally extended to the operator level and reformulated for the expansion of the time-ordered product of two composite operators at short distances (large momenta q). For simplicity, let us continue to consider the scalar theory, instead of QCD, and analyse the expansion of the time-ordered product of two equal composite operators $J = (1/2)\phi^2$ in the massless ϕ^4 model with the interaction Lagrangian $\mathcal{L}_{\mathrm{I}} = -(g/4!)\phi^4$:

$$T J(x) J(0) \sim C_0(x) \mathbf{1} + C_2(x) J(0) + \ldots \, . \tag{1.18}$$

The expansion is performed in the limit where the difference between four-coordinates of the operators tends to zero, so that the terms are ordered according to the strength of their singular behaviour when $x \to 0$. It happens that in these first terms of the operator product expansion (OPE), only the unit operator and the operator J itself are present, as the operators with the lowest possible canonical dimensions (zero and two, respectively, in mass units). As is well known, the products of quantum field operators are singular at points where their arguments coincide, so that the first coefficient functions (Wilson coefficients) C_i are singular.

The limit $x \to 0$ can be translated into momentum space language as the limit $q \to \infty$ for Fourier transforms. Turning to momentum space, we have

$$T\tilde{J}(q)J(0) \equiv \int \mathrm{d}^4x\, \mathrm{e}^{\mathrm{i}q\cdot x} TJ(x)J(0) \sim \tilde{C}_0(q)\mathbf{1} + \tilde{C}_2(x)J(x) + \dots . \tag{1.19}$$

The Fourier transform $T\tilde{J}(q)J(0)$ is represented in perturbation theory, according to the Feynman rules, by diagrams which have two external vertices for the two operators and a set of ϕ^4 vertices (either external or internal) generated by the interaction Lagrangian – see Fig. 1.2.

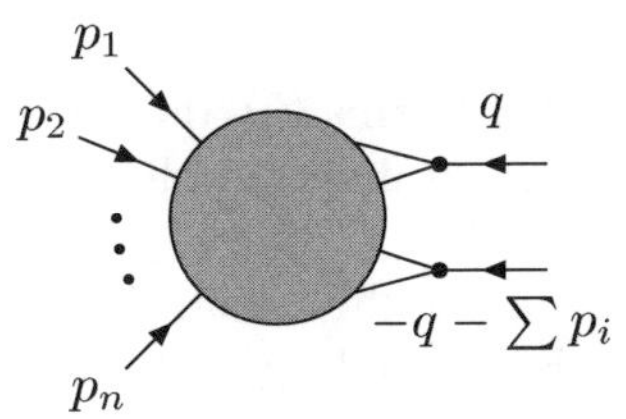

Fig. 1.2. The set of diagrams contributing to the product of two ϕ^2 operators

It is implied that all the quantities are renormalized. In the diagrammatic language, this means that we have rules for removing ultraviolet divergences from the diagrams. We shall usually assume a form of dimensional renormalization [144] which is based on subtracting poles in the parameter $\varepsilon = (4-d)/2$ of dimensional regularization. In addition to the rules for renormalizing diagrams constructed from the integration Lagrangian, we also have similar prescriptions for making finite one-particle-irreducible diagrams which involve one or both vertices corresponding to the composite operators. At the operator level, the renormalization is performed by inserting counterterms into the Lagrangian and redefining bare composite operators and their products. The renormalization of the Lagrangian (which reduces to a redefinition of the coupling constant) is irrelevant in our example up to one-loop order. The one-loop renormalization of the composite operator J reduces, in our massless case, to the substitution of the bare composite operator by a renormalized one: $J^{\mathrm{B}} \to Z^{-1}J$, where Z is a renormalization factor. Dia-

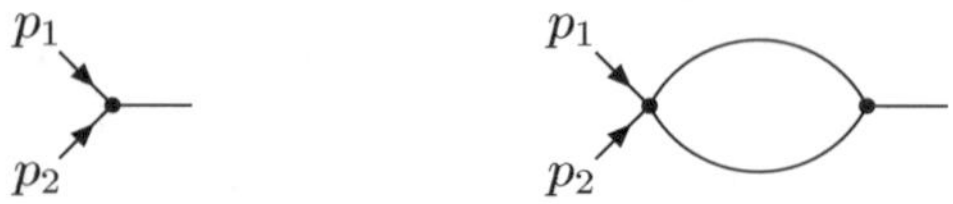

Fig. 1.3. Diagrams of zero and first order with two S-matrix external lines contributing to the operator J

grams of zero and first order in g with two S-matrix external lines and one vertex for the operator J are shown in Fig. 1.3.

Now, in accordance with the Feynman rules, we take into account a factor i for each propagator, $1/(2\pi)^d$ for the loop integration and $-\mathrm{i}g/4!$ for the S-matrix vertices. In particular, the value of the first diagram in Fig. 1.3 is one. The renormalization factor Z is generated, in one loop, by the second diagram,

$$\frac{\mathrm{i}g}{2(2\pi)^4} \int \frac{\mathrm{d}^d k}{k^2(q-k)^2} \, .$$

The $\overline{\mathrm{MS}}$ subtraction (where one subtracts the combination $1/\varepsilon + \ln(4\pi) - \gamma_{\mathrm{E}}$, rather than the pure pole in ε) gives the value

$$Z = 1 + \frac{g}{2(16\pi^2)} \frac{1}{\varepsilon} \, . \tag{1.20}$$

Multiplication of the sum of the two diagrams in Fig. 1.3 by this Z (equivalent to the renormalization of the operator J) then produces a finite result when $\varepsilon \to 0$.

The expansion of the product $T\tilde{J}(q)J(0)$ reduces to the expansion of the corresponding diagrams in the limit where q is much larger than all the other external momenta, in the Euclidean sense. The coefficient function C_0 can be evaluated through diagrams that depend only on the external momentum q and no other external momenta. The problem of the large-momentum expansion does not arise for C_0 and, since we are dealing with a massless theory, these diagrams have only power and logarithmic dependences on q. In this sense, the coefficient function C_0 is trivial, and we turn now to the Wilson coefficient C_2.

Consider diagrams with two ‘ S-matrix’ external vertices. In other words, consider the Green function

$$\begin{aligned} &\langle TJ(x)J(0)\phi(y_1)\phi(y_2)\rangle^{\mathrm{amp}} \\ &\qquad = \left\langle TJ(x)J(0) \exp\left(\mathrm{i}\int \mathrm{d}^4x \, \mathcal{L}_{\mathrm{I}}(x)\right) \phi(y_1)\phi(y_2)\right\rangle_0^{\mathrm{amp}} , \end{aligned}$$

where the symbol ‘amp’ denotes the contribution of one-particle-irreducible diagrams in which the external y lines have been amputated (i.e. the propagators corresponding to these lines are omitted). The vacuum expectation values are taken between vacuum states in the full and free theories, respectively. Turning to the Fourier transforms with respect to the variables x, y_1 and y_2 we have

$$\left\langle T\tilde{J}(q)J(0)\tilde{\phi}(p_1)\tilde{\phi}(p_2)\right\rangle^{\rm amp} = \left\langle T\tilde{J}(q)J(0)\exp\left({\rm i}\int {\rm d}^4x\,\mathcal{L}_{\rm I}(x)\right)\tilde{\phi}(p_1)\tilde{\phi}(p_2)\right\rangle_0^{\rm amp} .$$

The diagrams contributing in the zero and first orders in the coupling constant are shown in Fig. 1.4. The first two (zero-order) diagrams are expanded in the limit $q \to \infty$ just by Taylor expansion in the other external momenta, p_1 and p_2. The leading term of this expansion gives the value $2{\rm i}/q^2$. We can write down this term as the zero order contribution (equal to unity) to the Green function $\langle TJ(0)\phi(y_1)\phi(y_2)\rangle$ of the composite operator J times this value which thus gives the zero order contribution to the Wilson coefficient C_2 in (1.19).

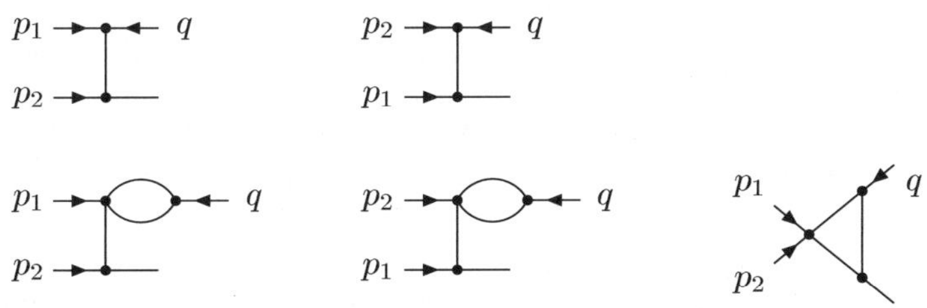

Fig. 1.4. Diagrams of zero and first order with two S-matrix external lines contributing to the product of two operators J

The three diagrams in the second row of Fig. 1.4 contribute in the first order of perturbation theory. The expansions of the first two diagrams in the limit $q \to \infty$ are given by Taylor expansion at zero values of $p_{1,2}$. Note that the UV divergences in the loops in these diagrams are connected with the initial composite operators and are removed by subtracting poles in ε. (This is equivalent to inserting the counterterm Z for the composite operator J.)

The third diagram in the second row of Fig. 1.4 is nothing but our diagram shown in Fig. 1.1, where we now have two external lines incident on the left vertex with a total external momentum $p = p_1 + p_2$, and p_i is the Fourier argument corresponding to y_i. This diagram has to be expanded in the limit where q is much larger than p. As we have seen, the leading behaviour of this diagram is given by (1.10). Observe now that the two terms admit a natural graph-theoretical interpretation, shown in Fig. 1.5. It turns out that the contribution of the two diagrams on the right-hand side of Fig. 1.5 and the first two diagrams in the second row of Fig. 1.4 can be represented in the form

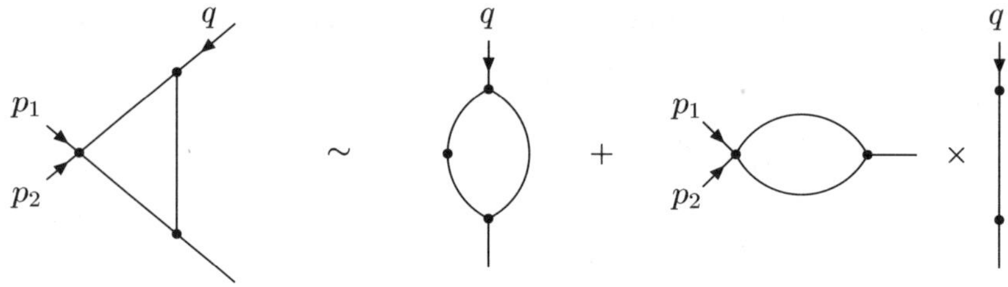

Fig. 1.5. Graph-theoretical description of the leading-order expansion of Fig. 1.1 when $q \to \infty$

$$\langle T\tilde{J}(q)J(0)\tilde{\phi}(0)\tilde{\phi}(0)\rangle^{\mathrm{amp}} \langle TJ^{\mathrm{B}}(0)\tilde{\phi}(p_1)\tilde{\phi}(p_2)\rangle \,, \tag{1.21}$$

where the factors are evaluated in either the zero or the first order in g. Indeed, the sum of the first term in Fig. 1.5 and the first two diagrams in the second row of Fig. 1.4 can be written as the contribution of the first order in g to $\langle T\tilde{J}(q)J(0)\tilde{\phi}(0)\tilde{\phi}(0)\rangle^{\mathrm{amp}}$ times the zero-order contribution to the Green function $\langle TJ^{\mathrm{B}}(0)\tilde{\phi}(p_1)\tilde{\phi}(p_2)\rangle$, equal to unity. Then the second term in Fig. 1.5 can be naturally recognized as the zero-order contribution to $\langle T\tilde{J}(q)J(0)\tilde{\phi}(0)\tilde{\phi}(0)\rangle^{\mathrm{amp}}$, equal to $2\mathrm{i}/q^2$ in agreement with what was obtained above, times the zero-order contribution to the Green function $\langle TJ^{\mathrm{B}}(0)\tilde{\phi}(p_1)\tilde{\phi}(p_2)\rangle$ of the 'unrenormalized' composite operator J^{B} given by the relevant integral.

The presence of J^{B} instead of J means that the UV divergence of the loop in the last term in Fig. 1.5 is not removed. We can obtain the renormalized operator by the substitution $J^{\mathrm{B}} \to Z^{-1}J$. After that, we obtain the following representation for the contribution of the zero- and first-order diagrams in Fig. 1.4 to C_2:

$$\tilde{C}_2 = \Pi[\tilde{J}(q)J(0)]Z^{-1} \,. \tag{1.22}$$

Here Π is the projector on the operator J:

$$\Pi[\tilde{J}(q)J(0)] = \langle T\tilde{J}(q)J(0)\tilde{\phi}(0)\tilde{\phi}(0)\rangle^{\mathrm{amp}} \,, \tag{1.23}$$

with the property $\Pi J = 1$ in the zero order in g.

Substituting the value of the second integral in (1.10), given by the second term in the large parentheses in (1.11) (with a factor i when going from Euclidean to Minkowski space), as well as the renormalization factor Z given by (1.20), we obtain, up to the first order in g,

$$\tilde{C}_2(q) = \frac{\mathrm{i}}{q^2}\left(2 - \frac{g}{2(16\pi^2)}\ln\left(-q^2/\mu^2\right)\right) \,, \tag{1.24}$$

where μ is a massive parameter that arises within dimensional regularization.

Our main goal here was to illustrate the origin of the very simple formula (1.22) rather than directly calculate the Wilson coefficient C_2. In this book, one will also find

- a generalization of (1.22) [128] applicable to any composite operators in an arbitrary order of perturbation theory;
- operator expansions and expansions of Green functions corresponding to other limits;
- various applications of asymptotic expansions in momenta and masses to problems of phenomenological importance.

The OPE is a typical example of the so-called factorization procedure, which separates factors responsible for the description of physical phenomena that take place on different scales. In the case of the OPE, the coefficient functions are responsible for the information about small distances (high energies), so that they can be treated perturbatively while the information about large distances (small energies) is encoded in the matrix elements of the composite operators present in the OPE.

The most typical form of the factorization is achieved by use of an effective theory described by an effective Lagrangian. The standard 'folklore' procedure is to include all operators having the necessary symmetries, with arbitrary coefficients. These 'matching' coefficients are then adjusted by solving a system of a sufficient number of equations which express the fact that the same results are obtained for the amplitudes of the full theory and the effective theory when expanded in the ratio of the small to the large scale. (The OPE is an example of an operator expansion where the matching conditions are explicitly solved in a general form, and the corresponding coefficient functions can be evaluated [128] by means of formulae of the type (1.22).) The effective Lagrangian consists of fields responsible for physics at lower energies and is typically written as a series in the expansion parameter, while the matching coefficients of the effective Lagrangian contain higher-energy information about the initial theory. The number and complexity of the terms increase with the order of the expansion. In physical slang, the transition to the effective theory is characterized as integrating out higher scales, although an explicit integration over the fields corresponding to higher scales in the functional integral appears to be an impossible task.

An essential point is that the strategy of expansion by regions turns out to be a universal technique for treating any asymptotic regime. This procedure gives a natural way of constructing effective field theories by providing adequate information about the form of the effective Lagrangian.

To fulfil our programme, a sketch of basic facts connected with Feynman integrals is first presented in Chap. 2. Chapter 3 then introduces the two basic strategies for expanding Feynman integrals in infinite series in powers and logarithms. Chapters 4 and 5 are devoted to typical Euclidean limits, while Chaps. 6–8 deal with typical pseudo-Euclidean on-shell and threshold limits. The threshold expansion in the case of one heavy (non-zero) mass in the threshold is studied in Chap. 6, the case of two non-zero masses is treated in Chap. 7 and limits of the Sudakov type are investigated in Chap. 8. The structure of each of Chaps. 4–8 is the same. We start with one-loop examples, then formulate prescriptions for a given limit, present two-loop examples and, finally, go up to the operator level and discuss applications to physical problems.

We shall (almost) always take scalar Feynman integrals as examples, not only because this choice simplifies our discussion, but also because, in real practical calculations with non-scalar numerators, one usually performs a tensor decomposition and reduces the problem to scalar diagrams.

In Chap. 9, I conclude by presenting some alternative approaches, characterizing the status of the methods described and conclude with some advice. In Appendix A, one can find a table of basic integrals and useful formulae, in particular the definitions and basic properties of some special functions that are used in the book. Basic notational conventions are presented below. The notation is described in more detail in the List of Symbols. In Appendix B, an analysis of the convergence of Feynman integrals and a proof of the prescriptions for the off-shell large-momentum limit are presented.

1.1 Notation

We use Greek and Roman letters for four-indices and spatial indices, respectively:

$$x^\mu = (x^0, \mathbf{x}) \, ,$$
$$q{\cdot}x = q^0 x^0 - \mathbf{q}{\cdot}\mathbf{x} \equiv g_{\mu\nu} q^\mu x^\nu \, .$$

The four-dimensional Fourier transform and its inverse are defined as

$$\tilde{f}(q) = \int \mathrm{d}^4 x \, \mathrm{e}^{\mathrm{i} q\cdot x} f(x) \, ,$$
$$f(x) = \frac{1}{(2\pi)^4} \int \mathrm{d}^4 q \, \mathrm{e}^{-\mathrm{i} x\cdot q} \tilde{f}(q) \, .$$

The parameter of dimensional regularization is

$$d = 4 - 2\varepsilon \, . \tag{1.25}$$

The operator denoted by

$$\mathcal{T}^n_{\dots} \equiv \sum_{j=0}^{n} \mathcal{T}^{(j)}_{\dots} \, , \tag{1.26}$$

where

$$\mathcal{T}_{\dots} \equiv \mathcal{T}^\infty_{\dots} \, , \tag{1.27}$$

performs a Taylor expansion of order n in the corresponding set of variables. For example, in the one-dimensional case,

$$\mathcal{T}^n_x f(x) = \sum_{j=0}^{n} \frac{1}{j!} f^{(j)}(0) x^j \, . \tag{1.28}$$

When one calculates Feynman integrals in the SU(N) non-Abelian gauge theory, the following standard invariants appear:

$$C_{\mathrm{F}} = \frac{N^2 - 1}{2N} \,, \quad C_{\mathrm{A}} = N \,, \quad T_{\mathrm{F}} = \frac{1}{2} \,. \tag{1.29}$$

These are the quadratic Casimir operator of the fundamental representation of the SU(N) group, the quadratic Casimir operator of the adjoint representation and the index of the fundamental representation, respectively.

The scalar product of a vector k and the four- (or d-) vector composed of the gamma matrices is denoted by

$$\not{k} = k^{\mu}\gamma_{\mu} \,.$$

2 Feynman Integrals: a Brief Review

In this chapter, it is briefly explained what Feynman integrals are, what properties they have, and how they are regularized, renormalized and evaluated. Special attention is paid to various divergences of Feynman integrals because their interplay is a characteristic feature of explicit prescriptions for asymptotic expansions in momenta and masses. A brief review of methods for analytic evaluation of Feynman integrals is presented. In addition to the method which is itself based on the use of our prescriptions for the asymptotic expansions, we need a number of methods for the evaluation of Feynman integrals that appear on the right-hand side of the expansions. Basic definitions connected with renormalization, in particular the forest formula, are also listed, because the remainder of an asymptotic expansion, as obtained by the method of expansion by subgraphs, has the structure of renormalization.

2.1 From Lagrangians to Feynman Integrals

In perturbation theory, any quantum field model is characterized by a Lagrangian, which is represented as a sum of a free-field part and an interaction part, $\mathcal{L} = \mathcal{L}_0 + \mathcal{L}_\mathrm{I}$. Amplitudes of the model, e.g. S-matrix elements and Green functions, are represented as power series in coupling constants. In a fixed perturbation order, the amplitudes are written as finite sums of Feynman diagrams which are constructed according to Feynman rules: lines correspond to $\mathcal{L}_0$ and vertices are determined by $\mathcal{L}_\mathrm{I}$.

The unrenormalized S-matrix is given by

$$S = T \exp\left(\mathrm{i} \int \mathrm{d}^4 x\, \mathcal{L}_\mathrm{I}(x)\right) \tag{2.1}$$

$$= \sum_{n=0}^{\infty} \frac{\mathrm{i}^n}{n!} \int \mathrm{d}^4 x_1 \ldots \mathrm{d}^4 x_n\, T \mathcal{L}_\mathrm{I}(x_1) \ldots \mathcal{L}_\mathrm{I}(x_n)\, . \tag{2.2}$$

The Wick theorem is then applied to time-ordered products of the free fields that enter the right-hand side through the interaction Lagrangian, i.e. any product is written as a sum of terms, where some pairs of fields of the same sort are replaced by the corresponding propagators, in accordance with the relation

$$T\phi_i(x_1)\phi_i(x_2) = :\phi_i(x_1)\phi_i(x_2): + D_i(x_1 - x_2) \,. \tag{2.3}$$

Here D_i is the propagator of the field of type i and the colons denote a normal product of the free fields. The Fourier transforms of the propagators have the form

$$\tilde{D}_i(p) \equiv \int \mathrm{d}^4x \, \mathrm{e}^{\mathrm{i}p\cdot x} D_i(x) = \frac{\mathrm{i}Z_i(p)}{(p^2 - m_i^2 + \mathrm{i}0)^{a_i}} \,, \tag{2.4}$$

where m_i is the corresponding mass, Z_i is a polynomial and $a_i = 1$ or 2 (for the gluon propagator in the general covariant gauge). This is not the most general form of the propagator. For example, in the axial or Coulomb gauge, the gluon propagator has another form. We shall usually deal in the book with the propagator (2.4), which we imply by default. We usually omit the causal i0 for brevity. Polynomials associated with vertices of graphs can be taken into account by means of the polynomials Z_l. We also omit the factors of i and $(2\pi)^4$ that enter in the standard Feynman rules (in particular, in (2.4)); these can be included at the end of a calculation.

Note that the products of the free fields in the Lagrangian are not required to be normal-ordered, so that products of fields of the same sort at the same point are allowed. The formal application of the Wick theorem therefore generates values of the propagators at zero. For example, in the case of the scalar free field, with the propagator

$$D(x) = \frac{\mathrm{i}}{(2\pi)^4} \int \mathrm{d}^4k \frac{\mathrm{e}^{-\mathrm{i}x\cdot k}}{k^2 - m^2} \,, \tag{2.5}$$

which satisfies $(\Box + m^2)D(x) = -\mathrm{i}\delta(x)$, we have

$$T\phi(x)\phi(x) = :\phi^2(x): + D(0) \,. \tag{2.6}$$

The value of $D(x)$ at $x = 0$ does not exist, because the propagator is singular at the origin:

$$D(x) = \frac{\mathrm{i}}{4\pi^2} \frac{1}{x^2 - \mathrm{i}0} + \ldots \,. \tag{2.7}$$

However, we imply the *formal* value at the origin rather than the 'honestly' taken value. This means that we set x to zero in some integral representation of this quantity. For example, using the inverse Fourier transformation, we can define $D(0)$ as the integral (2.5) with x set to zero *in the integrand.* Thus, by definition,

$$D(0) = \frac{\mathrm{i}}{(2\pi)^4} \int \frac{\mathrm{d}^4k}{k^2 - m^2} \,. \tag{2.8}$$

This integral is, however, divergent, as Feynman integrals typically are. We shall turn to divergences shortly in the next section.

Eventually, we obtain, for any fixed perturbation order, a sum of Feynman amplitudes labelled by Feynman graphs constructed from the given type of vertices and lines. In the commonly accepted physical slang, the graph, the corresponding Feynman amplitude and the integral are all often called the 'diagram'.

After the Wick theorem is applied, the S-matrix can be rewritten as

$$S = \sum_{l=0}^{\infty} \frac{\mathrm{i}^l}{[l]!} \int \mathrm{d}^4x_1 \dots \mathrm{d}^4x_l \, S_N(x_1, \dots, x_l) : \phi_1(x_1) \dots \phi_1(x_l) : , \tag{2.9}$$

where ϕ_i are the (free) fields present in the theory, and the coefficient functions S_l of the operator S are constructed from Feynman diagrams with l external lines. The vertices at the ends of these lines are external for any such diagram, and they carry a dependence on the four-coordinates x_i (or, after Fourier transforming, on the corresponding four-momenta). All the other vertices of any Feynman amplitude contributing to S_l are internal. The symbol $[l]!$ denotes the product of the $l_i!$ corresponding to fields of different sorts.

Any operator A can be written in a similar form in perturbation theory. If we replace normal products on the right-hand side by products of classical c-fields we obtain a representation of this operator in terms of a (non-linear) functional

$$A = \sum_{n=0}^{\infty} \frac{\mathrm{i}^N}{N!} \int \mathrm{d}^4x_1 \dots \mathrm{d}^4x_N \, A_N(x_1, \dots, x_N) \phi_1(x_1) \dots \phi_1(x_N) \, . \tag{2.10}$$

This language is useful in many situations. For example, the application of the Wick theorem is performed by use of the operator

$$\mathcal{W} = \exp\left(\frac{1}{2\mathrm{i}} \int \mathrm{d}x_1 \, \mathrm{d}x_2 \, \frac{\delta}{\delta\phi(x_1)} D(x_1 - x_2) \frac{\delta}{\delta\phi(x_2)} \right) , \tag{2.11}$$

where the scalar case is assumed and $D(x)$ is the corresponding propagator.

We shall now give some examples of Lagrangians. The Lagrangian of the scalar ϕ^4 theory is

$$\mathcal{L} = \frac{1}{2}\partial_\mu\phi\partial^\mu\phi - \frac{1}{2}m^2\phi^2 - \frac{g}{4!}\phi^4 \, . \tag{2.12}$$

The first two terms generate the free field theory. The corresponding propagator is (2.4) with $Z = 1, a = 1$. The interaction term generates, according to the Feynman rules, 4-vertices with factors $-\mathrm{i}g$.

The Lagrangian of quantum electrodynamics (QED) is

$$\mathcal{L} = \bar{\psi}(\mathrm{i}\partial\!\!\!/ - m)\psi - \frac{1}{4}F_{\mu\nu}F^{\mu\nu} - e\bar{\psi}\gamma^\mu\psi A_\mu \, , \tag{2.13}$$

where ψ is the electron field, A_μ is the electromagnetic vector potential, $F_{\mu\nu} = \partial_\mu A_\nu - \partial_\nu A_\mu$ is the electromagnetic field tensor, e is the electron charge and γ_μ are the gamma matrices. In the h-loop order of perturbation theory, a natural factor, $e^2/(16\pi^2)$, arises in the calculation of Feynman h-loop integrals. The standard way is to write this factor as $\alpha/(4\pi)$, where

$$\alpha = \frac{e^2}{4\pi} \tag{2.14}$$

is the fine-structure constant.

The Lagrangian of quantum chromodynamics (QCD) in the general covariant gauge has the form

$$\mathcal{L} = \sum_i \bar{q}_i(\mathrm{i}\not{D} - m_i)q_i - \frac{1}{4}G^a_{\mu\nu}G^{a\mu\nu} + \mathcal{L}^{\mathrm{GF}} + \mathcal{L}^{\mathrm{FP}} , \tag{2.15}$$

where $G^a_{\mu\nu}$ is the gluon field strength tensor

$$G^a_{\mu\nu} = \partial_\mu A^a_\nu - \partial_\nu A^a_\mu + g f^{abc} A^b_\mu A^c_\nu , \tag{2.16}$$

and f^{abc} are the structure constants of the colour SU(3) group. The gauge-fixing term and the ghost term are

$$\mathcal{L}^{\mathrm{GF}} = -\frac{1}{2\xi}(\partial_\mu A^\mu)^2 , \quad \mathcal{L}^{\mathrm{FP}} = (\partial_\mu \bar{c}^a)(D^\mu c^a) , \tag{2.17}$$

respectively. The Lagrangian includes the quark fields q_i, the gluon field A^a_μ and the ghost field c^a. The covariant derivative is defined by

$$D_\mu = \partial_\mu - \mathrm{i}g A^a_\mu t^a , \tag{2.18}$$

where t^a are SU(3) generators. In addition to the flavour index i, the quark field q_i has also spinor and colour indices, which are usually suppressed. It is useful to perform practical calculations in QCD in the case of the general colour gauge group SU(N). This is done to obtain better control on individual terms with different colour structures such as $C_{\mathrm{F}}, C_{\mathrm{A}}$ and T_{F} given by (1.29). At the end of the calculation one can insert the number of colours $N = 3$. The perturbative expansion is performed in the coupling constant g. Similarly to QED, one defines the strong coupling

$$\alpha_{\mathrm{s}} = \frac{g^2}{4\pi} , \tag{2.19}$$

which is the QCD analogue of the fine-structure constant. The QCD Feynman rules can be found in [148, 74, 190]. In particular, the gluon propagator in the ξ-gauge is

$$\frac{\mathrm{i}}{p^2}\left(-g^{\mu\nu} + \xi\frac{p^\mu p^\nu}{p^2}\right) , \tag{2.20}$$

where ξ is defined by (2.17).

Thus quantities that can be computed perturbatively are written, in any given order of perturbation theory, through a sum over Feynman graphs. For a given graph Γ, the corresponding Feynman amplitude

$$G_\Gamma(q_1, \ldots, q_n) = (2\pi)^4 \delta\left(\sum_i q_i\right) F_\Gamma(q_1, \ldots, q_{n-1}) \tag{2.21}$$

can be written in terms of an integral over loop momenta

$$F_\Gamma(q_1, \ldots, q_{n-1}) = \int \mathrm{d}^4 k_1 \ldots \int \mathrm{d}^4 k_h \prod_{l=1}^{L} \tilde{D}_l(p_l) , \tag{2.22}$$

where $\mathrm{d}^4k_i = \mathrm{d}k_i^0\,\mathrm{d}\mathbf{k}_i$. The Feynman integral F_Γ depends on $n-1$ linearly independent external momenta $q_i = (q_i^0, \mathbf{q}_i)$; the corresponding integrand is a function of L internal momenta p_i, which are certain linear combinations of the external momenta and $h = L - V + 1$ chosen loop momenta k_i, where L, V and h are the numbers of lines, vertices and (independent) loops, respectively, of the given graph.

2.2 Divergences

As has been known from early days of quantum field theory, Feynman integrals suffer from divergences. This word means that, taken naively, these integrals are ill-defined because the integrals over the loop momenta generally diverge. The *ultraviolet* (UV) *divergences* manifest themselves through divergence of the Feynman integrals at large loop momenta. Consider, for example, the Feynman integral corresponding to the one-loop graph Γ of Fig. 2.1 with scalar propagators, i.e. with $Z_{1,2} = 1$ and $a_{1,2} = 1$. This integral can be written as

$$F_\Gamma(q) = \int \frac{\mathrm{d}^4k}{(k^2 - m_1^2)[(q-k)^2 - m_2^2]}\,, \tag{2.23}$$

where the loop momentum k is chosen as the momentum of the first line. Introducing four-dimensional (generalized) spherical coordinates $k = r\hat{k}$ in (2.23), where $\hat{k}$ is on the unit (generalized) sphere and is expressed by means of three angles, and counting powers of propagators, we obtain, in the limit of large r, the following divergent behaviour: $\int_\Lambda^\infty \mathrm{d}r\, r^{-1}$. For a general diagram, a similar power counting at large values of the loop momenta gives $4h(\Gamma) - 1$ from the Jacobian that arises when one introduces generalized spherical coordinates in the $(4\times h)$-dimensional space of h loop four-momenta, plus a contribution from the powers of the propagators and the degrees of its polynomials, and leads to an integral $\int_\Lambda^\infty \mathrm{d}r\, r^{-\omega-1}$, where

$$\omega = 4h - \sum_l (2a_l - n_l) \tag{2.24}$$

is the (UV) *degree of divergence* of the graph. (Here n_l are the degrees of the polynomials Z_l.)

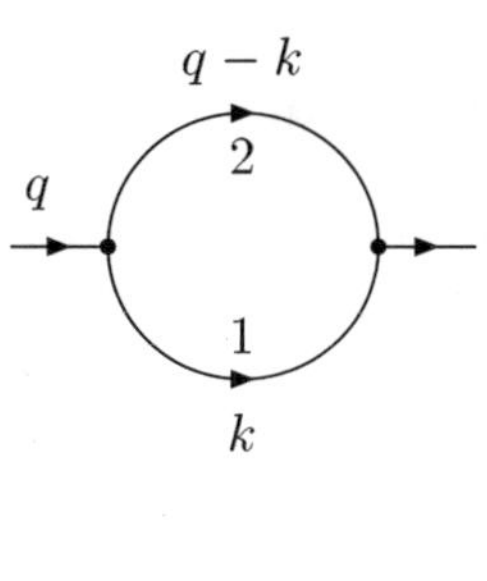

Fig. 2.1. One-loop self-energy diagram

This estimate shows that the Feynman integral is UV convergent overall (no divergences arise from the region where all the loop momenta are large) if the degree of divergence is negative. We say that the Feynman integral has a logarithmic, linear, quadratic, etc. overall divergence when $\omega = 0, 1, 2, \ldots$, respectively. To ensure a complete absence of UV divergences it is necessary to check convergence in various regions where some of the loop momenta become large, i.e. to satisfy the relation $\omega(\gamma) < 0$ for all the subgraphs γ of the graph. We call a subgraph UV divergent if $\omega(\gamma) \geq 0$. In fact, it is sufficient to check these inequalities only for *one-particle-irreducible* (1PI) subgraphs (which cannot be made disconnected by cutting a line). It turns out that these rough estimates are indeed true – more comments will be made on this later.

If we turn from momentum space integrals to some other representation of Feynman diagrams the UV divergences will manifest themselves in other ways. For example, in coordinate space, the Feynman amplitude (i.e. the inverse Fourier transform of (2.21)) is expressed in terms of a product of the Fourier transforms of propagators

$$\prod_{l=1}^{L} D_l(x_{l_{\rm i}} - x_{l_{\rm f}}) \,, \tag{2.25}$$

where $l_{\rm i}$ and $l_{\rm f}$ are the beginning and the end, respectively, of a line l. The propagators in coordinate space,

$$D_l(x) = \frac{1}{(2\pi)^4} \int {\rm d}^4 p \, \tilde{D}_l(p) {\rm e}^{-{\rm i}x\cdot p} \,, \tag{2.26}$$

are singular at small values of coordinates $x = (x_0, \mathbf{x})$. For example, the leading singularity of the scalar propagator at $x = 0$ is $1/(x^2 - {\rm i}0)$ (see (2.7)). In this case the coordinate space version of (2.23) can be expressed in terms of the square of the scalar propagator and involves the singularity $(x^2 - {\rm i}0)^{-2}$. Power-counting shows that this singularity produces integrals that are divergent in the vicinity of the point $x = 0$, and this is the coordinate space manifestation of the UV divergence.

The divergences caused by singularities at small loop momenta are called *infrared* (IR) *divergences*. First we distinguish IR divergences that arise at general values of the external momenta. A typical example of such a divergence is given by the graph of Fig. 2.1 when one of the lines contains the second power of the corresponding propagator, so that $a_1 = 2$. If the mass of this line is zero we obtain a factor $1/(k^2)^2$ in the integrand, where k is chosen as the momentum of this line. Then, keeping in mind the introduction of generalized spherical coordinates and performing power-counting at small k (i.e. when all the components of the four-vector k are small), we again encounter a divergent behaviour $\int_0^\Lambda {\rm d}r \, r^{-1}$ but now at small values of r. There is a similarity between the properties of IR divergences of this kind and those of UV divergences. One can define, for such off-shell IR divergences, an IR degree of divergence, in a similar way to the UV case. A reasonable choice is provided by the value

$$\tilde{\omega}(\gamma) = -\omega(\Gamma/\overline{\gamma}) \equiv \omega(\overline{\gamma}) - \omega(\Gamma) , \tag{2.27}$$

where $\overline{\gamma} \equiv \Gamma \backslash \gamma$ is the completion of the subgraph γ in a given graph Γ. The absence of off-shell IR divergences is guaranteed if the IR degrees of divergence are negative for all subgraphs γ whose completions $\overline{\gamma}$ include all the external vertices in the same connectivity component. (See details in [61, 208].) The off-shell IR divergences are the worst but they are in fact absent in physically meaningful theories. However, they play an important role in asymptotic expansions of Feynman diagrams, as we shall see shortly.

The other kinds of IR divergences arise when the external momenta considered are on a surface where the Feynman diagram is singular: either on a mass shell or at a threshold. Consider, for example, the graph Fig. 2.1, with the indices $a_1 = 2$ and $a_2 = 1$, with the masses, $m_1 = m \neq 0$ and $m_2 = 0$, on the mass shell, $q^2 = m^2$. With k as the momentum of the second line, the corresponding Feynman integral is of the form

$$F_\Gamma(q; d) = \int \frac{\mathrm{d}^d k}{k^2(k^2 - 2q \cdot k)^2} . \tag{2.28}$$

At small values of k, the integrand behaves like $1/[4k^2(q \cdot k)^2]$, and, with the help of power counting, we see that there is an *on-shell IR divergence* which would not be present for $q^2 \neq m^2$.

If we consider Fig. 2.1 with equal masses and indices $a_1 = a_2 = 2$ at the threshold, i.e. at $q^2 = 4m^2$, it might seem that there is a *threshold IR divergence* because, choosing the momenta of the lines as $q/2 + k$ and $q/2 - k$, we obtain the integral

$$\int \frac{\mathrm{d}^d k}{(k^2 + q \cdot k)^2 (k^2 - q \cdot k)^2} , \tag{2.29}$$

with an integrand that behaves at small k as $1/(q \cdot k)^4$ and is formally divergent. However, the divergence is in fact absent. (The threshold singularity at $q^2 = 4m^2$ is, of course, present – see Chap. 7.) Nevertheless, threshold IR divergences do exist. For example, the sunset diagram of Fig. 2.2 with general masses at threshold, $q^2 = (m_1 + m_2 + m_3)^2$, is divergent in this sense when the sum of the integer powers of the propagators is greater than or equal to five (see, e.g. [93]).

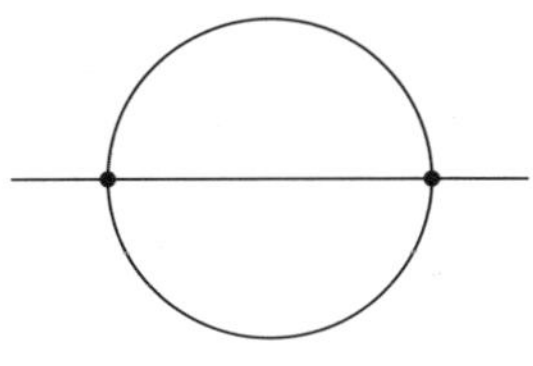

Fig. 2.2. Sunset diagram

The IR divergences characterized above are local in momentum space, i.e. they are connected with special points of the loop integration momenta. *Collinear* divergences arise at lines parallel to certain light-like four-vectors. A typical example of a collinear divergence is provided by the massless triangle graph of Fig. 2.3.

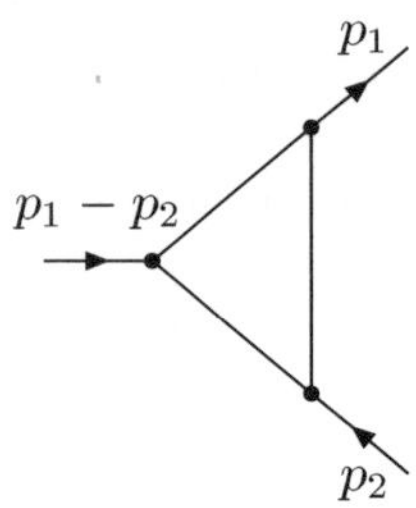

Fig. 2.3. One-loop triangle diagram

Let us take $p_1^2 = p_2^2 = 0$ and all the masses equal to zero. The corresponding Feynman integral is

$$\int \frac{\mathrm{d}^d k}{(k^2 - 2p_1 \cdot k)(k^2 - 2p_2 \cdot k)k^2} \,. \tag{2.30}$$

At least an on-shell IR divergence is present, because the integral is divergent when $k \to 0$ (componentwise). However, there are also divergences at non-zero values of k that are collinear with p_1 or p_2 and where $k^2 \sim 0$. This follows from the fact that the product $1/[(k^2 - 2p \cdot k)k^2]$, where $p^2 = 0$ and $p \neq 0$, generates collinear divergences. To see this, let us take residues in the upper complex half plane when integrating this product over k_0. For example, taking the residue at $k_0 = -|\mathbf{k}| + \mathrm{i}0$ leads to an integral containing $1/(p \cdot k) = 1/[p^0 |\mathbf{k}|(1 - \cos\theta)]$, where θ is the angle between the spatial components $\mathbf{k}$ and $\mathbf{p}$. Thus, for small θ, we have a divergent integration over angles because of the factor $\mathrm{d}\cos\theta/(1 - \cos\theta) \sim \mathrm{d}\theta/\theta$. The second residue generates a similar divergent behaviour – this can be seen by making the change $k \to p - k$.

2.3 How They Are Regularized

The standard of way of dealing with divergent Feynman integrals is to introduce a *regularization*. This means that, instead of the original ill-defined Feynman integral, we consider a quantity which depends on a regularization parameter, λ, and formally tends to the initial, meaningless expression when this parameter takes some limiting value, $\lambda = \lambda_0$. This new, regularized, quantity turns out to be well-defined, and the divergence manifests itself as a singularity with respect to the regularization parameter. Experience tells us that this singularity can be of a power or logarithmic type, i.e. $\ln^n(\lambda - \lambda_0)/(\lambda - \lambda_0)^i$.

An obvious way of regularizing Feynman integrals is to introduce a cut-off at large values of the loop momenta. Another well-known regularization procedure is the Pauli–Villars regularization [187], which is described by the replacement

$$\frac{1}{p^2 - m^2} \to \frac{1}{p^2 - m^2} - \frac{1}{p^2 - M^2}$$

and its generalizations. For finite values of the regularization parameter M, this procedure clearly improves the UV asymptotics of the integrand. Here the limiting value is $M = \infty$.

If we replace the integer powers a_l in the propagators by general complex numbers λ_l we obtain an *analytically regularized* Feynman integral [219], where the divergences of the diagram are encoded in the poles of this regularized quantity with respect to the analytic regularization parameters λ_l. For example, power counting at large values of the loop momentum in the analytically regularized version of (2.23) leads to the divergent behaviour $\int_\Lambda^\infty \mathrm{d}r\, r^{\lambda_1+\lambda_2-3}$, which results in a pole $1/(\lambda_1+\lambda_2-2)$ at the limiting values of the regularization parameters $\lambda_i = 1$.

The analysis of the divergences can be greatly simplified by the use of parametric integral representations of Feynman diagrams. The well-known alpha representation is obtained by writing down each propagator (2.4) as

$$\tilde{D}_l(p) = Z_l \left(\frac{1}{2\mathrm{i}}\frac{\partial}{\partial u_l}\right) \mathrm{e}^{2\mathrm{i}u_l\cdot p}\bigg|_{u_l=0} \frac{(-\mathrm{i})^{a_l}}{\Gamma(a_l)} \int_0^\infty \mathrm{d}\alpha_l\, \alpha_l^{a-1} \mathrm{e}^{\mathrm{i}(p^2-m^2)\alpha_l} . \tag{2.31}$$

Then the order of the integration over the loop momenta and the alpha parameters α_l is interchanged, and all the Gaussian momentum integrations are explicitly performed using the formula

$$\int \mathrm{d}^4k\, \mathrm{e}^{\mathrm{i}(\alpha k^2+2q\cdot k)} = -\mathrm{i}\pi^2\alpha^{-2}\mathrm{e}^{-\mathrm{i}q^2/\alpha} , \tag{2.32}$$

which is nothing but a product of four one-dimensional Gauss integrals:

$$\begin{aligned} \int_{-\infty}^{\infty} \mathrm{d}k_0\, \mathrm{e}^{\mathrm{i}(\alpha k_0^2+2q_0k_0)} &= \sqrt{\frac{\pi}{\alpha}}\, \mathrm{e}^{-\mathrm{i}q^2/\alpha+\mathrm{i}\pi/4} , \\ \int_{-\infty}^{\infty} \mathrm{d}k_j\, \mathrm{e}^{-\mathrm{i}(\alpha k_j^2+2q_jk_j)} &= \sqrt{\frac{\pi}{\alpha}}\, \mathrm{e}^{\mathrm{i}q_j^2/\alpha-\mathrm{i}\pi/4} , \quad j=1,2,3 . \end{aligned} \tag{2.33}$$

For example, in the case of the analytically regularized integral of Fig. 2.1, we obtain

$$\begin{aligned} F_\Gamma(q;\lambda_1,\lambda_2) = {}& \frac{\mathrm{e}^{-\mathrm{i}\pi(\lambda_1+\lambda_2+1)/2}\pi^2}{\Gamma(\lambda_1)\Gamma(\lambda_2)} \int_0^\infty\int_0^\infty \mathrm{d}\alpha_1\,\mathrm{d}\alpha_2\, \frac{\alpha_1^{\lambda_1-1}\alpha_2^{\lambda_2-1}}{(\alpha_1+\alpha_2)^2} \\ & \times \exp\left(\mathrm{i}q^2\frac{\alpha_1\alpha_2}{(\alpha_1+\alpha_2)} - \mathrm{i}(m_1^2\alpha_1+m_2^2\alpha_2)\right) . \end{aligned} \tag{2.34}$$

After the change of variables $\eta = \alpha_1 + \alpha_2$, $\xi = \alpha_1/(\alpha_1+\alpha_2)$ and explicit integration over η, we arrive at

$$F_\Gamma(q;\lambda_1,\lambda_2) = \mathrm{e}^{\mathrm{i}\pi(\lambda_1+\lambda_2)} \frac{\mathrm{i}\pi^2 \Gamma(\lambda_1+\lambda_2-2)}{\Gamma(\lambda_1)\Gamma(\lambda_2)} \times \int_0^1 \mathrm{d}\xi \frac{\xi^{\lambda_1-1}(1-\xi)^{\lambda_2-1}}{[m_1^2\xi + m_2^2(1-\xi) - q^2\xi(1-\xi) - \mathrm{i}0]^{\lambda_1+\lambda_2-2}} . \quad (2.35)$$

Thus the UV divergence manifests itself through the first pole of the gamma function $\Gamma(\lambda_1+\lambda_2-2)$ in (2.35), which results from the integration over small values of η due to the power $\eta^{\lambda_1+\lambda_2-3}$.

Let us write down the representation (2.34) for the unregularized integral of Fig. 2.1 (i.e. for $\lambda_1 = \lambda_2 = 1$):

$$F_\Gamma(q) = \mathrm{i}\pi^2 \int_0^\infty \int_0^\infty \mathrm{d}\alpha_1\, \mathrm{d}\alpha_2\, (\alpha_1+\alpha_2)^{-2} \times \exp\left(\mathrm{i}q^2 \frac{\alpha_1\alpha_2}{\alpha_1+\alpha_2} - \mathrm{i}(m_1^2\alpha_1 + m_2^2\alpha_2)\right) . \quad (2.36)$$

This representation turns out to be very useful for the introduction of *dimensional* regularization, which is a commonly accepted computational technique successfully applied in practice and which will serve as the main kind of regularization in this book. Let us imagine that the number of space–time dimensions differs from four. To be more precise, the number of space dimensions is considered to be $d-1$, rather than three. (But, of course, we still think of an integer number of dimensions!) The derivation of the alpha representation does not change much in this case. The only essential change is that, instead of (2.32), we need to apply its generalization to an arbitrary number of dimensions, d:

$$\int \mathrm{d}^d k\, \mathrm{e}^{\mathrm{i}(\alpha k^2 + 2q\cdot k)} = \mathrm{e}^{\mathrm{i}\pi(1-d/2)/2} \pi^{d/2} \alpha^{-d/2} \mathrm{e}^{-\mathrm{i}q^2/\alpha} . \quad (2.37)$$

So, instead of (2.36), we have the following in d dimensions:

$$F_\Gamma(q;d) = \mathrm{e}^{-\mathrm{i}\pi(1+d/2)/2} \pi^{d/2} \int_0^\infty \int_0^\infty \mathrm{d}\alpha_1\, \mathrm{d}\alpha_2\, (\alpha_1+\alpha_2)^{-d/2} \times \exp\left(\mathrm{i}q^2 \frac{\alpha_1\alpha_2}{\alpha_1+\alpha_2} - \mathrm{i}(m_1^2\alpha_1 + m_2^2\alpha_2)\right) . \quad (2.38)$$

The only two places where something has been changed are the exponent of the combination $(\alpha_1+\alpha_2)$ in the integrand and the exponents of the overall factors.

Now, in order to introduce dimensional regularization, we want to consider the dimension d as a complex number. So, by definition, the dimensionally regularized Feynman integral for Fig. 2.1 is given by (2.38) and is a function of q^2 as given by this integral representation. We choose $d = 4-2\varepsilon$, where the value $\varepsilon = 0$ corresponds to the physical number of the space–time dimensions. By the same change of variables as used after (2.34), we obtain

$$F_\Gamma(q;d) = \mathrm{e}^{-\mathrm{i}\pi(1+d/2)/2}\pi^{d/2} \int_0^\infty \mathrm{d}\eta\, \eta^{\varepsilon-1} \int_0^1 \mathrm{d}\xi \\ \times \exp\left\{\mathrm{i}q^2\xi(1-\xi)\eta - \mathrm{i}[m_1^2\xi + m_2^2(1-\xi)]\eta\right\} . \quad (2.39)$$

This integral is absolutely convergent for $0 < \mathrm{Re}\,\varepsilon < \Lambda$ (where $\Lambda = \infty$ if both masses are non-zero and $\Lambda = 1$ otherwise; this follows from an IR analysis of convergence, which we omit here) and defines an analytic function of ε, which is extended from this domain to the whole complex plane as a meromorphic function.

After evaluating the integral over η, we arrive at the following result:

$$F_\Gamma(q;d) = \mathrm{i}\pi^{d/2}\Gamma(\varepsilon) \int_0^1 \frac{\mathrm{d}\xi}{\left[m_1^2\xi + m_2^2(1-\xi) - q^2\xi(1-\xi) - \mathrm{i}0\right]^\varepsilon} . \quad (2.40)$$

The UV divergence manifests itself through the first pole of the gamma function $\Gamma(\varepsilon)$ in (2.40), which results from the integration over small values of η in (2.39).

This procedure of introducing dimensional regularization is easily generalized [28, 35, 61] to an arbitrary Feynman integral. We first follow the standard derivation (see, e.g., [181]) of the alpha representation (in an integer number of dimensions d) using (2.31), taking dh-dimensional Gauss integrals by means of a generalization of (2.37) to the case of an arbitrary number of loop integration momenta:

$$\int \mathrm{d}^d k_1 \ldots \mathrm{d}^d k_h \exp\left[\mathrm{i}\left(\sum_{i,j} A_{ij} k_i \cdot k_j + 2\sum_i q_i \cdot k_i\right)\right] \\ = \mathrm{e}^{\mathrm{i}\pi h(1-d/2)/2}\pi^{hd/2}(\det A)^{-d/2} \exp\left[-\mathrm{i}\left(\sum_{i,j} A_{ij}^{-1} q_i \cdot q_j\right)\right] . \quad (2.41)$$

Here A is an $h \times h$ matrix and A^{-1} its inverse.[1]

The elements of the inverse matrix involved here are rewritten in graph-theoretical language (see details in [181, 27]), and the resulting alpha representation takes the form [28, 35]

$$F_\Gamma(q_1,\ldots,q_n;d) = (-1)^a \frac{\mathrm{e}^{\mathrm{i}\pi[a+h(1-d/2)]/2}\pi^{hd/2}}{\prod_l \Gamma(a_l)} \\ \times \int_0^\infty \mathrm{d}\alpha_1 \ldots \int_0^\infty \mathrm{d}\alpha_L \prod_l \alpha_l^{a_l-1} D^{-d/2} Z \mathrm{e}^{\mathrm{i}A/D - \mathrm{i}\sum m_l^2 \alpha_l} , \quad (2.42)$$

[1] In fact, the matrix A involved here equals $e\beta e^+$ with the elements of an arbitrarily chosen column and row with the same number deleted. Here e is the incidence matrix of the graph, i.e. $e_{il} = \pm 1$ if the vertex i is the beginning/end of the line l, e^+ is its transpose and β consists of the numbers $1/\alpha_l$ on the diagonal – see, e.g., [181].

where $a = \sum a_l$, and D and A are the well-known functions (sometimes called Symanzik polynomials)

$$D = \sum_{T \in T^1} \prod_{l \notin T} \alpha_l \,, \tag{2.43}$$

$$A = \sum_{T \in T^2} \prod_{l \notin T} \alpha_l \left(q^T\right)^2 \,. \tag{2.44}$$

In (2.43), the sum runs over *trees* of the given graph, i.e. maximal connected subgraphs without loops, and, in (2.44), over *2-trees*, i.e. subgraphs that do not involve loops and consist of two connectivity components; $\pm q^T$ is the sum of the external momenta that flow into one of the connectivity components of the 2-tree T. (It does not matter which component is taken because of the conservation law for the external momenta.) The products of the alpha parameters involved are taken over the lines that do not belong to the given tree T.

The factor Z is responsible for the non-scalar structure of the diagram:

$$Z = \prod_l Z_l \left(\frac{1}{2\mathrm{i}} \frac{\partial}{\partial u_l} \right) \mathrm{e}^{\mathrm{i}(2B-K)/D} \Bigg|_{u_1 = \ldots u_L = 0} \,, \tag{2.45}$$

where (see, e.g., [246, 208])

$$B = \sum_l \sum_{T \in T_l^1} q_T \prod_{l \notin T} \alpha_l \,, \tag{2.46}$$

$$K = \sum_{T \in T^0} \prod_{l \notin T} \alpha_l \left(\sum_l \pm u_l \right)^2 \,. \tag{2.47}$$

In (2.46), the sum is taken over trees T_l^1 that include a given line l, and q_T is the total external momentum that flows through the line l (in the direction of its orientation). In (2.47), the sum is taken over pseudotrees T^0 (a *pseudotree* is obtained from a tree by adding a line), and the sum in l is performed over the loop (circuit) of the pseudotree T, with a sign dependent on the coincidence of the orientations of the line l and the pseudotree T.

The alpha representation of a general h-loop Feynman integral is useful for general analyses. In practical calculations, e.g. at the two-loop level, one can derive the alpha representation for concrete diagrams by hand, rather than deduce it from the general formulae presented above. Still, even in practice, such general formulae can provide advantages because the evaluation of the functions of the alpha representation can be performed on a computer. It will be described later how non-scalar Feynman integrals can be evaluated with the help of the general formulae.

Let us stress that this terrible-looking machinery for evaluating the determinant of the matrix A that arises from Feynman integrals, as well as for evaluating the elements of the inverse matrix, together with interpreting these results from the graph-theoretical point of view, is exactly the same as that used in the problem of solution of Kirchhoff's laws, a problem typical of the nineteenth century. Recall, for example, that the parameters α_l play the role of ohmic resistances and that the expression (2.43) for the function D as a sum over trees is a Kirchhoff result.

Let us now define[2] the *dimensionally regularized* Feynman integral by means of (2.42), treating the quantity d as a complex number. This is a function of kinematical invariants constructed from the external momenta and contained in the function A. In addition to this, we have to take care of polynomials in the external momenta and the auxiliary variables u_l hidden in the factor Z. We treat these objects q_i and u_l, as well as the metric tensor $g_{\mu\nu}$, as elements of an algebra of covariants, where we have, in particular,

$$\left(\frac{\partial}{\partial u_l^\mu}\right) u_{l'}^\nu = g_\mu^\nu \delta_{l,l'} \;, \quad g_\mu^\mu = d \;.$$

This algebra also includes the γ-matrices with anticommutation relations $\gamma_\mu\gamma_\nu + \gamma_\nu\gamma_\mu = 2g_{\mu\nu}$ so that $\gamma^\mu\gamma_\mu = d$, the tensor $\varepsilon_{\kappa\mu\nu\lambda}$, etc.

Thus the dimensionally regularized Feynman integrals are defined as linear combinations of tensor monomials in the external momenta and other algebraic objects with coefficients that are functions of the scalar products $q_i \cdot q_j$. However, this is not all, because we have to see that the α-integral is well-defined. Remember that it can be divergent, for various reasons. Let us consider first a Feynman diagram with *Euclidean* external momenta q_i, i.e. when

$$\left(\sum_{i\in I} q_i\right)^2 < 0 \tag{2.48}$$

for any subset I of external lines. This is the domain where the Feynman integral does not have singularities as a function of the external momenta. We exclude, in particular, on-shell and threshold configurations. Therefore we can encounter only off-shell IR divergences, which have a deep analogy with the UV ones.

[2] An alternative definition of algebraic character [242, 221] (see also [75]) exists and is based on certain axioms for integration in a space with non-integer dimension. It is unclear how to perform analysis within such a definition, for example, how to apply the operations of taking a limit, differentiation, etc. to algebraically defined Feynman integrals in d dimensions, in order to say something about the analytic properties with respect to momenta and masses and the parameter of dimensional regularization. After evaluating a Feynman integral according to the algebraic rules, one arrives at some concrete function of these parameters but, *before integration*, one is dealing with an abstract algebraic object. Let us remember, however, that, in practical calculations, one usually does not bother about precise definitions. From the purely pragmatic point of view, it is useless to think of a diagram when it is not calculated. On the other hand, from the theoretical point of view, such a position is beneath criticism.

The alpha representation is a very convenient tool for the analysis of convergence of Feynman integrals. This analysis is presented in detail in Appendix B.1; here only some of the steps and conclusions are presented. The UV divergences come from the region of small values of the α-parameters (2.42), where the factor $D^{\varepsilon-2}$ and the negative integer powers of the function D arising from the non-scalar structure of the diagram are singular, while the off-shell IR divergences arise from the integration over large α_l. The alpha representation is a very convenient tool for studying the divergences by resolving the structure of the singularities of the integrand. The standard way to perform the analysis of the UV convergence is to divide the integration region into so-called 'sectors'. These sectors can be of the type $\alpha_1 \leq \ldots \leq \alpha_L$ (applied, for example, in [137, 219]) plus similar sectors obtained by permutation of the indices, or more complicated sectors associated with 1PI subgraphs [194, 222]. Then, in each sector, one introduces new integration variables in which the singularities of the integrand factorize, i.e. the integral takes the form of a product of some powers of the sector variables with a non-zero function. Using this analysis, the above mentioned statement about the UV convergence of a Feynman diagram with negative degrees of divergence of all subgraphs can be proven in a natural way.

Suppose that all the masses are non-zero so that the diagram is IR finite. In this case we can choose a sufficiently large real part of ε to make the α-integral absolutely convergent. For such values of the regularization parameter, the alpha representation defines the dimensionally regularized Feynman integral as an analytic function of ε. This integral is uniquely extended from this domain to the whole complex plane as a meromorphic function of ε. (In other words, the only possible singularities of this function are poles.)

If there are zero masses we need IR power counting. To perform simultaneously the analysis of UV and IR convergence one uses more general decomposition of the α-integral and a subsequent introduction of sector variables [220, 222, 194, 208]. Here the similarity of UV and IR (off-shell) properties plays an important role. Eventually, in the new variables, the integrand is factorized and the analysis of convergence reduces to power counting (for both UV and IR convergence) in one-dimensional integrals.

This analysis of the alpha representation shows that if there are both UV and IR divergences in a diagram then, for any ε, the integral is divergent. In this case, we can introduce an auxiliary analytic regularization into all the lines. The Feynman integrals, regularized both analytically and dimensionally, are obtained from the dimensionally regularized expressions by inserting factors $\mathrm{i}^{\lambda_l}\alpha^{\lambda_l-1}/\Gamma(\lambda_l)$ for each line. Let us assume one more condition: that the diagram does not have massless detachable subgraphs. (According to a 'physical' definition, a subgraph is *detachable* if its external momenta are zero. For example, a *tadpole*, i.e. a line with coincident end points, is a detachable subgraph.) A general theorem [222] says that if a diagram without massless detachable subgraphs is considered for Euclidean values of the external momenta there exists a non-empty domain of parameters $(\varepsilon; \lambda_1, \ldots, \lambda_L)$ where the alpha representation is absolutely convergent and defines an analytic function of these parameters. The dimensionally regularized Feynman integral is then by definition [61] obtained as the analytic continuation of this function to the point $(\varepsilon; 1, \ldots, 1)$ where the analytic regularization is switched off.

And what does one do with the massless detachable subdiagrams? To provide a definition for them, let us observe that we can separate the UV and IR contributions of every propagator by dividing the integration region for each α_l into UV and IR regions, $0 \leq \alpha_l \leq \mu^{-2}$ and $\mu^{-2} \leq \alpha_l < \infty$. The α-parametric integral is then divided into 2^L pieces, which can be characterized by subsets of UV contributions (for example). For a given subset, the corresponding contribution can certainly be made into an absolutely convergent integral over α_l by choosing the real parts of the regularization parameters λ_l of the UV/IR lines sufficiently large/small. Then each piece is defined as an analytic function of the regularization parameters. Collecting all the pieces extended to the point $(\varepsilon; 1, \ldots, 1)$ gives, by definition [61], the dimensionally regularized Feynman integral (still considered for Euclidean external momenta) for an *arbitrary* graph.

2.4 Properties of Dimensionally Regularized Feynman Integrals

We can formally write down dimensionally regularized Feynman integrals as integrals in d-dimensional vectors k_i:

$$F_\Gamma(q_1, \ldots, q_n; d) = \int \mathrm{d}^d k_1 \ldots \int \mathrm{d}^d k_h \prod_{l=1}^{L} \tilde{D}_l(p_l) \,. \tag{2.49}$$

In order to obtain dimensionally regularized integrals with their dimension independent of ε, a factor of $\mu^{-2\varepsilon}$ per loop, where μ is a massive parameter, is introduced. This parameter serves as a renormalization parameter for schemes based on dimensional regularization. However, we shall usually omit these factors for brevity.

We have reasons for using the notation (2.49), because dimensionally regularized Feynman integrals as defined above possess the standard properties of integrals of the usual type in integer dimensions. (We still assume that the external momenta are Euclidean.) In particular,

- the integral of a linear combination of integrands equals the same linear combination of the corresponding integrals;
- one may contract the same factors in the numerator and denominator of integrands.

These properties follow directly from the above definition. A less trivial property is that

- a derivative of an integral with respect to a mass or momentum equals the corresponding integral of the derivative.

This is also a consequence (see [61, 208]) of the above definition. To prove this statement, one uses standard algebraic relations between the functions entering the alpha representation [181, 35]. (We note again that these are relations quite similar to those encoded in the solutions of Kirchhoff's laws for a circuit defined by the given graph.) A corollary of the last property is the possibility of integrating by parts and always neglecting surface terms:

- $$\int \mathrm{d}^d k_1 \ldots \int \mathrm{d}^d k_h \frac{\partial}{\partial k_i^\mu} \prod_{l=1}^{L} \tilde{D}_l(p_l) = 0 \;, \quad i = 1, \ldots, h \;. \tag{2.50}$$

This property is the basis of a powerful method of evaluation of Feynman integrals – see the next section.

The next property says that

- any diagram with a detachable massless subgraph is zero.

Setting massless detachable subdiagrams to zero turns out to be a specific case of a more general prescription (which we shall discuss in the next chapter and use throughout the book).

This property can also be shown to be a consequence of the above definition [61, 208], by use of an auxiliary analytic regularization using pieces of the α-integral considered in different domains of the regularization parameters. Let us consider, for example, the massless tadpole diagram, which can be reduced by means of alpha parameters to a scaleless one-dimensional integral:

$$\int \frac{\mathrm{d}^d k}{k^2} = -\mathrm{i}^\varepsilon \pi^{d/2} \int_0^\infty \mathrm{d}\alpha \, \alpha^{\varepsilon - 2} \;. \tag{2.51}$$

We divide this integral into two pieces, from 0 to 1 and from 1 to ∞, integrate these two integrals and find results that are equal except for opposite signs, which lead to the zero value. (These arguments can be found, for example, in [163].) It should be stressed here that the two pieces that contribute to the right-hand side of (2.51) are convergent in *different* domains of the regularization parameter ε, namely, $\mathrm{Re}\,\varepsilon > -1$ and $\mathrm{Re}\,\varepsilon < -1$, with no intersection, and that this procedure here is equivalent to introducing analytic regularization and considering its parameter in different domains for different pieces.

But let us distinguish between two qualitatively different situations: the first when we have to deal with a massless Feynman integral which arises from the Feynman rules, and the second when we obtain such integrals after some manipulations: after using partial fractions, differentiation, integration by parts, etc. We also include in this second class all such integrals that appear on the right-hand side of explicit formulae for (off-shell) asymptotic expansions in momenta and masses. In the first situation, the only possibility is to use the ad hoc prescription of setting the integral to zero. In the second situation, we start with an α-integral which is convergent in some non-empty domain of the regularization parameters. Then we can split the α-integral into our 2^L pieces and improve, in advance, its convergence as much as possible (increase $\mathrm{Re}\,\lambda_l$ for the UV α_l and decrease $\mathrm{Re}\,\lambda_l$ for the IR α_l). A very important point here is that all the properties of dimensionally regularized integrals given above, apart from the last one, can be justified in a purely algebraic way [61, 208], through identities between functions in the alpha representation. Therefore, we may safely set to zero all the resulting massless Feynman integrals with zero external momenta and need not bother about the subtlety of the non-existence of an appropriate domain of the regularization parameters for them. We shall come back later to a discussion of this point when we consider scaleless integrals that arise on the right-hand side of asymptotic expansions.

Let us now remind ourselves of reality and observe that it is necessary to deal in practice with diagrams on a mass shell or at a threshold. What about the properties of dimensionally regularized Feynman integrals in this case? At least the algebraic proof of the basic properties of dimensionally regularized Feynman integrals is not sensitive to putting the external momenta in any particular place. However, the bad news here is that a general analysis of the convergence of such integrals, even in specific cases, is still absent,[3] so that we do not have any control over convergence. Technically, this means that the sectors used for the analysis of the convergence in the off-shell case are no longer sufficient for the resolution of the singularities of the integrand of the alpha representation. These singularities are much more complicated and can even appear (e.g. at a threshold) at non-zero, finite values of the α-parameters. Thus we cannot guarantee that the standard analytic regularization can be used to improve convergence.

However, the good news is that numerous practical applications have shown that there is no sign of breakdown of these properties for on-shell or threshold Feynman integrals. In particular, when the Feynman integrals are evaluated on a mass shell or at a threshold, the convergence can be put under control. We now switch to various methods of evaluation of Feynman integrals and then come back to a discussion of Feynman integrals on a mass shell or at a threshold.

[3] We should, however, mention [23], where an algorithm for resolving singularities in ε for any *concrete* diagram in the situation where all kinematical invariants have the same sign has been developed. There is no graph-theoretical interpretation of the desired kind in this algorithm and it has not been formulated for a *general*, rather than a specific diagram. This algorithm has been successfully applied to numerical evaluation of multiloop diagrams with strong IR and collinear divergences both for checks of analytic results [230, 215] and when no analytic results were available.

2.5 How They Are Evaluated

There exists a great variety of methods for evaluation of Feynman integrals. Here is a brief summary of them, with special attention to the methods that are applied in calculations in this book.

Let us note first that the asymptotic expansions which are the subject of the book can be regarded, in a narrow sense, as a method of evaluation of Feynman diagrams. Still, even when we apply an asymptotic expansion and substitute a given Feynman integral by the first terms of its expansion, it is necessary to compute these terms. As we shall see later, the functions that are encountered on the right-hand side of the expansions are nothing but Feynman integrals obtained from the primary Feynman integrals by Taylor expansions of their integrands with respect to certain sets of variables. It turns out that one can apply the same techniques as used for the usual integrals to the evaluation of such more general integrals.

2.5.1 Alpha Representation as a Method of Evaluation

The alpha representation itself can serve as a method of evaluating Feynman integrals. Let us evaluate, for example, the dimensionally regularized massless Feynman diagram of Fig. 2.1 with general powers of propagators,

$$F_\Gamma(q;\lambda_1,\lambda_2,d) = \int \frac{\mathrm{d}^d k}{(-k^2+m_1^2)^{\lambda_1}[-(q-k)^2+m_2^2]^{\lambda_2}} \,. \tag{2.52}$$

From now on, we shall use the following convention: when powers of propagators are integers we use them with $+k^2+\mathrm{i}0$, but when they are non-integral or complex, we take the opposite sign, i.e. $-k^2-\mathrm{i}0$. The second choice is more natural if we wish to obtain a Euclidean, $-q^2$, dependence of the results (see, e.g., (2.54)). We also prefer to use a_l for integer and λ_l for general complex indices. Starting from its alpha representation, we obtain the analytically regularized version of (2.40), i.e.

$$F_\Gamma(q;\lambda_1,\lambda_2,d) = \mathrm{i}\pi^{d/2}\Gamma(\lambda_1+\lambda_2+\varepsilon-2) \times \int_0^1 \frac{\mathrm{d}\xi\, \xi^{\lambda_1}(1-\xi)^{\lambda_2}}{[m_1^2\xi+m_2^2(1-\xi)-q^2\xi(1-\xi)-\mathrm{i}0]^{\lambda_1+\lambda_2+\varepsilon-2}} \,. \tag{2.53}$$

Suppose that the masses are zero. In this case the integral over ξ can be evaluated in terms of gamma functions, and we arrive at the following result:

$$\int \frac{\mathrm{d}^d k}{(-k^2)^{\lambda_1}[-(q-k)^2]^{\lambda_2}} = \mathrm{i}\pi^{d/2}\frac{G(\lambda_1,\lambda_2)}{(-q^2)^{\lambda_1+\lambda_2+\varepsilon-2}} \,, \tag{2.54}$$

where

$$G(\lambda_1,\lambda_2) = \frac{\Gamma(\lambda_1+\lambda_2+\varepsilon-2)\Gamma(2-\varepsilon-\lambda_1)\Gamma(2-\varepsilon-\lambda_2)}{\Gamma(\lambda_1)\Gamma(\lambda_2)\Gamma(4-\lambda_1-\lambda_2-2\varepsilon)} \,. \tag{2.55}$$

In the case where the powers of propagators are equal to one, we have

$$\int \frac{\mathrm{d}^d k}{k^2(q-k)^2} = \mathrm{i}\pi^{d/2} \frac{\Gamma(\varepsilon)\Gamma(1-\varepsilon)^2}{\Gamma(2-2\varepsilon)(-q^2)^{\varepsilon}} \; . \tag{2.56}$$

Note that although the indices of the diagrams are integral at the beginning, non-integral indices shifted by amounts proportional to ε appear after intermediate integration, e.g. after the use of (2.56).

Let us present another example of evaluation of Feynman diagrams by means of alpha parameters: consider Fig. 2.1 with $m_1 = 0$, $m_2 = m$, $a_1 = 1$, $a_2 = 2$. We shall use this example in the next chapter to illustrate prescriptions for asymptotic expansions. Using (2.53), we have

$$F_\Gamma(q;d) = -\mathrm{i}\pi^{d/2}\Gamma(1+\varepsilon) \int_0^1 \frac{\mathrm{d}\xi \, \xi^{-\varepsilon}}{[m^2 - q^2(1-\xi) - \mathrm{i}0]^{1+\varepsilon}} \; . \tag{2.57}$$

In fact, the initial Feynman integral is finite. If we want to evaluate it for $d = 4$, we put $\varepsilon = 0$ in (2.57) and easily obtain an explicit result,

$$F_\Gamma(q) = \mathrm{i}\pi^2 \frac{\ln\left(1 - q^2/m^2\right)}{q^2} \; . \tag{2.58}$$

Suppose we now want to perform evaluation for general d. There could be various reasons for doing this. First, when using the method of integration by parts (see below), one obtains expressions with coefficients involving negative powers of ε. Second, when one performs renormalization, insertion of counterterms also generates diagrams with coefficients singular in ε. In any case, it is often necessary to know not only the pole and the finite part of a diagram but also some initial terms of its Laurent expansion in ε.

Starting from (2.57) and referring to [195] or evaluating this expression by computer, e.g. in MATHEMATICA [243], we see (on a page of [195] or on the screen of the computer) an explicit result:

$$F_\Gamma(q;d) = -\mathrm{i}\pi^{d/2} \frac{\Gamma(1+\varepsilon)}{(1-\varepsilon)(m^2)^{1+\varepsilon}} {}_2F_1\left(1, 1+\varepsilon; 2-\varepsilon; q^2/m^2\right) \; , \tag{2.59}$$

where ${}_2F_1(a,b;c;z)$ is the Gauss hypergeometric function [103] (see (A.52)) so that we have obtained a result in the form of a series in q^2/m^2.

A lot of one-loop integration formulae can be derived by use of the alpha representation and subsequent straightforward integration. A collection of such formulae is presented in Appendix A.1.

2.5.2 Recursively One-Loop Feynman Integrals

Massless integrals are often evaluated with the help of successive application of the one-loop formula (2.54). Consider, for example, the two-loop diagram shown in Fig. 2.4. The internal one-loop integral can be evaluated by use of (2.56) and is effectively replaced by a line with index ε. Then the sequence of two lines with indices 1 and ε is replaced by one line with index $1+\varepsilon$, and the one-loop diagram so obtained, which has indices 2 and $1+\varepsilon$, is

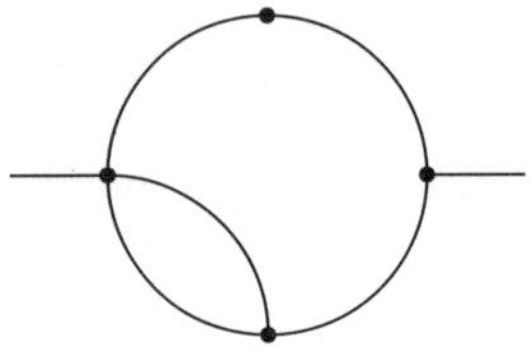

Fig. 2.4. A recursively one-loop diagram

evaluated by means of the one-loop formula (2.54), with the following result expressed in terms of gamma functions: $G(1,1)G(2,1+\varepsilon)/(-q^2)^{1+2\varepsilon}$. The class of Feynman diagrams that can be evaluated in this way by means of (2.54) is called *recursively one-loop.*

2.5.3 Partial Fractions

When evaluating dimensionally regularized Feynman integrals one uses their properties, in particular the possibility of manipulations based on the first two of the properties listed earlier. Here the following standard decomposition proves to be useful:

$$\frac{1}{(x+x_1)^{a_1}(x+x_2)^{a_2}} = \sum_{i=0}^{a_1-1} \binom{a_2-1+i}{a_2-1} \frac{(-1)^i}{(x_2-x_1)^{a_2+i}(x+x_1)^{a_1-i}} + \sum_{i=0}^{a_2-1} \binom{a_1-1+i}{a_1-1} \frac{(-1)^{a_1}}{(x_2-x_1)^{a_1+i}(x+x_2)^{a_2-i}} , \qquad (2.60)$$

where

$$\binom{n}{j} = \frac{n!}{j!(n-j)!}$$

is a binomial coefficient.

2.5.4 Feynman Parameters

Let us now present the alpha representation of scalar dimensionally regularized integrals in a modified form by making the change of variables $\alpha_l = \eta\alpha_l'$, where $\sum \alpha_l' = 1$. Performing the integration over η from 0 to ∞ explicitly and omitting primes from the new variables, we obtain

$$F_\Gamma(q_1,\ldots,q_n;d) = (-1)^a \frac{\left(\mathrm{i}\pi^{d/2}\right)^h \Gamma(a-hd/2)}{\prod_l \Gamma(a_l)} \times \int_0^\infty \mathrm{d}\alpha_1 \ldots \int_0^\infty \mathrm{d}\alpha_L\, \delta\left(\sum \alpha_i - 1\right) \frac{D^{a-(h+1)d/2} \prod_l \alpha_l^{a_l-1}}{\left(A - D\sum m_l^2 \alpha_l\right)^{a-hd/2}} . \qquad (2.61)$$

As is well known [45] (see also [24]), one can choose the sum of an arbitrary subset of $\alpha_l\,, i = l,\ldots,L$, in the argument of the delta function in (2.61). This choice can greatly simplify the evaluation.

In addition to alpha parameters, the closely related Feynman parameters can also be used. For a product of two propagators, one writes down the following relation:

$$\frac{1}{(m_1^2 - p_1^2)^{\lambda_1}(m_2^2 - p_2^2)^{\lambda_2}} = \frac{\Gamma(\lambda_1+\lambda_2)}{\Gamma(\lambda_1)\Gamma(\lambda_2)} \int_0^1 \frac{\mathrm{d}\xi\, \xi^{\lambda_1-1}(1-\xi)^{\lambda_2-1}}{[(m_1^2-p_1^2)\xi + (m_2^2-p_2^2)(1-\xi)]^{\lambda_1+\lambda_2}} \,. \tag{2.62}$$

This relation is usually applied to a pair of appropriately chosen propagators if an explicit integration over a loop momentum then becomes possible. Then new Feynman parameters for other factors in the integral can be chosen, etc. In fact, any choice of the Feynman parameters can be achieved by starting from the alpha representation and making certain changes of variables. However, the possibility of an intermediate explicit loop integration of the kind mentioned above can be hidden in the alpha integral.

The generalization of (2.62) to an arbitrary number of propagators is of the form

$$\frac{1}{\prod A_i^{\lambda_i}} = \frac{\Gamma(\sum \lambda_i)}{\prod \Gamma(\lambda_i)} \int_0^1 \mathrm{d}\xi_1 \ldots \int_0^1 \mathrm{d}\xi_L \prod_l \xi_l^{\lambda_l - 1} \frac{\delta\left(\sum \xi_i - 1\right)}{(\sum A_i \xi_i)^{\sum \lambda_i}} \,, \tag{2.63}$$

where $A_i = m_i^2 - p_i^2$.

For the evaluation of diagrams with a small number of loops, the choice of applying either alpha or Feynman parameters is usually just a matter of taste. In particular, if we apply (2.62) to a two-loop diagram and then integrate over two loop momenta, with the help of (A.1) and its generalizations to integrals with numerators, we obtain the same result as that obtained starting from (2.61).

2.5.5 Reduction of Tensor Integrals

A lot of one-loop Feynman integrals with numerators can be found in Appendix A.1. One can use these tabulated integrals for traceless monomials, e.g. (A.8). Sometimes it is reasonable to convert general monomials to traceless monomials or vice versa. Here formulae (A.40a) and (A.40b) are of use.

In the case of a general h-loop Feynman integral, let us observe that the function (2.45) can be taken into account by shifting the space–time dimension d and indices a_l of a given diagram, because any factor that arises after the differentiation with respect to the auxiliary parameters u_l is a sum of products of positive integer powers of the α-parameters and negative integer powers of the function D. In particular, the factor $1/D^n$ is taken into account by the shift $d \to d + 2n$. This observation enables us to express any given Feynman integral with numerators through a linear combination of scalar integrals with shifted indices and shifted dimensions. At the one-loop level, this property has been used in [88] to derive general formulae for integrals with numerators. Systematic algorithms oriented towards realization on a computer, with a demonstration up to two-loop level, have been constructed in [228].

2.5.6 Integration by Parts

The next method in our list is based on integration by parts[4] (IBP) [68] within dimensional regularization, i.e. (2.50). The idea is to write down various equations (2.50) and obtain a set of relations between Feynman integrals. These relations can then be used for two purposes:

(a) to reduce any Feynman integral from a given family to a set of 'master integrals', i.e. with the lowest powers of propagators, in addition to integrals that are easily evaluated, for example, in terms of gamma functions;
(b) to evaluate the master integrals.

If both parts of this programme are fulfilled then the evaluation of the given family of integrals reduces to algebraic manipulations and substitutions of values of the master and 'boundary' integrals.

Let us illustrate how the IBP method works using the massless diagram of Fig. 2.5 with arbitrary integer powers of the propagators:

$$J(a_1, a_2, a_3, a_4, a_5) = \int\int \frac{\mathrm{d}^d k\,\mathrm{d}^d l}{(k^2)^{a_1}[(q-k)^2]^{a_2}(l^2)^{a_3}[(q-l)^2]^{a_4}[(k-l)^2]^{a_5}} \,. \tag{2.64}$$

First we note that if $a_5 = 0$ the integrals in k and l decouple and can be evaluated in terms of gamma functions by use of (2.54). When some other index a_l is zero, the integral can also be evaluated in terms of gamma functions by successively applying the same one-loop formula. When some index is negative, a generalization of this formula works.

Suppose now that all the indices are positive integers. Let us write down the following IBP identity [68]:

$$\int\int \frac{\mathrm{d}^d k\,\mathrm{d}^d l}{(l^2)^{a_3}[(q-l)^2]^{a_4}} \frac{\partial}{\partial k_\mu}\left(\frac{k_\mu}{(k^2)^{a_1}[(q-k)^2]^{a_2}[(k-l)^2]^{a_5}}\right) = 0 \,. \tag{2.65}$$

Taking derivatives and recognizing terms on the left-hand side as integrals (2.64), we arrive at the following relation:

$$(a_1 + a_2 + 2a_5 - d)J = \left[a_1\mathbf{1}^+\left(\mathbf{3}^- - \mathbf{5}^-\right) + a_2\mathbf{2}^+\left(\mathbf{4}^- - \mathbf{5}^-\right)\right] J \,, \tag{2.66}$$

where the standard notation for increasing and lowering operators has been used, e.g. $\mathbf{1}^+\mathbf{3}^- J(a_1, a_2, a_3, a_4, a_5) = J(a_1+1, a_2, a_3-1, a_4, a_5)$. Equation

[4] For one loop, IBP was used in [146]. A crucial step – an appropriate modification of the *integrand* before differentiation, with an application at the two-loop level (to massless propagator diagrams) — was taken in [68] and, in a coordinate-space approach, in [235]. The case of three-loop massless propagators has been treated in [68].

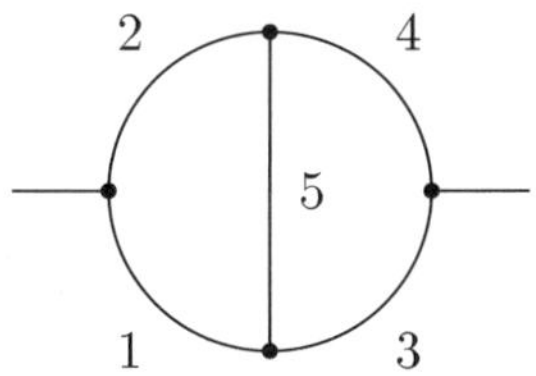

Fig. 2.5. The master two-loop self-energy diagram

(2.66) can be used as a recurrence relation for the given family of integrals because, on the right-hand side, we encounter integrals where the sum $a_3 + a_4 + a_5$ is less by one than that on the left-hand side. Thus, successive application of this relation reduces any given integral to integrals with some index equal to zero.

We observe at this point that (2.66) gives a solution to both problems (a) and (b) simultaneously. For example, the master integral, i.e. the integral with all indices equal to one, is evaluated by means of the same equation (2.66):

$$J(1,1,1,1,1) = \frac{1}{\varepsilon}\left[J(2,1,0,1,1) - J(2,1,1,1,0)\right] . \tag{2.67}$$

The Feynman integral $J(2,1,1,1,0)$ is recursively one-loop and has already been considered in Sect. 2.5.3, and $J(2,1,0,1,1)$ is a product of two one-loop integrals that can be evaluated by means of (2.54). After expanding the gamma functions as Laurent series in ε, we arrive at the well-known result [198, 53]

$$\begin{aligned} J(1,1,1,1,1) &= \frac{1}{\varepsilon}G(1,1)\left[G(2,1) - G(2,1+\varepsilon)\right]\frac{1}{(-q^2)^{1+2\varepsilon}} \\ &= \left(\mathrm{i}\pi^2\right)^2 \frac{6\zeta(3)}{q^2} + O(\varepsilon) . \end{aligned} \tag{2.68}$$

In this simple example, it was sufficient to use only one IBP relation, which, in fact, follows from an IBP identity for the triangle massless diagram of Fig. 2.6 with general indices. This 'triangle' rule takes the form

$$1 = \frac{1}{d - a_1 - a_2 - 2a_3}\left[a_1\mathbf{1}^+\left(\mathbf{3}^- - p_1^2\right) + a_2\mathbf{2}^+\left(\mathbf{3}^- - p_2^2\right)\right] . \tag{2.69}$$

When applied to the left triangle in Fig. 2.5 it provides (2.66).

Generally, one tries to use all possible IBP relations. For example, for a two-loop Feynman integral over the loop momenta k and l depending on n external momenta q_i, all possible IBP relations for derivatives $(\partial/\partial k_\mu)p_\mu$ and $(\partial/\partial l_\mu)p_\mu$ are used, where $p = k, l, q_i$.

Step (a), i.e. reduction of a Feynman integral from a given family to master integrals, can be a rather non-trivial problem even at the two-loop level. Recently, there have been some attempts to make this step systematic. One of them is based on the possibility, mentioned in the previous subsection,

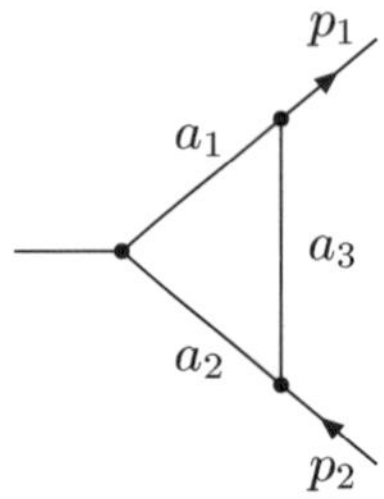

Fig. 2.6. Triangle diagram with general integer indices

of reducing any integral with a numerator to scalar integrals with shifted dimension and indices. As was demonstrated in [228], this property can provide a more general set of recurrence relations. Applications of this technique of shifting dimensions for one and two loops have been presented in [228, 229].

Another new approach [122] applies so-called Lorentz invariance identities together with IBP relations. This method is primarily oriented towards Feynman integrals with four or more external lines and is based on the fact that when the total dimension of the denominator and numerator in the Feynman integrals associated with a given graph is increased the total number of IBP and Lorentz invariance equations grows faster than the number of independent Feynman integrals (labelled by the powers of propagators and the powers of independent scalar products in the numerators). Therefore this system of equations sooner or later becomes overconstrained, and one obtains a possibility of performing a reduction to master integrals. This method has been successfully applied to reduction of massless double box diagrams with one leg off shell [123].

The IBP relations are consequences of the property (2.50), which is valid at least for off-shell Feynman integrals. Let us now see how the IBP method works for an on-shell Feynman integral with two scales. Consider the triangle diagram shown in Fig. 2.6, with $m_3 = 0$, $p_1^2 = m_1^2$, $p_2^2 = m_2^2$ and $(p_1 - p_2)^2 = 0$:

$$F_\Gamma(m_1, m_2; d) = \int \frac{\mathrm{d}^d k}{(k^2 - 2p_1\cdot k)(k^2 - 2p_2\cdot k)k^2} \,. \tag{2.70}$$

Using the IBP identity

$$\int \mathrm{d}^d k \frac{\partial}{\partial k_\mu} \left(\frac{k_\mu}{(k^2 - 2p_1\cdot k)(k^2 - 2p_2\cdot k)k^2} \right) = 0 \tag{2.71}$$

and changing the loop integration momentum, we obtain

$$F_\Gamma(m_1, m_2; d) = \frac{1}{2\varepsilon} \left(\int \frac{\mathrm{d}^d k}{(k^2 - m_1^2)^2 (k^2 - m_2^2)} + \int \frac{\mathrm{d}^d k}{(k^2 - m_1^2)(k^2 - m_2^2)^2} \right) \,. \tag{2.72}$$

The integrals on the right-hand side are then evaluated by means of partial fractions (2.60) and the integral (A.1), with the following simple result:

$$F_\Gamma(m_1, m_2; d) = -\mathrm{i}\pi^{d/2} \frac{\Gamma(\varepsilon)}{2\varepsilon} \frac{(m_1^2)^{-\varepsilon} - (m_2^2)^{-\varepsilon}}{m_1^2 - m_2^2} . \tag{2.73}$$

Let us stress that the method of IBP is successfully applied in practice, without any hesitation, not only to off-shell Feynman integrals, where the convergence is certainly under control, but also to Feynman integrals on the mass shell or at threshold. Examples where IBP breaks down are unknown at present. In particular, we shall systematically apply the method in what follows to integrals that appear on the right-hand side of asymptotic expansions. As one of the first applications of IBP in such 'non-standard' situations, let us refer to [41], where IBP was applied to Feynman integrals arising in the Heavy Quark Effective Theory.

2.5.7 Mellin–Barnes Representation

One also often uses the Mellin transformation and, in particular, the Mellin–Barnes representation of a factor which is decomposed into two pieces:

$$\frac{1}{(X+Y)^\lambda} = \frac{1}{\Gamma(\lambda)} \frac{1}{2\pi \mathrm{i}} \int_{-\mathrm{i}\infty}^{+\mathrm{i}\infty} \mathrm{d}z \frac{Y^z}{X^{\lambda+z}} \Gamma(\lambda+z)\Gamma(-z) . \tag{2.74}$$

Here the contour of integration is chosen in the standard way: the poles with a $\Gamma(\ldots + z)$ dependence (let us call them IR poles) are to the left of the contour and the poles with a $\Gamma(\ldots - z)$ dependence (UV poles) are to the right of it.

Such factors can be chosen as functions in parametric integrals. But the simplest way to apply this representation is to write down a massive propagator in terms of massless ones:

$$\frac{1}{(m^2-k^2)^\lambda} = \frac{1}{\Gamma(\lambda)} \frac{1}{2\pi \mathrm{i}} \int_{-\mathrm{i}\infty}^{+\mathrm{i}\infty} \mathrm{d}z \frac{(m^2)^z}{(-k^2)^{\lambda+z}} \Gamma(\lambda+z)\Gamma(-z) . \tag{2.75}$$

We consider once again Fig. 2.1 with $m_1 = 0$, $m_2 = m$, $a_1 = 1$, $a_2 = 2$, insert (2.75) with $\lambda = 2$ into the Feynman integral, apply the formula (2.54) and obtain the following result:

$$F_\Gamma(q; d) = -\mathrm{i}\pi^{d/2} \frac{\Gamma(1+\varepsilon)}{(-q^2)^{1+\varepsilon}}$$
$$\times \frac{1}{2\pi \mathrm{i}} \int_{-\mathrm{i}\infty}^{+\mathrm{i}\infty} \mathrm{d}z \left(\frac{m^2}{-q^2}\right)^z \frac{\Gamma(1+\varepsilon+z)\Gamma(-\varepsilon-z)\Gamma(-z)}{\Gamma(1-2\varepsilon-z)} . \tag{2.76}$$

By closing the integration contour to the left and taking a series of residues at the points $z_n = -1 - \varepsilon - n$, where $n = 0, 1, \ldots$, we reproduce (2.59).

2.6 Properties of Dimensionally Regularized Feynman Integrals (Continued)

Although on-shell and threshold Feynman integrals have been already mentioned many times, let us now be more precise in our definitions. We must realize that, generally, an on-shell or threshold Feynman integral is *not* the value of the given Feynman integral $F_\Gamma(q^2,\ldots)$, defined as a function of q^2 and other kinematical variables, at a value of q^2 on a mass shell or at a threshold. Consider, for example, the Feynman integral corresponding to Fig. 2.1, with $m_1 = 0$, $m_2 = m$, $a_1 = 1$, $a_2 = 2$. We know an explicit result for the diagram, given by (2.58). There is a logarithmic singularity at threshold, $q^2 = m^2$ so that we cannot strictly speak about the value of the integral there. Still, we can certainly define the threshold Feynman integral by putting $q^2 = m^2$ *in the integrand* of the integral over the loop momentum or over the alpha parameters. And this is what was really meant and will be meant by 'on-shell' and 'threshold' integrals. In this example, we obtain an integral which can be evaluated by means of (A.13):

$$\int \frac{\mathrm{d}^d k}{k^2(k^2-2q\cdot k)^2} = \mathrm{i}\pi^{d/2}\frac{\Gamma(\varepsilon)}{2(m^2)^{1+\varepsilon}} \,. \tag{2.77}$$

This integral is divergent, in contrast to the original Feynman integral defined for general q^2.

Thus on-shell or threshold dimensionally regularized Feynman integrals are defined by the alpha representation or by integrals over the loop momenta with restriction of some kinematical invariants to appropriate values in the corresponding integrands. In this sense, these regularized integrals are 'formal' values of general Feynman integrals at the chosen variables.

Another example of this sort is provided by the value $D(0)$ of the scalar propagator $D(x)$ at the origin – see (2.8). According to (2.7), there is a singularity, so that such a quantity does not exist in the naive sense. We can understand this object as a dimensionally regularized formal value when we put $x = 0$ in the Fourier integral and obtain, using (A.1),

$$\int \frac{\mathrm{d}^d k}{k^2-m^2} = \mathrm{i}\pi^{d/2}\Gamma(\varepsilon-1)(m^2)^{1-\varepsilon} \,. \tag{2.78}$$

This Feynman integral in fact corresponds to the tadpole ϕ^4 theory graph shown in Fig. 2.7. The corresponding quadratic divergence manifests itself through an UV pole in ε – see (2.78).

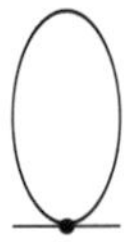

Fig. 2.7. Tadpole

Observe that we successfully dealt with an on-shell Feynman integral in Sect. 2.5.6 by applying IBP. Note also that one can trace the derivation of the integrals tabulated in Appendix A.1 and see that the integrals are convergent in some non-empty domains of the complex parameters λ_i and ε and that the results are analytic functions of these parameters with UV, IR and collinear poles.

Before continuing our discussion of setting scaleless integrals to zero, let us present an analytic result for the one-loop massless triangle integral with two on-shell external momenta, $p_1^2 = p_2^2 = 0$. Using (A.28), we obtain

$$\int \frac{\mathrm{d}^d k}{(k^2 - 2p_1\cdot k)(k^2 - 2p_2\cdot k)k^2} = \mathrm{i}\pi^{d/2} \frac{\Gamma(1+\varepsilon)\Gamma(-\varepsilon)^2}{\Gamma(1-2\varepsilon)(-q^2)^{1+\varepsilon}} \,. \tag{2.79}$$

A double pole at $\varepsilon = 0$ arises from the IR and collinear divergences.

A similar formula with a monomial in the numerator is also straightforwardly obtained:

$$\int \frac{\mathrm{d}^d k \; k^\mu}{(k^2 - 2p_1\cdot k)(k^2 - 2p_2\cdot k)k^2} = \mathrm{i}\pi^{d/2} \frac{\Gamma(\varepsilon)\Gamma(1-\varepsilon)^2}{\Gamma(2-2\varepsilon)} \frac{p_1^\mu + p_2^\mu}{(-q^2)^{1+\varepsilon}} \,. \tag{2.80}$$

Now only a simple pole is present, because the factor k^μ kills the IR divergence.

Consider now a massless one-loop integral with the external momentum on the massless mass shell, $p^2 = 0$:

$$\int \frac{\mathrm{d}^d k}{(p-k)^2 k^2} \,. \tag{2.81}$$

If we write down the alpha representation for this integral we obtain the same expression (2.51) as for $p = 0$ because only p^2, equal to zero in both cases, is involved there. In spite of this obvious fact, there is still a qualitative difference: for $p = 0$, there are UV and IR poles which enter with opposite signs and, for $p^2 = 0$ (but with $p \neq 0$ as a d-dimensional vector), there is a similar interplay of UV and collinear poles.

Now we follow the arguments presented in [182] and write down the following identity for (2.81), with $p = p_1$:

$$\int \frac{\mathrm{d}^d k}{(k^2 - 2p_1\cdot k)k^2} = \int \frac{\mathrm{d}^d k}{(k^2 - 2p_1\cdot k)(k^2 - 2p_2\cdot k)} - \int \frac{\mathrm{d}^d k \; 2p_2\cdot k}{(k^2 - 2p_1\cdot k)(k^2 - 2p_2\cdot k)} \,, \tag{2.82}$$

where $p_2^2 = 0$ and $p_1 \cdot p_2 \neq 0$. We then evaluate the integrals on the right-hand side by means of (2.56) and (2.80), respectively, and obtain a zero value. This fact again exemplifies the consistency of our rules.

Thus we are going to systematically apply the properties of dimensionally regularized Feynman integrals in any situation, no matter where the external momenta are considered to be.

2.7 How They Are Renormalized

Although a regularization makes it possible to deal with divergent Feynman integrals, it does not actually remove UV divergences, because this operation is of an auxiliary character so that sooner or later it will be necessary to switch off the regularization. To provide finiteness of physical observables evaluated through Feynman diagrams, another operation, called *renormalization*, is used. This operation is described, at the Lagrangian level, as a redefinition of the bare parameters of a given Lagrangian by inserting counterterms. A very important class of theories is described by interaction Lagrangians with dimensionless coupling constants, where the counterterms have the same form as the terms of the original total Lagrangian. These theories are also called *exactly renormalizable.*

It is usually rather difficult to carry out the programme of choosing the renormalized parameters of the Lagrangian in order to make finite the sum of diagrams of a given order. This problem is greatly simplified by introducing renormalization at the diagrammatic level, which is called *R-operation* and removes the UV divergence from individual Feynman integrals. We denote this procedure symbolically by $F_\Gamma \to RF_\Gamma$. Thus, for any R-operation, the quantity RF_Γ is UV finite.

Of course, not all operations that make Feynman integrals finite can be described as renormalization. The requirement for the R-operation to be implemented by inserting counterterms into the Lagrangian leads to the following structure [26, 137, 247, 35]:

$$RF_\Gamma = \sum_{\gamma_1,\ldots,\gamma_j} \Delta(\gamma_1)\ldots\Delta(\gamma_j)F_\Gamma \equiv R'\,F_\Gamma + \Delta(\Gamma)\,F_\Gamma\;, \tag{2.83}$$

where $\Delta(\gamma)$ is the corresponding counterterm operation, and the sum is over all sets $\{\gamma_1,\ldots,\gamma_j\}$ of disjoint divergent 1PI subgraphs, with $\Delta(\emptyset)=1$. The 'incomplete' R-operation R', by definition, includes all the counterterms except the overall counterterm $\Delta(\Gamma)$. For example, if a graph is primitively divergent, i.e. does not have divergent subgraphs, the operation is of the form $RF_\Gamma = [1+\Delta(\Gamma)]\,F_\Gamma$.

The action of the counterterm operations is described by

$$\Delta(\gamma)\,F_\Gamma = F_{\Gamma/\gamma}\circ P_\gamma\;, \tag{2.84}$$

where $F_{\Gamma/\gamma}$ is the Feynman integral corresponding to the reduced graph Γ/γ, and the right-hand side of (2.84) denotes the Feynman integral that differs from $F_{\Gamma/\gamma}$ by insertion of the polynomial P_γ in the external momenta and internal masses of γ into the vertex v_γ to which the subgraph γ was reduced. The degree of each P_γ equals the degree of divergence $\omega(\gamma)$. It is implied that a UV regularization is present in (2.83) and (2.84) because these quantities are UV divergent. The coefficients of the polynomial P_γ are connected in a straightforward manner with the counterterms of the Lagrangian.

A specific choice of the counterterm operations for the set of the graphs of a given theory defines a *renormalization scheme*. In the framework of dimensional renormalization, i.e. renormalization schemes based on dimensional regularization, the polynomials P_γ have coefficients that are linear combinations of pure poles in $\varepsilon = (4-d)/2$. In the minimal subtraction (MS) scheme [143], these polynomials are defined recursively by equations of the form

$$P_\gamma \equiv \Delta(\gamma)\, F_\gamma = -\hat{K}_\varepsilon R'\, F_\gamma \tag{2.85}$$

for the graphs γ of the given theory. Here $\hat{K}_\varepsilon$ is the operator that picks up the pole part of the Laurent series in ε. The modified MS scheme [13] ($\overline{\mathrm{MS}}$) is obtained from the MS scheme by the replacement $\mu^2 \to \mu^2 \mathrm{e}^{\gamma_{\mathrm{E}}}/(4\pi)$ for the massive parameter of dimensional regularization that enters through the factors of $\mu^{-2\varepsilon}$ per loop.

Note that the most important part of the basic theorem about the R-operation in the framework of dimensional renormalization [35] is just the above polynomial dependence of the diagrammatic counterterms $P_\gamma(\underline{q}, \underline{m})$ on the masses and external momenta. Although a requirement for this polynomial dependence on the masses [71] is not obligatory from the Lagrangian point of view, it turns out to be rather natural. For example, it leads to mass-independent renormalization group equations.

Another well-known renormalization scheme is the BPHZ renormalization [26, 137, 247]. In its original variant [26, 137], it is based on counterterms determined only for *complete* subgraphs (which include sets of vertices together with all lines between them). This kind of renormalization is natural for theories with a normal-ordered Lagrangian (and composite operators), while renormalization in all 1PI subgraphs is designed for Lagrangian and composite operators without normal order. After the BPH renormalization scheme had been reformulated for the latter case, the fourth letter in the acronym appeared. This scheme is based on (2.83) and (2.84), with the counterterms given recursively by

$$P_\gamma \equiv \Delta(\gamma)\, F_\gamma = -M R'\, F_\gamma \, , \tag{2.86}$$

where the subtraction operator M of the BPHZ scheme performs Taylor expansion of order $\omega(\gamma)$ of the subdiagram γ in its external momenta (but not in the masses!) at their zero values.

The BPHZ subtraction procedure is not applicable to theories with massless particles, because Taylor expansion at zero external momenta generates IR divergences. An example of its safe generalization to such theories is the auxiliary-mass renormalization scheme [169, 170], where an auxiliary massive parameter is introduced into subtracted terms in order to avoid these IR divergences. One can also use momentum subtractions for various off-shell configurations. This class of renormalization schemes is referred to as MOM schemes – see, for example, [44].

Note that the introduction of a regularization is not always necessary for defining renormalized Feynman integrals. Examples of renormalization without regularization are BPHZ renormalization, in its last version [247], and the recently developed differential renormalization [117] (see also [212] for prescriptions formulated for an arbitrary diagram).

There exists a great variety of renormalization schemes. Let us also mention a form of renormalization based on some normalization conditions where the renormalization is fixed by requirements imposed on Green functions at certain values of the external momenta. Another example is analytic renormalization, which was originally defined by the action of some operator on analytically regularized Feynman integrals [219] but can be also formulated in the style of the minimal subtractions, when one uses the same analytic parameter for all the lines in the same manner as ε.

When one is proving the basic properties of the R-operation, different points are non-trivial for different schemes: in contrast to dimensional renormalization, the most non-trivial point for the BPHZ renormalization and its massless generalizations is the finiteness of renormalized Feynman integrals (the Bogoliubov–Parasiuk theorem [26, 137]).

It is possible to resolve the recurrence structure of the R-operation and express it in terms of the subtraction operator M that enters through (2.86) (and can be chosen as the BPHZ subtraction or some other operator that provides the necessary properties of the renormalization). A well-known solution is given by the forest formula [247] (written earlier in equivalent language in [245])

$$R = \sum_f \prod_{\gamma \in f} (-M_\gamma) \;, \tag{2.87}$$

where the sum runs over *forests* (i.e. families of non-overlapping subgraphs), consisting of 1PI divergent subgraphs. By definition, the empty forest is also implied in the sum, with the corresponding term equal to unity. For example, for the one-loop graph Γ of Fig. 2.1, there are two forests, $\emptyset$ and Γ, so that the R-operation takes the form $R = 1 - M$. For the two-loop scalar graph shown in Fig. 2.8, we have the following forests: $\{\emptyset\}$, $\{\gamma\}$, $\{\Gamma\}$ and $\{\gamma, \Gamma\}$, where γ is the one-loop subgraph. The corresponding R-operation takes the form

$$R = 1 - M_\gamma - M_\Gamma + M_\Gamma M_\gamma \equiv (1 - M_\Gamma)(1 - M_\gamma) \;. \tag{2.88}$$

Moreover, for the two-loop QED graph shown in Fig. 2.9, with two logarithmically divergent partially overlapping one-loop vertex subgraphs γ_1 and γ_2, the complete set of forests is $\{\emptyset\}$, $\{\gamma_1\}$, $\{\gamma_2\}$, $\{\Gamma\}$, $\{\gamma_1, \Gamma\}$, $\{\gamma_2, \Gamma\}$. The corresponding R-operation is

$$\begin{aligned} R &= 1 - M_{\gamma_1} - M_{\gamma_2} - M_\Gamma + M_\Gamma M_{\gamma_1} + M_\Gamma M_{\gamma_2} \\ &\equiv (1 - M_\Gamma)(1 - M_{\gamma_1} - M_{\gamma_2}) \;. \end{aligned} \tag{2.89}$$

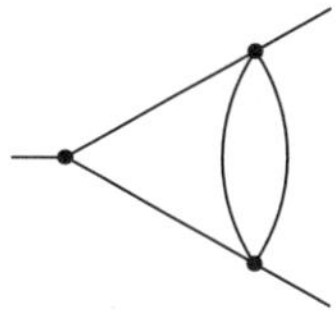

Fig. 2.8. Two-loop scalar vertex graph with nested UV divergent subgraphs

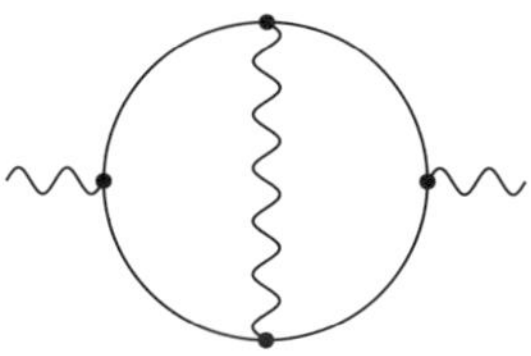

Fig. 2.9. Two-loop QED diagram

Although dimensional renormalization can be also described through an appropriate subtraction operator [72], this is achieved by a cumbersome procedure, at the cost of introducing different space–time dimensions for 1PI subgraphs, so that it is more reasonable to apply the MS and $\overline{\text{MS}}$ schemes within the recursive definition. However, we shall certainly need the forest formula for another purpose: it happens that the structure of the R-operation represented by the forest formula arises when one solves the problem of asymptotic expansions in off-shell limits of momenta and masses. As we shall see, the remainder of the expansion is obtained from the original diagram by the action of an operation given by the forest formula with an appropriate (pre-)subtraction operator.

2.8 Back to the Operator Level

The renormalized S-matrix

$$S = RT \exp\left(\mathrm{i} \int \mathrm{d}^4 x \, \mathcal{L}_{\mathrm{I}}(x)\right) \tag{2.90}$$

can be obtained, on the one hand, by the action of an R-operation on the Feynman diagrams contributing to the coefficient functions S_N in (2.9). On the other hand, it can also be obtained from the unrenormalized S-matrix by replacing all the bare parameters in the Lagrangian by their renormalized values. For example, for MS renormalized QCD, this means the following replacement of the bare quantities present in (2.15), labelled here by the index B:

$$\begin{aligned} g^{\mathrm{B}} &= \mu^{\varepsilon} Z_g g \,, \quad m_i^{\mathrm{B}} = Z_m m_i \,, \quad \xi^{\mathrm{B}} - 1 = Z_3(\xi - 1) \,, \\ q_i^{\mathrm{B}} &= \sqrt{Z_2} q_i \,, \quad A_\nu^{\mathrm{B},a} = \sqrt{Z_3} A_\nu^a \,, \quad c^{\mathrm{B},a} = \sqrt{\tilde{Z}_3} c_a \,, \end{aligned} \tag{2.91}$$

where the factor μ^ε makes the dimension of the coupling constant independent of the parameter ε of dimensional regularization. Any counterterm Z_i within the MS or $\overline{\text{MS}}$ scheme is a linear combination of pure poles in ε:

$$Z_i = 1 + \sum_{n=0} \sum_{j=0}^{n} z_{nj} \left(\frac{\alpha_s}{\pi}\right)^n \frac{1}{\varepsilon^j} , \tag{2.92}$$

where z_{nj} are dimensionless constants.

The fact that the diagrammatic renormalization performed by the R-operation (2.83), with appropriate counterterm operations (2.84), can be implemented by modifying parameters of the Lagrangian looks (almost) obvious at the one-loop level. For a general perturbation order, this property is written as

$$RT \exp\left(\mathrm{i} \int \mathrm{d}^4x \, \mathcal{L}_\mathrm{I}^\mathrm{B}(x)\right) = T \exp\left(\mathrm{i} \int \mathrm{d}^4x \, \mathcal{L}_\mathrm{I}^\mathrm{R}(x)\right) , \tag{2.93}$$

where $\mathcal{L}_\mathrm{I}^\mathrm{R}$ is the renormalized Lagrangian, which consists of the interaction part of the bare Lagrangian, $\mathcal{L}_\mathrm{I}^\mathrm{B}$, and counterterms. The renormalized Lagrangian can be written explicitly in terms of the counterterm operation involved in (2.83) and (2.84) by means of the so-called counterterm technique developed in [4, 246] and generalized to the case of Lagrangians and composite operators without normal order in [63] (see also [234, 42]). In the latter case,

$$\mathcal{L}_\mathrm{I}^\mathrm{R} = \mathcal{W}\Delta \left[T \exp\left(\mathrm{i} \int \mathrm{d}x \, \mathcal{L}_\mathrm{I}^\mathrm{B}(x)\right) - 1\right] , \tag{2.94}$$

where the action of the counterterm operation Δ reduces to the action of the corresponding diagrammatic operation in (2.84) on Feynman integrals, and the operation $\mathcal{W}$ removes the symbol that defines the normal order in representations of operators of the form (2.9) (so that the Wick theorem is applied again after the action of this operation). In the language of functionals, when the quantum field operators are represented by (2.10), this operator is given by (2.11).

The normal-ordered variant [4, 246] differs from (2.83) by omission of the operator $\mathcal{W}$ and use of the R-operation (2.84), where the sum runs only over complete subgraphs. The proof of this formula is based on combinatorial arguments that have been used in many situations, in particular for the transition from full Green functions to connected Green functions (see, e.g., [147]). Here is a sketch of the proof. After expansion of the exponent on the left-hand side of (2.93) and application of the Wick theorem the diagrammatic R-operation given by (2.83) and (2.84) is applied to the resulting diagrams. From the order-N expansion term of the exponent, diagrams with N vertices arise. The set of vertices is decomposed, according to (2.84), into subsets consisting of a different number of vertices, and the number N is represented as $N_1 + 2N_2 + \ldots + kN_k + \ldots$, in $N!/[\prod N_k!(k!)^{N_k}]$ distinct ways. Then one performs summation over the numbers N_k and eventually arrives at (2.94).

A (bare) composite operator

$$J^{\mathrm{B}}(x) = Tj(x) \exp\left(\mathrm{i} \int \mathrm{d}^4x \, \mathcal{L}_{\mathrm{I}}^{\mathrm{R}}(x)\right) \tag{2.95}$$

is generated by a polynomial $j(x)$ composed of the fields present in the theory. As will be explained shortly (and, in detail, in Chap. 4), it is very convenient to include powers of the masses of the theory in the composite operators. The canonical dimension of the operator is defined in the standard way, in mass units, as $N_{\mathrm{b}} + N_{\mathrm{f}} + N_{\mathrm{m}}$, where N_{b} N_{f} and N_{m} are the number of boson fields, the number of fermion fields and the total power of the masses involved in the product j, respectively.

The bare composite operators are defined by means of the renormalized Lagrangian so that all the UV divergences present in purely S-matrix (sub)diagrams are removed. However, there are no prescriptions for removing UV divergences from diagrams that involve the vertex corresponding to the composite operator. The renormalized composite operators

$$J(x) = RJ^{\mathrm{B}}(x) \tag{2.96}$$

are obtained from the unrenormalized (i.e. incompletely renormalized) operators by including counterterms for these specific UV divergences. Usually, one chooses a basis of composite operators O_i generated by some polynomials o_i. Then the local property of the counterterms in (2.83) leads to the following expression for the renormalized composite operators:

$$O_i(x) \equiv RO_i^{\mathrm{B}}(x) = \sum_k Z_{ik} O_k^{\mathrm{B}}(x) \,, \tag{2.97}$$

where Z_{ik} is a renormalization matrix. One says that the operators are mixed by the renormalization. An example of mixing of operators (in QCD) can be found in Sect. 4.5.

In diagrams that contribute to bare products of two renormalized composite operators $[TO_i(x)O_k(0)]^{\mathrm{B}}$, all the UV divergences present in purely S-matrix 1PI (sub)diagrams and diagrams involving only one vertex corresponding to a composite operator are removed. But one needs additional prescriptions for removing UV divergences from diagrams that involve both of the vertices connected with the composite operators. The renormalized product of two composite operators is represented in a form similar to (2.97):

$$TO_i(x)O_k(0) = \sum_{i'k'} Z_{ik,i'k'} [TO_{i'}(x)O_{k'}(0)]^{\mathrm{B}} \,, \tag{2.98}$$

where $Z_{ik,i'k'}$ is a renormalization matrix of products of the composite operators.

Although the monomials of the fields present in the Lagrangian and the composite operators were traditionally[5] supposed, in the early days of quan-

[5] Even the very technique of definition of the composite operators and analysing their properties was called the 'Normal Product Algorithm' [248]. So, now, 'Abnormal Product Algorithm' is in use.

tum field theory, to be normal-ordered, it is very important to consider them as operators without normal order. This prescription results in the possibility of performing the factorization of contributions depending on phenomena connected with different scales. We shall come to back to discussion of this factorization in Chap. 4. Thus products of fields at the same point are present both in the Lagrangian and in the composite fields, so that Feynman diagrams can involve tadpoles. Note also that, within the method of path integrals, Feynman diagrams naturally arise without any prescription to exclude such diagrams. Anyway, the use of Lagrangians and composite operators without normal order is now commonly accepted.

3 Why, Where and How to Expand

Divide et impera

(A successful strategy)

Suppose that we have to analytically evaluate a two-loop Feynman diagram which depends on two parameters, e.g. a mass squared and a momentum squared. This is generally a rather complicated problem. It usually happens, however, that the parameters involved differ in scale so that it is reasonable to expand the diagram in their small ratio. We shall see in this book that an asymptotic expansion in an arbitrary limit with two scales can be written explicitly as an infinite series of products of certain one-scale Feynman integrals, with a power and logarithmic dependence on the expansion parameter, which can be evaluated analytically much more easily than the initial two-scale integral. The original Feynman integral can then be replaced by a sufficiently large number of terms of its expansion.

In this chapter, the limits we are going to study are characterized and the form of the asymptotic expansion we are aiming at is described. An example of a one-loop Feynman integral in the large-momentum limit and a related toy one-dimensional example are then used to formulate two basic strategies for expanding Feynman integrals that we are going to apply. The first of them, expansion by regions, is physically motivated and turns out to be more general. In this strategy, one divides the whole integration domain into various regions, then performs some simplifications and eventually obtains an expansion as a sum of contributions from the regions that can be handled in much easier way than the initial diagram. Although it is clearly a heuristic strategy, examples where it breaks down are unknown at present. The second strategy can, at present, be applied to off-shell limits of momenta and masses and to some specific on-shell limits typical of Minkowski space. In contrast to the first strategy, it has been mathematically proven. Since these strategies lead to the same prescriptions for the off-shell limits, we thereby obtain an indirect confirmation of the first, heuristic strategy for such a family of simpler limits.

3.1 Limits and Asymptotic Expansions

The problem of expanding Feynman integrals in momenta and masses arises quite naturally. Suppose that we consider phenomena at a given scale, Λ.

Then it is reasonable to consider small all the masses and kinematical invariants that are below this scale, and the remaining parameters as large. So, it is a decomposition of the given set of momenta and masses into these two subsets that determines a *limit* of momenta and masses. For simplicity, let us first consider an h-loop Feynman integral F_Γ corresponding to a graph Γ with two external vertices and depending on two parameters. To be specific, let us choose the small parameter to be the square of a mass and the large parameter as the external momentum squared, i.e. $m^2 \ll -q^2$. We have chosen q to be Euclidean so that $q^2 = -Q^2$, with $Q > 0$.

Experience tells us that, in all limits, we obtain expansions of Feynman integrals in powers and logarithms. If we pull out an overall dimensional factor, we expect that the corresponding expansion has the form

$$F_\Gamma(Q^2, m^2) \sim (Q^2)^\omega \sum_{n=n_0}^{\infty} \sum_{j=0}^{2h} C_{nj}\, x^n \ln^j x \, , \tag{3.1}$$

where $x = m^2/Q^2$ and ω is the degree of divergence of the graph Γ.

The sum over n runs from some minimal value. The index n can generally take, in some limits, not only integer but also half-integer values. The second index, j, is bounded, for any n, by twice the number of loops. It should be stressed that these expectations are, in general, a matter of experience rather than a corollary of a mathematical theorem. However we shall see that, in some simpler limits, these statements can indeed be justified.

According to a standard definition of an asymptotic expansion [185], (3.1) means that, for an arbitrary number N, the remainder

$$R_N(Q^2, m^2) = F_\Gamma(Q^2, m^2) - (Q^2)^\omega \sum_{n=N_0}^{N} \sum_{j=0}^{2h} C_{nj}\, x^n \ln^j x \tag{3.2}$$

has order $o(x^N)$, i.e., for Q in a finite interval $A < Q < B$, there exist $C > 0$ and $\varepsilon > 0$ such that

$$\left| x^{-N} R_N(Q^2, Q^2 x) \right| \leq C \tag{3.3}$$

for $0 < x < \varepsilon$. The above formulae are modified just a little when there are more than two parameters: we multiply each small parameter by a dimensionless variable x, and deal with the resulting function of x in the limit $x \to 0$.

Another 'experimental' fact is the existence of a non-zero radius of convergence of the asymptotic expansion of any Feynman integral in any limit. Normally, we shall just point out this phenomenon in subsequent chapters and shall not aim at a deep investigation of this problem. Nevertheless, we shall believe in this property because we want to apply our expansions by substituting given Feynman integrals by a sufficiently large number of initial terms of the corresponding expansion.

We have already specified the natural requirement for an expansion in powers and logarithms. Before formulating the next requirement let us note that, on the left-hand side of (3.1), we assumed a UV- and IR-finite quantity, so that a regularization was not introduced. However, even if the original Feynman integral is finite, it is necessary to introduce a regularization, which we choose to be dimensional. The dimensionally regularized analogue of (3.1) is of the form

$$F_\Gamma(Q^2, m^2; \varepsilon) \sim (Q^2)^{\omega - h\varepsilon} \sum_{n=n_0}^{\infty} \sum_{j=0}^{2h} \sum_{i=0}^{h} C_{nji}(\varepsilon)\, x^{n-j\varepsilon} \ln^i x \,. \tag{3.4}$$

The necessity for the regularization is explained by the fact that, without it, it is very difficult to provide simple explicit prescriptions for the terms of the expansion; it is a negligible price for obtaining a natural solution of the problem. Here the dimensional regularization enables us to separate various terms on the right-hand side of the expansion which would be ill-defined without it. Nevertheless, in all the limits that we are going to investigate, the sum of terms at a fixed power of the expansion parameter turns out to be finite, provided the original integral is finite. We shall also see in Chap. 8 that, in some cases, individual terms at a fixed power in the expansion become unregularized when considered separately, even in the presence of dimensional regularization. In this case a natural solution of this problem is to temporarily introduce an auxiliary analytic regularization, which is switched off immediately after these integrals have been evaluated and summation has been performed over every 'subset' of such dangerous terms. We can keep, at this step, dimensional regularization while the logarithms of the expansion parameter in (3.4) 'remember' the above problem.

The second requirement is that, starting from the Feynman integral for the given graph on the left-hand side, we would like to obtain Feynman integrals on the right-hand side also. This appears rather natural because Feynman integrals are the fundamental objects we are dealing with and it is better to have them in the expansion rather than, say, in some artificially introduced parametric integrals. Moreover we would like to formulate *universal* rules that make a correspondence between the original Feynman integral and the terms of its expansion, i.e. the rules should be formulated in the same way for an arbitrary graph. After we have set out such rules, we shall be able to apply them to any given diagram, with any number of loops, and not perform any analytic work at this point. As we shall see, the application of these rules will be just a task (which can in principle be done by computer) of looking through all relevant subgraphs or regions, from a known list, and writing down the corresponding contributions. Of course, the integrals that appear on the right-hand side (which are much simpler than the initial Feynman integral because they depend on a lower number of scales) have not yet been calculated at this stage, but their calculation is another problem, while the problem of expansion is already essentially solved.

Another very important reason for the second requirement is that, starting from the expansions of individual Feynman integrals and analysing their right-hand sides written in terms of Feynman integrals, it is then much simpler to arrive at the corresponding expansions at the operator level. At this step, the above universality plays a crucial role.

We shall see, in Chaps. 4–8, that we can fulfil both requirements and, in all the limits we are going to investigate, we shall encounter, on the right-hand side, certain Feynman integrals generated, according to some rules that are universal for the given limit, from the original integral by Taylor expansions of the integrand in certain parameters.

In a great variety of limits associated with physical problems, we can distinguish an important class of simpler limits which are typical of Euclidean space, i.e. can be formulated for Euclidean Feynman integrals. It is not so simple to characterize this class more explicitly. Ironically, this is the form of the general prescriptions which we are aiming at, which is characteristic of the classification of the limits. In fact, the limits for which prescriptions can be written in the form of a sum of a specific family of subgraphs belong to the Euclidean type of limit. Anyway, for typically Euclidean limits, all the large external momenta are considered large in the Euclidean sense, so that any scalar product of a large momentum Q and any other (either large or small) external momentum q is considered large.

We shall investigate such limits in Chaps. 4 and 5. We shall see that the corresponding prescriptions for expanding Feynman integrals in these limits can be expressed simply in a graph-theoretical language, within our second strategy. This is a 'safe' class of limits, in the sense that these results can be proven – see Appendix B.2. At the operator level, this class of limits covers the Wilson operator product expansion and the large-mass limit described by an effective Lagrangian of light fields.

The remaining limits are connected with typically Minkowskian configurations of the large momenta, which are located on some singular surface, either on a mass shell or at a threshold. Such limits cannot be formulated for Feynman integrals in Euclidean space. For example, for a typical pseudo-Euclidean limit, a momentum can be large, i.e. some its components are large, while its momentum squared can be small, or even equal to zero from the beginning (see Chap. 8). The method of expansion by subgraphs has been developed only for some special cases of such limits and, in situations where there is no mathematical confirmation of the methods used to deal with these limits, we are forced to use expansion by regions and confine ourselves to heuristic arguments and experimental computational confirmations. We shall study these limits in Chaps. 6–8.

3.2 Expansion by Regions

Consider the example of a one-loop Feynman diagram Γ shown in Fig. 3.1 in the large-momentum limit, $|q^2| \gg m^2$. We suppose that the upper line is massless. In order to deal with a convergent Feynman integral from the

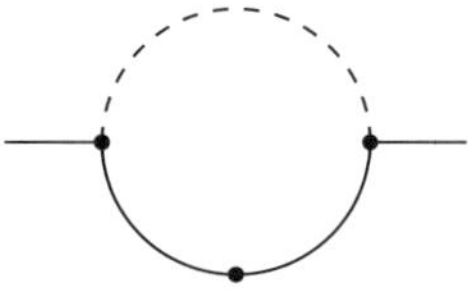

Fig. 3.1. One-loop self-energy graph. The *dashed line* denotes a massless propagator

beginning, we have chosen the second power of the lower propagator, which is indicated by a dot on the line. The corresponding dimensionally regularized Feynman integral takes the form

$$F_\Gamma(q^2, m^2; d) = \int \frac{\mathrm{d}^d k}{(k^2 - m^2)^2 (q-k)^2} \,. \tag{3.5}$$

Regularization has been consciously introduced for a finite initial quantity because we shall obtain various divergences in individual terms of our result for the expansion. This appearance of divergences is characteristic (well, almost) of all expansions in momenta and masses and is a negligible price paid in order to obtain explicit formulae for individual terms that are as simple as possible in the general order of the expansion.

In fact, we used this diagram in the previous chapter to illustrate methods of evaluating Feynman integrals, so that we already know the result (2.59). Since the argument of the hypergeometric functions is q^2/m^2, this result is naturally written as an expansion in powers of this ratio. However, we want an expansion in the opposite variable. To rewrite the result as an expansion in m^2/q^2, we use the identity (A.54) that relates the hypergeometric functions with arguments z and $1/z$ and obtain

$$F_\Gamma(q^2, m^2; d) = \mathrm{i}\pi^{d/2} \left[\frac{(\varepsilon-1)m^2}{\varepsilon q^2} {}_2F_1\left(1, \varepsilon; 1-\varepsilon; m^2/q^2\right) \right.$$
$$\left. + \frac{\Gamma(2-\varepsilon)\Gamma(\varepsilon)}{\Gamma(1-2\varepsilon)} \left(-\frac{m^2}{q^2}\right)^{1+\varepsilon} \left(1 - \frac{m^2}{q^2}\right)^{-2\varepsilon} \right] . \tag{3.6}$$

We can obtain the same result by the Mellin–Barnes technique, using (2.76), closing the integration contour to the right and taking two series of residues, at the points $z_{1,n} = -\varepsilon + n$ and $z_{2,n} = n$, where $n = 0, 1, \ldots$.

But we want to arrive at the terms of this expansion in an essentially easier way, and by following some *general* prescriptions which will avoid the need for us to evaluate this full result or to perform an analysis of MB integrals oriented towards *concrete* expressions for the given diagram. Note that if we proceed naively and begin to expand the corresponding integrand in a Taylor series in our small parameter m^2, we immediately obtain, in the leading order of the expansion, an IR divergence, which comes from the factor $1/(k^2)^2$.

To illustrate our strategies for expanding Feynman integrals let us simplify the situation as much as possible and consider first a toy example of the following one-dimensional analogue of (3.5):

$$F(q,m,\varepsilon)=\int_0^\infty \frac{k^{-\varepsilon}\,\mathrm{d}k}{(k+m)(k+q)}\,, \tag{3.7}$$

where q and m are numbers, with $0 < m \ll q$. The regularization factor $k^{-\varepsilon}$ plays the role of dimensional regularization in (3.5). Of course, this integral can be evaluated explicitly without any expansion (e.g. by use of partial fractions), with the following result:

$$F(q,m,\varepsilon)=-\frac{\pi}{\sin\pi\varepsilon}\,\frac{q^{-\varepsilon}-m^{-\varepsilon}}{q-m}\,, \tag{3.8}$$

which, in the limit $\varepsilon = 0$, gives

$$F(q,m,0)=\frac{\ln(q/m)}{q-m}\,. \tag{3.9}$$

Let us now reproduce this result by expanding (3.7) following two different strategies.

As with its four-dimensional prototype, we shall immediately obtain an integral that is divergent for $\varepsilon = 0$ if we begin to expand the integrand into a Taylor series in m:

$$\int_0^\infty \frac{k^{-\varepsilon-1}\,\mathrm{d}k}{k+q}+\ldots\,. \tag{3.10}$$

It is clear that this cannot be the true leading-order term of the expansion, since (3.7) is finite in ε. The breakdown of this 'naive' expansion is explained by the fact that the Taylor series, in m, of the factor $1/(k + m)$ converges only for $|k| < m$, rather than for arbitrary values of k. Still, let us continue to be naive and evaluate the whole series that results from this expansion:

$$F_{\text{large}} \sim \int_0^\infty \mathrm{d}k\,\frac{k^{-\varepsilon-1}}{k+q}-m\int_0^\Lambda \mathrm{d}k\,\frac{k^{-\varepsilon-2}}{k+q}+\ldots \tag{3.11}$$

$$=\sum_{n=0}^\infty(-1)^n m^n\int_0^\infty\frac{k^{-n-\varepsilon-1}\,\mathrm{d}k}{k+q}$$

$$=\frac{1}{q}\sum_{n=0}^\infty(-1)^n\Gamma(n+1+\varepsilon)\Gamma(-n-\varepsilon)\left(\frac{m}{q}\right)^n$$

$$=\frac{\pi}{q^{1+\varepsilon}\sin\pi\varepsilon}\sum_{n=0}^\infty\left(\frac{m}{q}\right)^n\,. \tag{3.12}$$

We attach the label 'large' in the sense that the expansion of the integrand is still legitimate at large values of k, of order q.

Now, suppose we were naive in the opposite way and looked at the region of small values of k, i.e. of order m. Then we would see that was legitimate to expand the second factor in k and would obtain the following series:

$$F_{\rm small} \sim \frac{1}{q}\int_0^\infty {\rm d}k\,\frac{k^{-\varepsilon}}{k+m} - \frac{1}{q^2}\int_0^\infty {\rm d}k\,\frac{k^{-\varepsilon+1}}{k+m} + \ldots \tag{3.13}$$

$$= \sum_{n=0}^\infty \frac{(-1)^n}{q^{n+1}}\int_0^\infty \frac{k^{n-\varepsilon}\,{\rm d}k}{k+m}$$

$$= \frac{1}{qm^\varepsilon}\sum_{n=0}^\infty (-1)^n \Gamma(n+1-\varepsilon)\Gamma(-n+\varepsilon)\left(\frac{m}{q}\right)^n$$

$$= -\frac{\pi}{qm^\varepsilon \sin\pi\varepsilon}\sum_{n=0}^\infty \left(\frac{m}{q}\right)^n . \tag{3.14}$$

A curious fact is that if we add together these two naive results and sum up the geometric progressions involved we obtain nothing but the true explicit result (3.8), so that the desired expansion is given by

$$F \sim F_{\rm small} + F_{\rm large} . \tag{3.15}$$

Let us now prove that such a way of expanding is indeed correct, in our example, and that we really can obtain the final result by expanding the integrand in two different ways. We note that the procedure for expanding in m would work if the integration was taken strictly above the value m. Let us therefore choose an intermediate scale Λ such that $m < \Lambda < q$ and divide our integral into two pieces:

$$F = f_{\rm small} + f_{\rm large} , \tag{3.16a}$$

$$f_{\rm small}(q,m,\Lambda) = \int_0^\Lambda \frac{k^{-\varepsilon}\,{\rm d}k}{(k+m)(k+q)} , \tag{3.16b}$$

$$f_{\rm large}(q,m,\Lambda) = \int_\Lambda^\infty \frac{k^{-\varepsilon}\,{\rm d}k}{(k+m)(k+q)} . \tag{3.16c}$$

We can safely expand the integrand in m in the second piece:

$$f_{\rm large} \sim \int_\Lambda^\infty {\rm d}k \frac{k^{-\varepsilon}}{k+q}\left(\frac{1}{k} - \frac{m}{k^2} + \ldots\right) . \tag{3.17}$$

On the other hand, since $k < \Lambda$ in the first piece, we may expand the second factor there:

$$f_{\rm small} \sim \int_0^\Lambda {\rm d}k\,\frac{k^{-\varepsilon}}{k+m}\left(\frac{1}{q} - \frac{k}{q^2} + \ldots\right) . \tag{3.18}$$

After this we rewrite the two contributions above in identical form by extending the integrations to infinite limits and compensating this by subtracting the corresponding additional pieces:

$$f_{\rm small} \sim F_{\rm small} - \left(\frac{1}{q}\int_\Lambda^\infty {\rm d}k\,\frac{k^{-\varepsilon}}{k+m} - \frac{1}{q^2}\int_\Lambda^\infty {\rm d}k\,\frac{k^{-\varepsilon+1}}{k+m} + \ldots\right) , \tag{3.19a}$$

$$f_{\rm large} \sim F_{\rm large} - \left(\int_0^\Lambda {\rm d}k\,\frac{k^{-\varepsilon-1}}{k+q} - m\int_0^\Lambda {\rm d}k\,\frac{k^{-\varepsilon-2}}{k+q} + \ldots\right) , \tag{3.19b}$$

where $F_{\rm small}$ and $F_{\rm large}$ are given by (3.13) and (3.11), respectively.

For each integral on the right-hand sides in (3.19a) and (3.19b) there exists a domain of the regularization parameter ε for which it is absolutely convergent. All these integrals can be defined for general ε by analytic continuation. Observe that we may now expand any given integral in the large parentheses in (3.19a) and (3.19b) into Taylor series in m and k, respectively. The resulting integrals are easily evaluated. In particular, the leading-order terms that arise from the terms in the large parentheses are

$$-\frac{1}{q}\int_{\Lambda}^{\infty} \mathrm{d}k\, k^{-\varepsilon-1} = \frac{1}{\varepsilon q \Lambda^{\varepsilon}} , \tag{3.20a}$$

$$-\frac{1}{q}\int_{0}^{\Lambda} \mathrm{d}k\, k^{-\varepsilon-1} = -\frac{1}{\varepsilon q \Lambda^{\varepsilon}} , \tag{3.20b}$$

respectively.

It should be stressed that the above two integrals are initially defined in different domains of the parameter ε, namely Re $\varepsilon > 0$ and Re $\varepsilon < 0$, where they are absolutely convergent and from where they are analytically continued to general values of ε. Note that, formally, the sum of these two integrals equals an integral with the same integrand but taken from zero to infinity. As we discussed in Chap. 2, it is natural to set such integrals to zero. But here we do not need to refer to such ad hoc recipes, because the two pieces of the integral arise quite naturally when they are evaluated with their own values of ε which provide convergence.

It is easy to see that results with opposite signs arise not only from the leading-order terms but from the whole series of the contributions from (3.19a) and (3.19b). In fact, in both cases we obtain an integrand with factors $1/(k+m)$ and $1/(k+q)$, expanded into Taylor series in m and k, respectively, and the only difference is that (3.19a) and (3.19b) generate integrals from Λ to ∞ and from 0 to Λ, respectively, when considered in their own initial domains of the regularization parameter ε where they are convergent. Therefore the terms in the large parentheses are mutually cancelled after this expansion and calculation, and we arrive at the expansion of the given integral (3.7), written as (3.15).

Thus the asymptotic expansion of the original integral is given by two contributions, F_{small} and F_{large}. All terms in both contributions are homogeneous with respect to the expansion parameter, and the corresponding one-scale integrals can be evaluated trivially. Note that the regularization provided by the parameter ε is necessary because, individually, these two contributions, which would be divergent when $\varepsilon = 0$, contain poles in ε. Nevertheless, these poles are cancelled in the sum of the two contributions for every power of the expansion parameter.

We use on purpose the same labels in F_{small} and F_{large} as in the integrals f_{small} and f_{large}, which are indeed integrals in the regions of small and large k. We want thereby to stress the fact that these two contributions originate from these regions and the corresponding Taylor expansions. Moreover, we see that we can obtain the result (3.15) using the following prescriptions for expanding integrals

$$\int_0^\infty f(q,m,k)\,\mathrm{d}k \tag{3.21}$$

of *arbitrary* rational functions $f(q,m,k)$ with positive denominators in the limit $q \gg m$:

- Divide the integration domain into various regions and, in every region, expand the integrand into a Taylor series with respect to the parameters that are considered small there.
- Integrate every expanded integrand over the whole integration domain and sum the resulting contributions from all the regions.

In our toy example, we already know two typical regions, namely those of small and large k, i.e. $k \sim m$ and $k \sim q$, respectively. Guided by the above prescriptions, we immediately arrive at the result (3.15), without the proof presented above. However, this is the first and the last example where a direct confirmation of these prescriptions will be presented, because it happens to be very difficult to provide mathematical proofs (even for one-loop Feynman integrals!) of the second step.

Note that, in our later considerations, we shall not really use the word 'region' to denote a domain determined by inequalities, as was done in the previous analysis, with a choice of some intermediate scale. Rather, we shall treat this word in the physical sense and imply relations between the quantities involved that determine an order of the integration variables, expressed in terms of the parameters of the problem.

Suppose now that we have not performed the above analysis and are equipped only with the above two prescriptions. In that case we could ask whether other regions contribute to the expansion. For example, why not consider a region of the type $k \sim m^2/q$? According to our prescriptions, we should expand both propagators in k because $k \ll m$ and $k \ll q$. As a result, we obtain a contribution that consists of scaleless integrals. In particular, in the leading order, we have

$$\frac{1}{mq}\int_0^\infty k^{-\varepsilon}\,\mathrm{d}k\,. \tag{3.22}$$

This is a scaleless integral, so that it is tempting to set it to zero, as we did in the case of the scaleless integrals we met above. And this appears to be a reasonable decision because we know that, in our example, there are no other contributions in addition to those present in (3.15). By the way, if we choose some more exotic region, e.g. $k \sim m^{7/2}/q^{5/2}$, we end up with the same scaleless integrals.

Thus, it appears necessary to supplement the above prescriptions with a third one, which provides zero values for scaleless integrals. But let us now come back to the Feynman integrals and formulate some prescriptions which we shall use systematically in the rest of the book. Suppose that we want to expand a Feynman integral in some limit of momenta and masses. Let us follow the rules formulated in [18]:

- Divide the space of the loop momenta into various regions and, in every region, expand the integrand into a Taylor series with respect to the parameters that are considered small there.
- Integrate the integrand, expanded in the appropriate way in every region, over the whole integration domain of the loop momenta.
- Set to zero any scaleless integral.

Remember that we have agreed to understand the term 'region' in the sense of relations between the quantities involved. It is possible to formalize these relations mathematically, and we shall do this in Sect. 9.3, but, in the main part of the book, we shall still use our basic term in the heuristic sense and indeed think of some regions also because this language certainly has an unambiguous physical orientation.

Note that, in the last prescription, we do not mean only massless vacuum integrals. We shall meet various examples of scaleless integrals in later chapters. In particular, sometimes the presence of a scaleless integral is manifest only after several intermediate integrations rather than from the very beginning.

Let us apply this strategy to our one-loop d-dimensional Feynman integral (3.5). We have the regions of large and small loop momenta k:

$$\text{large,} \quad k \sim q \; ; \tag{3.23a}$$

$$\text{small,} \quad k \sim m \; . \tag{3.23b}$$

In the region of large k, the second propagator is unexpanded, while the first propagator is expanded into a Taylor series in m. Here the Euclidean sense is implied, i.e. the corresponding relations for any components of the d-dimensional quantities involved are satisfied. In particular, when k is considered large, we consider its scalar products also to be large. For example, k^2 is considered large and we do not bother about possible cancellation of k_0^2 and $\mathbf{k}^2$ in the combination $k^2 = k_0^2 - \mathbf{k}^2$.

In the region of small k, the second propagator is expanded into a Taylor series in k because $|k^2|, |2q{\cdot}k| \ll |q^2|$ in the combination $(q-k)^2 \equiv q^2 - (2q{\cdot}k - k^2)$. Thus, the first propagator is unexpanded there, while the expansion of the second propagator is given by a geometric progression:

$$\frac{1}{(q-k)^2} = \frac{1}{q^2} + \frac{2q{\cdot}k - k^2}{(q^2)^2} + \frac{(2q{\cdot}k - k^2)^2}{(q^2)^3} + \ldots \; . \tag{3.24}$$

Here the degree of the terms, measured in powers of k, is not strictly ordered: for example, $k^2/(q^2)^2$ from the second term on the right-hand side has the same degree as $(2q{\cdot}k)^2/(q^2)^3$ from the third one.

Combining both contributions together, we arrive at the following expansion:

$$F_\Gamma(q^2,m^2;d) \sim \int \frac{\mathrm{d}^d k}{(k^2)^2(q-k)^2} - 2m^2 \int \frac{\mathrm{d}^d k}{(k^2)^3(q-k)^2} + \ldots$$
$$+\frac{1}{q^2}\int \frac{\mathrm{d}^d k}{(k^2-m^2)^2} + \frac{1}{(q^2)^2}\int \frac{(2q\cdot k - k^2)\,\mathrm{d}^d k}{(k^2-m^2)^2} + \ldots\,. \quad (3.25)$$

Evaluating the integrals on the right-hand side by use of (A.6) and (A.7), we obtain the following result:

$$F_\Gamma(q^2,m^2;d) \sim \frac{\mathrm{i}\pi^{d/2}}{(-q^2)^{1+\varepsilon}} \frac{\Gamma(1-\varepsilon)^2\Gamma(\varepsilon)}{\Gamma(1-2\varepsilon)} \left(1 + 2\varepsilon\frac{m^2}{q^2} + \ldots\right)$$
$$+\frac{\mathrm{i}\pi^{d/2}}{q^2(m^2)^\varepsilon}\Gamma(\varepsilon)\left(1+\frac{\varepsilon}{1+\varepsilon}\frac{m^2}{q^2}+\ldots\right)\,. \quad (3.26)$$

The pole in ε in the contribution from the region of large k is of IR nature, while the pole in the contribution from the region of small k is UV. In fact, poles are present only in the leading-order terms. We observe that the poles are cancelled to produce a result finite at $\varepsilon = 0$:

$$F_\Gamma(q^2,m^2;4) \sim \frac{\mathrm{i}\pi^2}{q^2}\left[\ln\left(\frac{-q^2}{m^2}\right) - \frac{m^2}{q^2} + \ldots\right]\,. \quad (3.27)$$

An arbitrary, nth, term of the expansion can be easily evaluated, with a subsequent summation to obtain the explicit analytic results (2.58) and (2.59).

3.3 Expansion by Subgraphs

Now we turn to the second strategy. According to the definition of the asymptotic expansion (see Sect. 3.1), we can take a sufficiently large number of initial terms of the expansion in order to provide the desired asymptotic behaviour of the remainder. Let us define the remainder of order n for our toy example by

$$\mathcal{R}^n F(q,m;\varepsilon) = \int_0^\infty \mathrm{d}k\, k^{-\varepsilon}\left(\frac{1}{k+m} - \frac{1}{k} + \frac{m}{k^2} - \ldots - (-1)^n\frac{m^n}{k^{n+1}}\right)$$
$$\times\left(\frac{1}{k+q} - \frac{1}{q} + \frac{k}{q^2} - \ldots - (-1)^n\frac{k^n}{q^{n+1}}\right)\,. \quad (3.28)$$

We can rewrite this expression more concisely as

$$\mathcal{R}^n F(q,m;\varepsilon) = \int_0^\infty \mathrm{d}k\, k^{-\varepsilon}\left((1-\mathcal{T}^n_m)\frac{1}{k+m}\right)\left((1-\mathcal{T}^n_k)\frac{1}{k+q}\right)\,, \quad (3.29)$$

where $\mathcal{T}^n$ is the operator of Taylor expansion of order n in the corresponding variables, i.e., in the one-dimensional case,

$$\mathcal{T}^n_x f(x) = \sum_{j=0}^n \frac{1}{j!} f^{(j)}(0) x^j\,. \quad (3.30)$$

Since the remainder involves the operator $(1 - \mathcal{T}_m^n)$, we see that it has an asymptotic behaviour of order m^{n+1} (modulo logarithms), provided the corresponding integral is convergent when $\varepsilon \to 0$. To show this convergence, let us consider first the region of small k. Here the terms originating from $\mathcal{T}_m^n$ are dangerous because they involve negative powers of k, the last term, containing m^n/k^{n+1}, being the most dangerous of them. However this 'potential' divergence is removed by the expression in the other parentheses in the remainder. To see this, it suffices to apply a standard formula for the remainder of a Taylor expansion (e.g. (B.40)) and conclude that the operator $(1 - \mathcal{T}_k^n)$ generates an additional factor k^{n+1} in the numerator. Counting powers, we can convince ourselves that there is convergence at small k.

The situation in the region of large k is quite opposite: the operator $\mathcal{T}_k^n$ generates 'dangerous' positive powers of k there. However, the combination $(1 - \mathcal{T}_m^n)$ successfully cancels these powers. These are the arguments that will be generalized further to the analysis of general Feynman integrals in off-shell limits of momenta and masses. It should be pointed out, however, that in this toy example, a more explicit estimate is possible: using a well-known formula for the sum of the initial terms of a geometric progression, we obtain a very simple relation,

$$\mathcal{R}^n F(q, m; \varepsilon) = \left(\frac{m}{q}\right)^{n+1} F(q, m; \varepsilon) \,, \tag{3.31}$$

which makes trivial the above statements about the remainder.

The operation that determines the remainder can be written as

$$\mathcal{R}^n = (1 - \mathcal{M}^n_{\text{large}})(1 - \mathcal{M}^n_{\text{small}}) \,, \tag{3.32}$$

where $\mathcal{M}^n_{\text{large}}$ corresponds to $\mathcal{T}_m^n$ and $\mathcal{M}^n_{\text{small}}$ to $\mathcal{T}_k^n$. Starting from (3.29), let us rewrite the original integral as

$$\begin{aligned} F &= (1 - \mathcal{R}^n)\, F + \mathcal{R}^n F \\ &\equiv (\mathcal{M}^n_{\text{large}} + \mathcal{M}^n_{\text{small}} - \mathcal{M}^n_{\text{large}}\mathcal{M}^n_{\text{small}}) F + \mathcal{R}^n F \,. \end{aligned} \tag{3.33}$$

The last term is the remainder. The term containing the product of two subtraction operators produces scaleless integrals, which are naturally set to zero here. However we do not need to use such ad hoc prescriptions here. Rather, we again divide the integral into two pieces, with separate terms corresponding to the right-hand side of (3.33), so that these scaleless integrals arise by definition as combinations of two integrals evaluated in their own domains of the regularization parameter ε, where they are convergent and then produce zero in the sum. Now we let n tend to infinity and observe that the remaining two terms in the brackets generate nothing but the terms of the asymptotic expansion (3.15), so that we have obtained the same result in another way.

For our Feynman integral, the corresponding procedure looks quite similar. The remainder is defined as

$$\mathcal{R}^n F_\Gamma(q^2, m^2; d) = \int \mathrm{d}^d k \left((1 - \mathcal{T}^n_m) \frac{1}{(k^2 - m^2)^2} \right) \left((1 - \mathcal{T}^n_k) \frac{1}{(q-k)^2} \right)$$
$$\equiv (1 - \mathcal{M}^n_{\text{large}})(1 - \mathcal{M}^n_{\text{small}}) F_\Gamma(q^2, m^2; d) \; ; \qquad (3.34)$$

the only difference is that the operator $\mathcal{T}^n_k$ now performs Taylor expansion in a d-dimensional rather than a one-dimensional variable. To show the convergence and obtain the asymptotic estimate, we can simply repeat the arguments presented after (3.30).

The operators involved can be associated naturally with subgraphs of the given graph. The operator $\mathcal{M}^n_{\text{large}}$ can be labelled as $\mathcal{M}_\Gamma$ because it performs Taylor expansion in the small parameter m of the whole integrand of the Feynman integral. (It acts only on the propagator containing the mass m but would act on the second propagator if the corresponding (small) mass were non-zero.) On the other hand, the second operator, $\mathcal{M}^n_{\text{small}}$, acts on the expression $1/(q-k)^2$ associated with the subgraph γ consisting of a single massless line. This action reduces to expanding the expression into a Taylor series in the loop momentum k of the whole graph which is external for the subgraph. Let us now adopt the following convention:

- we consider all the loop momenta to be small *by definition* when associating subtraction operators with subgraphs.

The second operator involved is then naturally identified as $\mathcal{M}_\gamma$ because it is the Taylor expansion operator with respect to the small parameters of the subgraph γ. Here we introduce one more convention:

- when considering an operator corresponding to a subgraph, we choose a routing of the external momenta in such a way that they all flow through the given subgraph.

Let us now ask why there is no contribution from the second subgraph, consisting of two successive massive lines. According to the second of the above conventions, we choose the loop momentum by letting the external momentum flow through the massive lines. We then obtain the following contribution from this subgraph:

$$\int \mathrm{d}^d k \, \frac{1}{k^2} \mathcal{T}^n_{k,m} \frac{1}{[(q-k)^2 - m^2]^2} = \frac{1}{(q^2)^2} \int \mathrm{d}^d k \, \frac{1}{k^2} + \ldots \; . \qquad (3.35)$$

Here the corresponding Taylor operator acts on m and (the new loop momentum) k, which are considered as small parameters with respect to this subgraph. However, every resulting term is a scaleless integral, which we set to zero. So the answer is that there is indeed a contribution from the second subgraph but it is zero. However, it would be non-zero if this subgraph had a non-zero mass – see Example 4.2 in Sect. 4.1.

Thus we have the following correspondence: $\mathcal{M}^n_\Gamma = \mathcal{M}^n_{\text{large}}$ and $\mathcal{M}^n_\gamma = \mathcal{M}^n_{\text{small}}$. We can rewrite the remainder (3.34) as

$$\mathcal{R}^n F_\Gamma = \left(1 - \mathcal{M}^n_\Gamma\right)\left(1 - \mathcal{M}^n_\gamma\right) F_\Gamma \; . \qquad (3.36)$$

We then use the same manipulations as in (3.33) and obtain

$$F_\Gamma = (\mathcal{M}^n_\Gamma + \mathcal{M}^n_\gamma - \mathcal{M}^n_\Gamma \mathcal{M}^n_\gamma) F_\Gamma + \mathcal{R}^n F_\Gamma \,. \tag{3.37}$$

We want to consider individual terms separately. It turns out that they possess divergences, so that we introduce a regularization (which we always choose to be dimensional) even if the original Feynman integral is UV and IR finite. Let us also observe that the (dimensionally regularized) Feynman integral expanded by the operator $\mathcal{M}_\gamma$,

$$\mathcal{M}_\gamma F_\Gamma = \int \mathrm{d}^d k \, \frac{1}{(k^2-m^2)^2} \mathcal{T}^n_k \frac{1}{(q-k)^2} \tag{3.38}$$

(equal to the last line of (3.25)), can be interpreted as an integral for the reduced graph Γ/γ (which is obtained from Γ by contracting the subgraph γ to a point and is nothing but a tadpole consisting of the two massive lines), where the expansion of the expression corresponding to the subgraph $\mathcal{M}_\gamma F_\gamma$ in its small external momentum k is inserted. We formally write down this insertion as

$$\mathcal{M}_\gamma F_\Gamma = F_{\Gamma/\gamma} \circ \mathcal{M}_\gamma F_\gamma \,, \tag{3.39}$$

where we use the same symbol for insertion of a polynomial into a reduced vertex as in (2.84).

For both parts on the right-hand side in (3.37), there is a domain of the regularization parameter ε where they are convergent. We can introduce (in advance) an auxiliary analytic regularization in an appropriate way into various UV/IR sectors in the alpha representation, as explained in Sect. 2.4, and consider separately each term of the first part. The third term is then set to zero according to one the properties of dimensional regularization. By this process, we arrive at the asymptotic large-momentum expansion in the form

$$F_\Gamma \sim \mathcal{M}_\Gamma F_\Gamma + F_{\Gamma/\gamma} \circ \mathcal{M}_\gamma F_\gamma \,, \tag{3.40}$$

and thereby reproduce results (3.25) and (3.26) obtained earlier within the strategy of expansion by regions.

The operation (3.32) clearly has the structure of an R-operation and corresponds to the situation where a given graph and one of its subgraphs are divergent – see (2.88). It turns out that the similarity is rather deep and applies to an arbitrary Feynman integral in an off-shell limit. Although the R-operation and the operation that determines the remainder of the asymptotic expansion have the same structure, the corresponding subtraction operators serve different purposes. The subtraction operators entering the R-operation remove UV divergences, while the goal of the operators in the remainders of the asymptotic expansions is to remove a sufficiently large number of initial terms of the expansion in the small parameters of the problem.

We shall see in the next two chapters how the remainder for a general diagram in an off-shell limit of momenta and masses can be represented, by the forest formula, through appropriate subtractions in a certain family of subgraphs of the given graph. This will enable us to obtain, in Chap. 4, a straightforward generalization of (3.40) to any diagram with an *arbitrary* number of loops. On the other hand, this very representation of the remainder is adequate for the proof of the asymptotic expansion presented in Appendix B.2, which resembles very much proofs of the so-called Bogoliubov–Parasiuk theorem on renormalization.

4 Off-Shell Large-Momentum Expansion

In this and the next chapter, we investigate two limits where the method of expansion by subgraphs can be applied and justified. First we consider the off-shell large-momentum limit, i.e. when all the large momenta are large in the Euclidean sense and there are no large masses. We have already seen an example of a one-loop diagram in this limit in Chap. 3. After investigating two more one-loop examples in Sect. 4.1 we formulate, in Sect. 4.2, general prescriptions for the expansion and compare them with the prescriptions for expansion by regions. We continue with two-loop examples and then go to the operator level by investigating the concept of the operator product expansion (OPE), which is the operator analogue of the diagrammatic off-shell large-momentum expansion. The chapter is concluded by presenting typical applications of OPE. We shall discuss two-point quark current correlators in the large-momentum limit and the calculation of the total cross-section for hadron production in e^+e^- annihilation.

4.1 One-Loop Examples

Let us continue to consider instructive one-loop examples of the off-shell large-momentum expansion. The next example to be considered is the following.

Example 4.1. The one-loop triangle diagram of Fig. 1.1 with all masses equal to zero, and one large and one small external momentum, q and p, respectively.

This is the diagram discussed in the introduction, where we derived its leading asymptotics and illustrated the two basic strategies. The corresponding Feynman integral is

$$F_{4.1}(q,p;d) = \int \frac{\mathrm{d}^d k}{k^2(q-k)^2(p+k)^2} \,. \tag{4.1}$$

Guided by our first example in Chap. 3, let us construct an appropriate remainder of the expansion in the limit $q \to \infty$. As in Chap. 3, we introduce dimensional regularization for the finite initial quantity because we know in advance that divergences can appear in individual terms of the expansion. In fact, the situation looks quite similar to the previous case: the factor $1/(k^2-m^2)^2$ is replaced here by $1/[k^2(p+k)^2]$, and the Taylor expansion of the integrand in p here generates IR divergences, starting from the zero-order term as the previous expansion in m. In the previous example, the remainder was constructed by including the operator $(1 - \mathcal{T}_k^n)$, together with $(1 - \mathcal{T}_m^n)$. After that there appeared a balance between subtraction operators that provided the convergence and the desired asymptotic behaviour. So we construct the new remainder in the same way:

$$\mathcal{R}^n F_{4.1}(q,p;d) = \int \frac{\mathrm{d}^d k}{k^2} \left(\left(1 - \mathcal{T}_p^n\right) \frac{1}{(p+k)^2} \right) \left(\left(1 - \mathcal{T}_k^n\right) \frac{1}{(q-k)^2} \right) . \tag{4.2}$$

The arguments which show the convergence and the asymptotic behaviour of order $n+1$ are exactly as in the previous case: in the region of small k, the operator $\mathcal{T}_p^n$ is dangerous but the factor $(1 - \mathcal{T}_k^n)$ cures these potential divergences and, in the region of large k, the operator $\left(1 - \mathcal{T}_p^n\right)$ removes the divergences caused by $\mathcal{T}_k^n$.

We again identify the operators involved with subgraphs of the given graph. The operator $\mathcal{T}_p^n$ performs Taylor expansion of the integrand with respect to the small quantity, p, of the problem and corresponds to the whole graph. The operator $\mathcal{T}_k^n$ performs Taylor expansion of the factor $1/(q-k)^2$ associated with the subgraph consisting of this line and expands it in the momentum k which is the loop momentum for the whole graph and is considered small. Symbolically, we have

$$\mathcal{R}^n = \left(1 - \mathcal{M}_\Gamma^n\right) \left(1 - \mathcal{M}_\gamma^n\right) , \tag{4.3}$$

where the structure of the forest formula (2.87) can be recognized.

We then write down the given Feynman integral as

$$F_{4.1} = \left(1 - \mathcal{R}^n\right) F_{4.1} + \mathcal{R}^n F_{4.1} , \tag{4.4}$$

where the last term is the remainder of the expansion, and substitute $\mathcal{R}^n$, given by (4.2), into the first term on the right-hand side. Any product of two different subtraction operators $\mathcal{M}$ generates massless integrals with zero external momenta, which are zero within dimensional regularization. Using the same arguments as in the one-loop example in Chap. 3, we observe that these integrals vanish *automatically.* Therefore only two terms survive, and we arrive at the expansion

$$F_{4.1} = \mathcal{M}_\Gamma^n F_{4.1} + \mathcal{M}_\gamma^n F_{4.1} + \mathcal{R}^n F_{4.1} , \tag{4.5}$$

with terms up to order n with respect to p.

Letting n tend to infinity, we again obtain an asymptotic expansion of the form (3.40):

$$\begin{aligned} F_{4.1}(q,p;d) \sim & \int \frac{\mathrm{d}^d k}{k^2 (q-k)^2} \left(\frac{1}{k^2} - \frac{2p\cdot k + p^2}{(k^2)^2} + \ldots \right) \\ & + \int \frac{\mathrm{d}^d k}{k^2 (p+k)^2} \left(\frac{1}{q^2} + \frac{2q\cdot k - k^2}{(q^2)^2} + \ldots \right) . \end{aligned} \tag{4.6}$$

Each term in the first series (originating from $\mathcal{M}_\Gamma F_{4.1}$) and in the second series (from $\mathcal{M}_\gamma F_{4.1}$) can be analytically evaluated by means of (A.10):

$$
\begin{aligned}
F_{4.1}(q,p;d) \sim \mathrm{i}\pi^{d/2} \Biggl(& \frac{\Gamma(1-\varepsilon)^2\Gamma(\varepsilon)}{\Gamma(1-2\varepsilon)} \frac{1}{(-q^2)^{1+\varepsilon}} \\
& - \frac{\Gamma(1-\varepsilon)\Gamma(-\varepsilon)\Gamma(2+\varepsilon)}{\Gamma(1-2\varepsilon)} \frac{q\cdot p}{(-q^2)^{2+\varepsilon}} + \ldots \\
& + \frac{\Gamma(1-\varepsilon)^2\Gamma(\varepsilon)}{\Gamma(2-2\varepsilon)} \frac{1}{q^2(-p^2)^{\varepsilon}} \\
& - \frac{2\Gamma(1-\varepsilon)\Gamma(2-\varepsilon)\Gamma(\varepsilon)}{\Gamma(3-2\varepsilon)} \frac{q\cdot p}{(q^2)^2(-p^2)^{\varepsilon}} \Biggr) + \ldots . \qquad (4.7)
\end{aligned}
$$

We have, on the right-hand side of (4.7), an interplay of IR (in the first series) and UV (in the second series) divergences, which are cancelled in the sum. We obtain, at $\varepsilon = 0$,

$$
\begin{aligned}
F_{4.1}(q,p;4) \sim \mathrm{i}\pi^2 \Biggl(& \frac{2+\ln(q^2/p^2)}{q^2} - \frac{1+\ln(q^2/p^2)}{(q^2)^2} q\cdot p \\
& - \frac{2+3\ln(q^2/p^2)}{9(q^2)^3} \left[q^2p^2 - 4(q\cdot p)^2\right] \\
& + \frac{1+2\ln(q^2/p^2)}{2(q^2)^4} \left[q^2p^2 - 2(q\cdot p)^2\right] q\cdot p \\
& + \frac{2+5\ln(q^2/p^2)}{25(q^2)^5} \left[(q^2p^2)^2 - 12(q\cdot p)^2 q^2p^2 + 16(q\cdot p)^4\right] \Biggr) + \ldots . \qquad (4.8)
\end{aligned}
$$

This expansion is in agreement with the following explicit result [145, 233]:

$$
F_{4.1}(q,p;4) = \frac{\mathrm{i}\pi^2}{q^2} \Phi_1\left(\frac{p^2}{q^2}, \frac{(p+q)^2}{q^2}\right) , \qquad (4.9)
$$

where

$$
\begin{aligned}
\Phi_1(x,y) = \frac{1}{\lambda} \Biggl(& 2\left[\mathrm{Li}_2(-\rho x) + \mathrm{Li}_2(-\rho y)\right] \\
& + \ln\frac{y}{x} \ln\frac{1+\rho y}{1+\rho x} + \ln(\rho x)\ln(\rho y) + \frac{\pi^2}{3} \Biggr) , \qquad (4.10)
\end{aligned}
$$

$$
\lambda(x,y) = \sqrt{(1-x-y)^2 - 4xy} , \quad \rho(x,y) = \frac{2}{1-x-y+\lambda(x,y)} , \qquad (4.11)
$$

and $\mathrm{Li}_2(z)$ is the dilogarithm (see (A.57)).

Let us now expand, in the large-momentum limit, a more general one-loop diagram than the example treated in Chap. 3.

Example 4.2. The Feynman integral (2.23) with two non-zero masses, m_1 and m_2, in the limit $m_i^2/q^2 \to 0$.

Remember that the expansion of the diagram, with one non-zero mass, was determined by the remainder (3.34). The operator $\mathcal{T}_m$ performs expansion and improves convergence at large values of the loop momentum, while the operator $\mathcal{T}_k$ improves convergence at small values of the loop momentum and cures dangerous terms generated by $\mathcal{T}^n_m$. Let us generalize this construction to the new case. Since there are now non-zero (small) masses in both lines, the Taylor expansion in m_1^2 and m_2^2 generates IR divergences not only at $k = 0$ but also at $k = q$. Thus, in addition to the operator $\mathcal{M}_1^n$, which reduces to the Taylor expansion operator $\mathcal{T}^n_{k,m_1}$ acting on the propagator $1/[(q-k)^2 - m_1^2]$, we need to include a similar operator, $\mathcal{M}_2^n$, which reduces to the Taylor expansion operator $\mathcal{T}^n_{k-q,m_2}$ acting on the propagator $1/(k^2 - m_1^2)$. (One can also change the variable by means of the replacement $k \to q - k$, take the expression for $\mathcal{M}_1$ and then change the variable back.)

We define the remainder as

$$\mathcal{R}^n F_{4.2} = (1 - \mathcal{M}^{n+2})(1 - \mathcal{M}_1^n - \mathcal{M}_2^n) F_{4.2} \,. \tag{4.12}$$

As in the previous example, it has the structure of the forest formula (2.87). Using explicit representations for the operators $\mathcal{M}_i$ we obtain

$$\mathcal{R}^n F_{4.2}(q) = \int \mathrm{d}k \left(1 - \mathcal{T}^{n+2}_{m_1,m_2}\right) \left[\frac{1}{(q-k)^2 - m_1^2} \frac{1}{k^2 - m_2^2} \right.$$
$$- \frac{1}{k^2 - m_2^2} \left(\frac{1}{q^2} + \frac{m_1^2 + 2q\cdot k - k^2}{(q^2)^2} + \frac{(m_1^2 + 2q\cdot k - k^2)^2}{(q^2)^3} + \ldots \right)$$
$$\left. - \frac{1}{(q-k)^2 - m_1^2} \left(\frac{1}{q^2} + \frac{m_2^2 + q^2 - k^2}{(q^2)^2} + \frac{(m_2^2 + q^2 - k^2)^2}{(q^2)^3} + \ldots \right) \right] \,. \tag{4.13}$$

The operators $\mathcal{M}_i$ are naturally associated with the subgraphs γ_1 and γ_2, consisting of lines 1 and 2, respectively. We use here the shorthand notation $\mathcal{M}_i$ instead of $\mathcal{M}_{\gamma_i}$, and also $\mathcal{M}$ for $\mathcal{M}_\Gamma$.

An asymptotic behaviour of order $n+3$ up to logarithms will be achieved because of the operator $\left(1 - \mathcal{T}^{n+2}_{m_1,m_2}\right)$ if the remainder is UV and IR finite. Since we now have two overlapping subgraphs that contribute to the expansion, the power counting in the remainder (as in the BPHZ R-operation) is a little more complicated and we shall need some simple properties of Taylor expansion operators. Let $\mathcal{T}^{(n)}$ denote the terms of the nth order in the Taylor expansion operator $\mathcal{T}^n$ so that

$$\mathcal{T}^n = \sum_{j=0}^{n} \mathcal{T}^{(j)} \,. \tag{4.14}$$

In the case of one variable, we have

$$\mathcal{T}_x^{(j)} f(x) = \frac{f^{(j)}(0)}{j!} x^j \,, \tag{4.15}$$

and the following commutation relations:

$$\mathcal{T}_x^n x^j = x^j \mathcal{T}_x^{n-j}, \quad (1 - \mathcal{T}_x^n) x^j = x^j (1 - \mathcal{T}_x^{n-j}) \,. \tag{4.16}$$

In the case of two variables, the Leibniz rule gives

$$\mathcal{T}_{x,y}^{(n)} = \sum_{j=0}^{n} \mathcal{T}_x^{(j)} \mathcal{T}_y^{(n-j)} , \tag{4.17}$$

$$\mathcal{T}_{x,y}^{n} = \sum_{j=0}^{n} \mathcal{T}_x^{(j)} \mathcal{T}_y^{n-j} = \sum_{j=0}^{n} \mathcal{T}_x^{j} \mathcal{T}_y^{(n-j)} . \tag{4.18}$$

These relations are also valid when x and y denote two groups of variables.

To analyse the (UV) convergence in the region of large k, we consider separately the terms in $(1 - \mathcal{M}_1^n - \mathcal{M}_2^n)$. The first of them is UV convergent and the other two are similar. Let us choose $\mathcal{M}_1^n$ and consider the term $\mathcal{M}_1^{(n)}$ with the highest degree of expansion. Using the commutation relations (4.16), we transform the integrand as follows:

$$\begin{aligned} &\left(1 - \mathcal{T}_{m_1,m_2}^{n+2}\right) \frac{1}{k^2 - m_2^2} \sum_{j=0}^{n} \mathcal{T}_k^{(n-j)} \mathcal{T}_{m_1}^{(j)} \frac{1}{(q-k)^2 - m_1^2} \\ &= \sum_{j=0}^{n} \left(\mathcal{T}_k^{(n-j)} \mathcal{T}_{m_1}^{(j)} \frac{1}{(q-k)^2 - m_1^2} \right) \left(1 - \mathcal{T}_{m_2}^{n+2-j}\right) \frac{1}{k^2 - m_2^2} . \end{aligned} \tag{4.19}$$

Counting powers, we obtain $1/k^{5+n-j}$ and k^{n-j} from the last factor (where we use a formula for the remainder of the Taylor expansion, e.g. in the form (B.40)) and the first factor, respectively. Thus we have the asymptotic behaviour $1/k^5$, which guarantees convergence.

At finite values of k, there are only two dangerous points, $k = 0$ and $k = q$, where the Taylor expansion in the masses can generate IR divergences. The analysis of convergence at these points is quite similar, so that we may confine ourselves to the point $k = 0$. First, we observe that in this region the terms originating from $\mathcal{M}_2$ are not dangerous, because this operator puts the second denominator equal to q^2 and $\mathcal{T}_{m_2}$ cannot generate an IR divergence at $k = 0$ after that. We then consider $1 - \mathcal{M}_1$ and the most dangerous term in $\left(1 - \mathcal{T}_{m_1,m_2}^{n+2}\right)$, i.e. $-\mathcal{T}_{m_1,m_2}^{(n+2)}$. We have

$$- \sum_{j=0}^{n+2} \mathcal{T}_{m_1}^{(n+2-j)} \mathcal{T}_{m_2}^{(j)} \frac{1}{k^2 - m_2^2} \left[\left(1 - \mathcal{T}_{m_1,k}^{n}\right) \frac{1}{(q-k)^2 - m_1^2} \right] . \tag{4.20}$$

The expression in the square brackets can be represented as

$$\sum_{i=0}^{n+1} p_i(k) m_1^{n-i+1} f_i(k, m_1) , \tag{4.21}$$

where $p_i(k)$ is a monomial of degree i in k, and $f_i(k, m_1)$ is finite in the vicinity of the point $k = 0, m_1 = 0$. Using the commutation relations (4.16), we rewrite (4.20) as

$$\sum_{j=0}^{n+2} \sum_{i=0}^{n+1} p_i(k) m_1^{n-i+1} \left[\mathcal{T}_{m_1}^{(n+2-j)} f_i(k, m_1) \right] \left[\mathcal{T}_{m_2}^{(j)} \frac{1}{k^2 - m_2^2} \right] . \tag{4.22}$$

Counting powers of k at small k, we obtain k^i from $p_i(k)$ and $1/k^{2+j}$ from the last square brackets, so that the integral is convergent at $k=0$.

Now we again write down the given Feynman integral as (4.4), set to zero terms containing products of different operators and arrive at the expansion

$$F_{4.2} = \mathcal{M}_1^n F_{4.2} + \mathcal{M}_2^n F_{4.2} + \mathcal{M}^{n+2} F_{4.2} + \mathcal{R}^n F_{4.2} \,. \tag{4.23}$$

Letting n tend to infinity, we obtain

$$F_{4.2} \sim \mathcal{M}_{\gamma_1} F_{4.2} + \mathcal{M}_{\gamma_2} F_{4.2} + \mathcal{M} F_{4.2} \,. \tag{4.24}$$

All the terms in the series $\mathcal{M}F_{4.2}$ (corresponding to the whole graph) and $\mathcal{M}_i F_{4.2}$ (corresponding to the two subgraphs) can be analytically evaluated by means of (A.3) and (A.7), respectively. We obtain

$$\mathcal{M}F_{4.2} \sim \mathrm{i}\pi^{d/2} \left(\frac{\Gamma(1-\varepsilon)^2\Gamma(\varepsilon)}{\Gamma(2-2\varepsilon)(-q^2)^{\varepsilon}} + \frac{\Gamma(1-\varepsilon)^2\Gamma(\varepsilon)(m_1^2+m_2^2)}{\Gamma(1-2\varepsilon)(-q^2)^{1+\varepsilon}} \right.$$
$$\left. + \frac{[2m_1^2m_2^2 + \varepsilon(m_1^2+m_2^2)^2]\Gamma(-\varepsilon)^2\Gamma(1+\varepsilon)}{2\Gamma(-2\varepsilon)(-q^2)^{2+\varepsilon}} \right) + \ldots \,, \tag{4.25}$$

$$\mathcal{M}_1 F_{4.2} + \mathcal{M}_2 F_{4.2} \sim \mathrm{i}\pi^{d/2} \left[-\left(m_1^{1-\varepsilon} + m_2^{1-\varepsilon}\right) \frac{\Gamma(\varepsilon-1)}{q^2} \right.$$
$$- \left(m_1^2 m_2^2 \left(m_1^{-\varepsilon} + m_2^{-\varepsilon}\right) + \frac{\varepsilon}{2-\varepsilon}\left(m_1^{2-\varepsilon} + m_2^{2-\varepsilon}\right) \right)$$
$$\left. \times \frac{\Gamma(\varepsilon-1)}{(q^2)^2} \right] + \ldots \,. \tag{4.26}$$

The pole in the first term in (4.25) is of UV nature and is present in the original Feynman integral. All the other poles in the above series are artificial: these are IR poles in (4.25) and UV poles in (4.26). They are cancelled in the result when it is written as a Laurent expansion up to ε^0:

$$F_{4.2} \sim \mathrm{i}\pi^2 \left(2 - \ln(-q^2) + \frac{1}{\varepsilon} \right.$$
$$+ \left[m_1^2 + m_2^2 - m_1^2 \ln\left(-m_1^2/q^2\right) - m_2^2 \ln\left(-m_2^2/q^2\right) \right] \frac{1}{q^2}$$
$$\left. - \left\{ m_1^4 + m_2^4 + 2m_1^2m_2^2 \left[\ln\left(-m_1^2/q^2\right) + \ln\left(-m_2^2/q^2\right) \right] \right\} \frac{1}{2(q^2)^2} \right)$$
$$+ O\left(\frac{\ln(-q^2)}{(q^2)^3} \right) \,. \tag{4.27}$$

4.2 General Prescriptions

Let us now formulate explicit prescriptions for the asymptotic expansion in the off-shell large-momentum limit. We consider a general Feynman diagram

F_Γ which depends on the large Euclidean external momenta $Q_1, \ldots, Q_{n_1}$ and the small external momenta $q_1, \ldots, q_{n_2}$. All the masses are supposed to be small. Let us consider first a convergent diagram.

As in the previous examples, we start from an appropriate remainder. In accordance with experience obtained from these examples, we define the remainder by means of the forest formula (2.87):

$$\mathcal{R}F_\Gamma = \sum_f \prod_{\gamma \in f} \left(-\mathcal{M}_\gamma^{a(\gamma)}\right) F_\Gamma \,. \tag{4.28}$$

We now need to characterize

- the set of subgraphs in this sum,
- the corresponding subtraction operators,
- how the order of expansion given by the subtraction degrees $a(\gamma)$ is chosen.

Remember that the subtraction operators used in the previous examples provided factors that tended to zero sufficiently fast at zero values of the loop momenta. This was necessary to compensate the singular behaviour generated by Taylor expansion in the masses and the small external momenta. Suppose that we are dealing with an h-loop Feynman integral F_Γ. Let γ be a subgraph of Γ and let $\bar{\gamma}$ be composed of the lines that do not belong to γ. We can choose[1] loop momenta starting from γ and continuing with the rest of the lines so that the rest of the loop momenta correspond to those of the reduced graph $\Gamma/\bar{\gamma}$ [181]:

$$\begin{aligned} F_\Gamma &= \int \mathrm{d}k_1 \ldots \mathrm{d}k_h \, \Pi_\Gamma \\ &= \int \mathrm{d}k_1 \ldots \mathrm{d}k_{h(\gamma)} \, \Pi_\gamma \int \mathrm{d}k_{h(\gamma)+1} \ldots \mathrm{d}k_h \, \Pi_{\bar{\gamma}} \,, \end{aligned} \tag{4.29}$$

where the $\Pi_{\ldots}$ denote products of propagators. Note that the number $h-h(\gamma)$ is nothing but the number of independent loop momenta of the reduced graph Γ/γ and that we can label the product $\Pi_{\bar{\gamma}}$ as $\Pi_{\Gamma/\gamma}$ also.

Suppose that we want to remove divergences in the loop momenta of Γ/γ that are caused by expanding in the masses and the small external momenta of the whole graph. Our examples tell us that we may do this by Taylor expanding the rest of the integrand, Π_γ, in these variables $k_{h(\gamma)+1}, \ldots, k_h$. It is sufficient to take care only of those products where no dependence on the large external momenta Q_i is present because, otherwise, the corresponding propagators never become singular when they are expanded in the masses and the small momenta. We come to the conclusion that we have to perform subtractions only in those γ that can involve the flow of all the large external momenta. Moreover, it should be possible to distribute the flow of the large

[1] One can choose a family of independent loop momenta by choosing first a tree and then put a loop in correspondence with each line that does not belong to this tree. One can follow this procedure starting from a tree of the subgraph γ.

external momenta through *all* the lines of γ. This means that we exclude, in γ, the existence of lines where only small momenta flow. If we do not do this and perform subtraction in such γ, we shall obtain, owing to momentum conservation, the corresponding propagator at zero values of the momentum and the mass. So such lines should be attributed to $\bar{\gamma}$. (Moreover, there is no sense in removing IR divergences in $\bar{\gamma}$ without such lines, so that these lines have to be included in $\bar{\gamma}$ to make these divergences stronger.)

We shall call subgraphs with the properties formulated above *asymptotically irreducible* (AI), and this is the class of subgraphs that is involved in (4.28).

To translate this 'physical' definition of the AI subgraphs into mathematical language, let us denote by $\hat{\gamma}$ the graph that is obtained from a given subgraph γ by identifying[2] all the external vertices associated with the large external momenta. All the graphs of this form are subgraphs of the graph $\hat{\Gamma}$. So, a subgraph γ is AI if

(*i*) it contains all the vertices with the large external momenta and
(*ii*) $\hat{\gamma}$ is 1PI.

Observe that these two properties correspond to the two physical requirements formulated above.

Now we define the subtraction operator corresponding to an AI subgraph γ as the Taylor expansion operator with respect to its masses and small external momenta:

$$\mathcal{M}_{\gamma}^{a(\gamma)} F_{\Gamma} = \int \mathrm{d}k_1 \ldots \mathrm{d}k_h \, \Pi_{\Gamma/\gamma} \mathcal{T}_{q_1,\ldots,m_1,\ldots,k_{h(\gamma)+1},\ldots,k_h} \Pi_{\gamma} \; . \tag{4.30}$$

As we have agreed earlier, the large external momenta flow through γ, so that there is no dependence on them in $\Pi_{\Gamma/\gamma}$. Let us also remember that the loop momenta $k_{h(\gamma)+1}, \ldots, k_h$ are external for γ and, by definition, are considered small. Finally, we fix the order of the subtraction operators as

$$a(\gamma) = \omega(\gamma) + \bar{a} \; , \tag{4.31}$$

where ω is the UV degree of divergence, and the number of oversubtractions, $\bar{a}$, is chosen to be the same for all the AI subgraphs.

Note that, according to the accepted definition of AI subgraphs, they always have common vertices corresponding to the large external momenta so that they certainly intersect each other. This means that the sum in (4.28) runs over *nests* of AI subgraphs, i.e. families which can be ordered with respect to the inclusion, $\gamma^1 \subset \gamma^2 \subset \ldots$. (Of course, any nest is a forest.)

A general mathematical theorem [207] says that the remainder (4.28), with $a(\gamma)$ given by (4.31), i.e.

$$\mathcal{R}^{\bar{a}} F_{\Gamma} \left(Q_1/\varrho, \ldots, Q_{n_1}/\varrho, q_1, \ldots, q_{n_2}, m_1, \ldots, m_L \right) \; ,$$

---

[2]Another possibility is to introduce a new vertex and connect it with each external vertex by a new line.

has order $\varrho^{\bar{a}+1}$ modulo logarithms when $\varrho \to 0$. A simplified version of the proof of this theorem is presented in Appendix B.2.

To derive the expansion we start, as in our examples, from the remainder with a given number of oversubtractions $\bar{a}$, write down the identity

$$F_\Gamma = \left(1 - \mathcal{R}^{\bar{a}}\right) F_\Gamma + \mathcal{R}^{\bar{a}} F_\Gamma \ , \tag{4.32}$$

substitute $\mathcal{R}$ in $(1 - \mathcal{R})$ written as in (4.28), and consider each term separately. Note that, before this separation, we are dealing with Feynman integrals without scaleless diagrams (i.e. without massless detachable subgraphs). Dimensional regularization can be unambiguously introduced for these Feynman integrals, with the help of an auxiliary analytic regularization, as explained in Sect. 2.4 and Appendix B.1, by analytic continuation from a non-empty domain of regularization parameters where the integrals are convergent. At this point we can switch to an equivalent extended definition, by dividing the whole alpha integral into 2^L pieces where the integration over each α_l is performed over either small or large values, when different contributions to the alpha integral are considered in different domains of the parameters λ_l of analytic regularization and one chooses $\mathrm{Re}\,\lambda_l \gg 0$ and $\mathrm{Re}\,\lambda_l \ll 0$ for the integrations at small and large values, respectively, of α_l.

Then, using the properties of dimensionally regularized Feynman integrals (see Chap. 2), we can separate contributions corresponding to various products of the subtraction operators in $(1 - \mathcal{R}^{\bar{a}})$. In particular, we can set to zero any massless integral that does not depend on the external momenta. But this is a consequence of self-consistent rules for dealing with dimensionally regularized integrals, rather than an additional assumption.

Thus we set to zero each term in the forest formula where a product of at least two subtraction operators is involved, because it involves massless integrals independent of the external momenta. As a result, only the contribution from forests with one element survives. As in our examples, the action of a subtraction operator is described graphically by (3.39), and we obtain the following prescription for the expansion in the off-shell large-momentum limit:

$$F_\Gamma = \sum_\gamma F_{\Gamma/\gamma} \circ \mathcal{M}_\gamma^{a(\gamma)} F_\gamma + \mathcal{R}^{\bar{a}} F_\Gamma \ , \tag{4.33}$$

where the sum runs over all AI subgraphs of Γ and the remainder is explicitly included. Letting the number of oversubtractions $\bar{a}$ tend to infinity, we obtain

$$F_\Gamma \sim \sum_\gamma F_{\Gamma/\gamma} \circ \mathcal{M}_\gamma F_\gamma \ . \tag{4.34}$$

Note that we have fulfilled the two requirements formulated in Sect. 3.1: we have Feynman integrals on the right-hand side and the expansion is in powers and logarithms. Moreover, when ε is non-zero, the expansion is in powers only. Indeed, if $\mathcal{M}_\gamma^{(j)}$ is the contribution from the terms of the jth order of the Taylor expansion involved in $\mathcal{M}_\gamma$ then it has a homogeneity degree equal to $\omega(\gamma) - j - 2h(\gamma)\varepsilon$ with respect to the set of the large external momenta Q_i. Thus, in the contribution from the sum over subgraphs with a given number of loops $h(\gamma) = h_0 \leq h(\Gamma)$, the dependence on ε is of the form $1/(Q^2)^{h_0\varepsilon}$, where Q^2 is a typical scale for the large momenta, i.e. each $Q_i = \xi_i Q$ with fixed numbers ξ_i.

The terms with a given number $\omega(\gamma) - j_\gamma$ generate contributions with the same power dependence.[3] (For example, in the situation with two scales, we arrive at an expansion of the form (3.4) without logarithms.) It should be stressed that such a contribution is finite in ε if the original Feynman integral is finite. In the limit $\varepsilon \to 0$, we obtain an expansion in powers and logarithms, the maximal power of the logarithm being no greater than the number of loops. This estimate, as well as the absence of logarithms when $\varepsilon \neq 0$, are essentially properties of asymptotic expansions typical of Euclidean space.

The expansion of a renormalized Feynman integral RF_Γ (by some R-operation R) in the off-shell large-momentum limit can be obtained by starting from an identity similar to (4.32),

$$RF_\Gamma = \left(R - \mathcal{R}^{\bar{a}}\right) F_\Gamma + \mathcal{R}^{\bar{a}} F_\Gamma \,. \tag{4.35}$$

One then applies a diagrammatic Zimmermann identity for the difference $(R - \mathcal{R}^{\bar{a}})$ and arrives at the following expansion (see, e.g., [208]):

$$R\, F_\Gamma \sim \sum_\gamma \bar{R} F_{\Gamma/\gamma} \circ \mathcal{M}_\gamma R F_\gamma \,, \tag{4.36}$$

where $\bar{R}F_{\Gamma/\gamma}$ is an 'incompletely' renormalized reduced Feynman diagram:

$$\bar{R}\, F_{\Gamma/\gamma} = \sum_{\gamma_1,\ldots,\gamma_j \not\ni v_\gamma} \Delta(\gamma_1)\ldots\Delta(\gamma_j) F_{\Gamma/\gamma} \,. \tag{4.37}$$

Here v_γ is the vertex to which the subgraph γ is collapsed when Γ/γ is produced. In other words, $\bar{R}$ is an incomplete R-operation (as compared with (2.83)) in the sense that the counterterms for subgraphs with this vertex are absent in $\bar{R}$.

The expansion in the form (4.37) was first derived within the method of gluing [46, 47]. Alternatively, one can informally obtain (4.37) from (4.34) by saying that the renormalization can be taken into account by adjusting the bare parameters of the Lagrangian. Then, any reduced diagram on the right-hand side of the expansion obtained will be renormalized with the exception of those subdiagrams which have a vertex of a new type (absent in the initial Lagrangian) generated by collapsing a subgraph to a point.

Since, in the large-momentum limit, the graph itself contributes to the sum in (4.34), there is always the term

$$\mathcal{T}_{q_1,\ldots,m_1,\ldots} F_\Gamma \,, \tag{4.38}$$

[3] Note that the leading-power behaviour in the large-momentum limit, which is described by the Weinberg theorem [237], naturally follows from (4.34).

which is nothing but a formal Taylor expansion (i.e. under the integral sign) in the small parameters of the problem. This series involves IR divergences starting from some minimal order of the expansion. The rest of the terms, corresponding to various subgraphs γ, possess UV divergences as well. From a practical point of view, the cancellation of the artificial poles is a good check of the expansion procedure. It is possible to rewrite (4.34) in a modified form where no new divergences, as compared with the divergences of the original Feynman integral, appear on the right-hand side [47]. To do this, one applies the so-called R^*-operation, which is a generalization of the R-operation to the case of off-shell IR divergences [62].

Let us now remember our first strategy. We have two scales in the problem: the large momenta are of the same order, $Q_i \sim Q$, and the masses and the small momenta are much smaller, $q_i \sim m_l \sim q$, with $q \ll Q$. For a given loop momentum we define regions where this momentum is large or small:

$$\text{large}, \quad k \sim Q \,, \tag{4.39a}$$

$$\text{small}, \quad k \sim q \,. \tag{4.39b}$$

Let us define the set of regions labelled by 1PI subgraphs of the given graph:

$$\begin{aligned} &k_i \sim Q, \quad \text{if } k_i \text{ is a loop momentum of } \gamma \,, \\ &k_i \sim q, \quad \text{if } k_i \text{ is not a loop momentum of } \gamma \,. \end{aligned} \tag{4.40}$$

In the contribution from the region corresponding to a given γ, we can expand every propagator from γ not only in its masses and the small external momenta flowing through it but also in the rest of the loop momenta of the whole graph (which actually correspond to the reduced graph Γ/γ). We thus obtain nothing but the contribution of the subgraph γ within the method of expansion by subgraphs. So we reproduce, within expansion by regions, the general prescriptions (4.34). Observe that there is no need to hesitate and look for other, probably exotic, regions, because we have already reproduced the known, mathematically proven result.

Similar results for explicit prescriptions in the off-shell large-momentum limit exist for Feynman integrals regarded as tempered distributions in the large momenta. See [207, 209], where such a point of view is taken. The corresponding prescriptions are almost the same. Only the class of AI subgraphs is a little bit different: one has to consider subgraphs of the graph $\hat{\Gamma}$ obtained from Γ by identifying all the external vertices associated with the large external momenta. (The empty subgraph is also included.) A distinction between the two versions manifests itself for the so-called contact terms of type $\delta(q)$, which give zero asymptotics at infinity for Feynman integrals considered as functions and certainly contribute to the expansion when Feynman integrals are treated as functionals. This alternative treatment turns out to be adequate in some phenomenological analyses where the contact terms are important (see, e.g., [59]).

4.3 Two-Loop Example

Let us illustrate the general prescriptions through the following example.

Example 4.3. The master two-loop diagram (see Fig. 2.5) with non-zero masses, which are all considered small with respect to the external momentum.

The Feynman integral has the form

$$J_m(a_1,\ldots,a_5;m_1,\ldots,m_5) = \int \frac{\mathrm{d}^d k}{(k^2-m_1^2)^{a_1}[(q-k)^2-m_2^2]^{a_2}} \times \int \frac{\mathrm{d}^d l}{(l^2-m_3^2)^{a_3}[(q-l)^2-m_4^2]^{a_4}[(k-l)^2-m_5^2]^{a_5}} \,. \tag{4.41}$$

The corresponding set of AI subgraphs is shown in Fig. 4.1. The other subgraphs are not AI and do not contribute to (4.34). For example, the subgraph

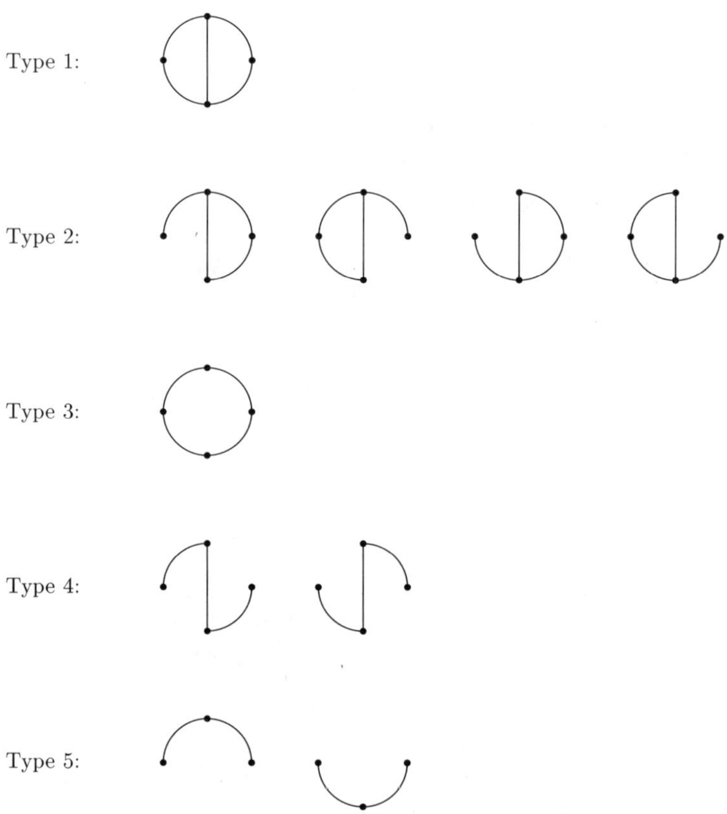

Fig. 4.1. The subgraphs contributing to the large-momentum expansion of the master two-loop diagram

$\{1,2,5\}$ does not have a path between external vertices, and the subgraph $\{1,3,5\}$ is one-particle reducible even after identifying the two external vertices. In 'physical' language, the flow of the external momentum cannot be distributed through line 5.

We shall discuss later the case of five general masses following [94], but let us now choose a specific case, where $m_1 = \ldots = m_4 = m$ and $m_5 = 0$ and all the indices a_i are equal to one.

The contribution of Type 1, i.e. when $\gamma = \Gamma$, is obtained by Taylor expanding the propagators in the masses. The resulting massless two-loop self-energy diagrams $J(a_1, \ldots, a_5)$ given by (2.64) are evaluated by means of the recurrence relations obtained by IBP, presented in Sect. 2.5.6:

$$\sum_{n_1,\ldots,n_4=0}^{\infty} (m^2)^{n_1+n_2+n_3+n_4} J(a_1+n_1, \ldots, a_4+n_4, 1)$$
$$= J(1,1,1,1,1) + 4m^2 J(2,1,1,1,1) + \ldots . \quad (4.42)$$

All four contributions of Type 2 (see Fig. 4.1) are equal to each other because of the symmetry of the diagram. The subgraph $\{2,3,4,5\}$ generates the following contribution:

$$\int \frac{\mathrm{d}^d k}{k^2 - m^2} \int \frac{\mathrm{d}^d l}{l^2 - m^2} \mathcal{T}_{k,m} \frac{1}{[(q-k)^2 - m^2](k-l)^2[(q-l)^2 - m^2]}$$
$$= \frac{1}{q^2} \int \frac{\mathrm{d}^d k}{k^2 - m^2} \int \frac{\mathrm{d}^d l}{(l^2)^2 (q-l)^2} + \ldots$$
$$= \left(\mathrm{i}\pi^{d/2}\right)^2 \frac{\Gamma(\varepsilon - 1) G(2,1)}{(-q^2)^{2+\varepsilon} (m^2)^{\varepsilon - 1}} + \ldots , \quad (4.43)$$

where the function G is given by (2.55). All the terms in the Type 2 contributions are products of massless one-loop integrals and massive vacuum integrals with numerators and can be analytically evaluated by means of (A.2) and (A.10), respectively.

In our simple case, the Type 3 contribution is zero because the fifth line is massless. It would be non-zero in the case $m_5 \neq 0$, when the corresponding series in the result would have the same structure as that of the Type 2 contribution and would be evaluated by means of the same formulae.

For Type 4, we have two equal contributions, from $\gamma = \{1,4,5\}$ and $\{2,3,5\}$. Let $\gamma = \{1,4,5\}$. According to our prescriptions, we choose the loop momenta in a different way and let the external momentum flow through all three lines of the subgraphs. We obtain the following contribution:

$$\int \frac{\mathrm{d}^d k}{k^2 - m^2} \int \frac{\mathrm{d}^d l}{l^2 - m^2} \mathcal{T}_{k,l,m} \frac{1}{[(q-l)^2 - m^2](q-k-l)^2[(q-k)^2 - m^2]}$$
$$= \frac{1}{(q^2)^3} \int \frac{\mathrm{d}^d k}{k^2 - m^2} \int \frac{\mathrm{d}^d l}{l^2 - m^2} + \ldots$$
$$= \left(\mathrm{i}\pi^{d/2}\right)^2 \frac{\Gamma(\varepsilon - 1)^2}{(q^2)^3 (m^2)^{2\varepsilon - 2}} + \ldots , \quad (4.44)$$

where all the terms on the right-hand side are products of tadpoles with numerators evaluated by means of (A.2).

There are two subgraphs of Type 5, $\{1,3\}$ and $\{2,4\}$, with equal contributions. The subgraph $\{2,4\}$ gives

$$\begin{aligned}&\int\int \frac{\mathrm{d}^d k\, \mathrm{d}^d l}{(k^2-m^2)(l^2-m^2)[(k-l)^2-m^2]}\\ &\qquad \times \mathcal{T}_{k,l,m} \frac{1}{[(q-k)^2-m^2][(q-l)^2-m^2]}\\ &\qquad = \frac{1}{(q^2)^2}\int\int \frac{\mathrm{d}^d k\, \mathrm{d}^d l}{(k^2-m^2)(l^2-m^2)[(k-l)^2-m^2]} + \ldots\\ &\qquad = \left(\mathrm{i}\pi^{d/2}\right)^2 \frac{\Gamma(\varepsilon)^2}{(1-\varepsilon)(1-2\varepsilon)(q^2)^2(m^2)^{2\varepsilon-1}} + \ldots\,,\end{aligned} \tag{4.45}$$

where all the terms on the right-hand side can be evaluated by means of (A.38) and its generalization to the case with numerators.

In the leading order (LO) of the expansion of our diagram, only the Type 1 contributes. In the next-to-leading order (NLO), m^2, we also have contributions of Types 2 and 5 because Type 3 gives zero and the Type 4 starts from the order m^4. Although the original diagram is finite, there are poles up to the second order in the individual contributions: IR poles in Type 1, products of UV and IR poles in Types 2–4 and UV poles in Type 5. Collecting the LO and NLO contributions we observe that the poles in ε are cancelled and we obtain the following result:

$$\begin{aligned}&J_m(1,\ldots,1;m,m,m,m,0)\\ &\sim \left(\mathrm{i}\pi^2\right)^2\left(\frac{6\zeta(3)}{q^2}+\frac{2m^2}{(q^2)^2}\left[\ln^2\left(-q^2/m^2\right)+4\ln\left(-q^2/m^2\right)+6\right]\right)+\ldots\,.\end{aligned} \tag{4.46}$$

Even if all five masses are different, every term in the corresponding large-momentum expansion can be analytically evaluated – see [94]. Each term in the contributions of Types 1–4 can be evaluated by means of the same tabulated formulae as in the previous specific case. For Type 5, vacuum two-loop Feynman integrals with three different masses appear:

$$\begin{aligned}&I(a_1,a_2,a_3;m_1,m_2,m_3)\\ &\qquad = \int\int \frac{\mathrm{d}^d k\, \mathrm{d}^d l}{(k^2-m_1^2)^{a_1}(l^2-m_2^2)^{a_2}[(k-l)^2-m_3^2]^{a_3}}\,.\end{aligned} \tag{4.47}$$

For example, in the case of three different masses and the indices $a_i = 1$, one has [95]

$$\begin{aligned}&I\left(1,1,1;m_1,m_2,m_3\right)\\ &= -\left(\mathrm{i}\pi^{d/2}\right)^2 (m_3^2)\frac{A(\varepsilon)}{2}\left(-\frac{1}{\varepsilon^2}(1+x+y)+\frac{2}{\varepsilon}(x\ln x+y\ln y)\right.\\ &\left.-x\ln^2 x-y\ln^2 y+(1-x-y)\ln x\ln y-\lambda(x,y)^2\Phi_2(x,y)\right)\,,\end{aligned} \tag{4.48}$$

where

$$x = \frac{m_1^2}{m_3^2}\,, \quad y = \frac{m_2^2}{m_3^2}\,, \quad \lambda(x,y) = \sqrt{(1-x-y)^2 - 4xy}\,,$$

$$A(\varepsilon) = \frac{\Gamma^2(1+\varepsilon)}{(1-\varepsilon)(1-2\varepsilon)}\,,$$

$$\Phi_2(x,y) = \frac{1}{\lambda}\{2\ln((1+x-y-\lambda)/2)\ln((1-x+y-\lambda)/2) - \ln x \ln y$$
$$-2\mathrm{Li}_2((1+x-y-\lambda)/2) - 2\mathrm{Li}_2((1-x+y-\lambda)/2) + \pi^2/3\}\,, \qquad (4.49)$$

and $\mathrm{Li}_2(z)$is the dilogarithm (see (A.57)). (Similar results were obtained earlier in [21].)

For example, in the case $a_i = 1, i = 1,\ldots,5$ and $\varepsilon = 0$ one has, for general masses [94],

$$J_m(1,\ldots,1;m_1,\ldots,m_5) \sim \left(\mathrm{i}\pi^2\right)^2 \left(\frac{6\zeta(3)}{q^2}\right.$$
$$+\frac{1}{(q^2)^2}\left\{\frac{m_1^2}{2}\left[\ln^2\left(-\frac{q^2}{m_1^2}\right) + 4\ln\left(-\frac{q^2}{m_1^2}\right) - \ln\frac{m_3^2}{m_1^2}\ln\frac{m_5^2}{m_1^2} + 6\right]\right.$$
$$+\left(\text{similar terms containing } m_2^2,\ m_3^2,\ m_4^2\right)$$
$$+\frac{m_5^2}{2}\left[2\ln^2\left(-\frac{q^2}{m_5^2}\right) + 4\ln\left(-\frac{q^2}{m_5^2}\right) - \ln\frac{m_1^2}{m_5^2}\ln\frac{m_3^2}{m_5^2} - \ln\frac{m_2^2}{m_5^2}\ln\frac{m_4^2}{m_5^2}\right]$$
$$\left.\left.+\frac{1}{2}\left[F(m_1^2,m_3^2,m_5^2) + F(m_2^2,m_4^2,m_5^2)\right]\right\}\right) + \ldots\,, \qquad (4.50)$$

where the symmetric function F is defined by

$$F(m_1^2,m_2^2,m_3^2) = m_3^2\lambda(x,y)^2\Phi_2\,(x,y) \qquad (4.51)$$

and (4.49).

Although the existence of the large-momentum expansion is guaranteed only for Euclidean large momenta, it is in fact valid in a larger domain. In particular, our example shows that we can consider the external momentum to be time-like. Note also that the general theorem does not give any information about the radius of convergence of the expansion. Our last example shows that the expansion converges above the highest threshold of the diagram – this can be seen by comparison with known explicit expressions in cases where some non-zero masses are equal [36, 38, 200] and with numerical calculations based on a twofold parametric representation [157] (see [94] for details).

4.4 Operator Product Expansion

The operator analogue of the off-shell large-momentum expansion is the operator product expansion (OPE), i.e. an expansion of time-ordered products of two or more composite operators $J_i(x_i)$ in the limit where the coordinates x_i tend to each other. The expansion[4] of the product of two composite operators is of the form

$$TJ_1(x)J_2(0) \sim \sum_i C_i(x)O_i(0) \, , \tag{4.52}$$

where $\{O_i\}$ is a basis of composite operators of a given theory and the $C_i(x)$ are coefficient functions (Wilson coefficients). The products of quantum field operators are in general singular when the difference between their coordinates vanish, so the first coefficient functions C_i are singular. The expansion is performed in the limit where the difference between the four-coordinates of the operators J_1 and J_2 tends to zero, so that the terms are ordered according to the strength of their singular behaviour when $x \to 0$. On the other hand, the operators O_i are ordered according to their dimension, starting from the unit operator $O_0 = \mathbf{1}$ of zero dimension.

The simplest basis is given by all possible monomials composed of the fields of the given theory and their derivatives taken at the same point. The general monomial is

$$j_{\{\lambda\}}(x) = \phi_1^{(\lambda_1)}(x) \ldots \phi_l^{(\lambda_l)}(x) \, ,$$

$$\phi_i^{(\lambda_i)}(x) = \left(\frac{\partial}{\partial x_{\lambda_{i1}}}\right) \cdots \left(\frac{\partial}{\partial x_{\lambda_{ir_i}}}\right) \phi_i(x) \, ,$$

where $\phi_1, \ldots, \phi_l$ are asymptotic free fields of the theory. The canonical dimension of this monomial, with a multi-Lorentz index $\{\lambda\} = \{(\lambda_1), \ldots, (\lambda_l)\}$, $(\lambda_i) = (\lambda_{i1}, \ldots, \lambda_{ir_i})$, equals $\sum_i (\delta_i + r_i)$ where δ_i is the dimension of the field ϕ_i. Since the composite operators are related to each other, owing to equations of motion, it is reasonable to include in the basis only independent operators so that the most general monomial basis of this kind is inconvenient in practice. On the other hand, in gauge theories, it is important to distinguish gauge-invariant combinations of the operators.

Turning to the Fourier transform with respect to the variable x, we have

$$T\tilde{J}_1(q)J_2(0) \equiv \int \mathrm{d}^4x \, \mathrm{e}^{\mathrm{i}q\cdot x} TJ_1(x)J_2(0) \sim \sum_i \tilde{C}_i(q)O_i(0) \, , \tag{4.53}$$

where the $\tilde{C}_i(q)$ are the Fourier transforms of $C_i(x)$. Now the expansion is performed in the limit[5] $q \to \infty$ and the terms are ordered according to their asymptotic behaviour at large four-momenta q.

[4]Suggested by Wilson [241].

[5]For a tempered distribution $F(x) = F(x_1, x_2, x_3, x_4)$, the behaviour at small x understood as the behaviour of the functional $F(\rho x)$ where the parameter ρ tends to zero, is in one-to-one correspondence with the asymptotic behaviour of its Fourier transform $\tilde{F}(q)$ at large values of q, understood as the behaviour of $\tilde{F}(\Lambda q)$ when $\Lambda \to \infty$. If we consider Feynman integrals as functions (in momentum space) rather than distributions, then the connection of asymptotics at small x to asymptotics at large values of the Fourier transforms is not so straightforward. Still, the power

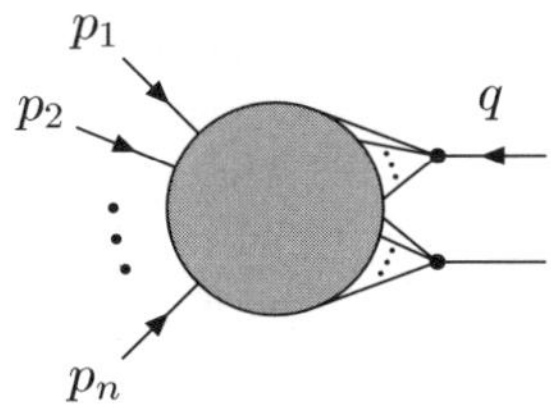

Fig. 4.2. The general diagram contributing to the product of two composite operators

The Fourier transform $T\tilde{J}(q)J(0)$ is represented in perturbation theory, according to the Feynman rules, through diagrams which have two external vertices for the two composite operators and a set of vertices generated by the interaction Lagrangian which can be either external or internal. The general diagram is shown symbolically in Fig. 4.2, where a Fourier transform with respect to all S-matrix external lines with external momenta $p_1, \ldots, p_n$ is implied. So, the limit $x \to 0$ is translated into momentum space language as the limit $|q| \ll |p_i|, m_l$ of the Feynman diagrams $F_\Gamma(q, p_1, \ldots, p_n, m_1, \ldots, m_L)$ contributing to Fig. 4.2.

The form of the asymptotic expansion (4.52) does not imply that the coefficient functions C_i are homogeneous in x (or, at most have logarithmic dependence). And, historically, in the OPE [248] based on the BPHZ renormalization, with subtraction only in complete subgraphs, where the general term of the expansion was explicitly written in terms of Green functions of the initial composite operators, the coefficient functions indeed were not homogeneous and involved a non-trivial dependence on the masses. Nevertheless, a requirement for at most a logarithmic dependence of the coefficient functions is rather natural, and the techniques described here certainly provide this property.

Another important property of the Wilson coefficients is connected with the possibility of providing a factorization of contributions from different scales. If it was possible to explicitly separate contributions from fields corresponding to large and small distances in functional integrals then the factorization of large and small scales would be manifestly achieved. However, there is no such possibility so that one has to apply indirect criteria. A sufficient factorization criterion is based on the polynomial dependence of the coefficient functions on the small momenta and masses. In the case of the OPE, this means a polynomial dependence on all the masses of a given theory. If a non-polynomial dependence on the masses is involved then there arises a new expansion parameter, $\ln(m^2/\mu^2)$, in addition to the coupling constant $g(\mu)$ and the renormalization group logarithm $\ln(Q^2/\mu^2)$. It turns out that, for sufficiently large $Q^2 = -q^2$ and small m^2, it is impossible to make all three expansion parameters small. In this situation, the coefficient functions involve contributions from large distances (small momenta) because

and logarithmic terms of the expansions of a function and its Fourier transform are certainly connected.

the logarithm $\ln(m^2/\mu^2)$ is generated by integration over regions of small loop momenta (of order m) in the corresponding Feynman integrals. On the other hand, if the coefficient functions depend only polynomially on the masses, the standard renormalization group strategy of choosing $\mu \sim Q$ can be used provided $g(\mu) \to 0$ when $\mu \to \infty$, i.e. in asymptotically free theories (see, e.g., [74]).

If this factorization criterion is satisfied then we can use the perturbative OPE beyond perturbation theory: the coefficient functions are evaluated perturbatively but one can then forget about the perturbative origin of the composite operators and treat their matrix elements as nonperturbative objects (for example condensates, i.e. their vacuum expectation values).

An essential point in the realization of the factorization procedure by means of OPE is the fact that the Lagrangian and all composite operators are generated by monomials without normal order, which therefore have non-zero perturbative vacuum expectation values.[6] To illustrate this fact, let us again consider the example of the OPE of two composite operators $J = (1/2)\phi^2$ in the ϕ^4 model with the interaction Lagrangian $\mathcal{L}_{\mathrm{I}} = -(g/4!)\phi^4$:

$$TJ(x)J(0) \sim C_0(x)\mathbf{1} + C_{m,2}(x)m^2\mathbf{1} + C_{J,2}(x)J(0) + \ldots . \tag{4.54}$$

In contrast to Chap. 1 (see (1.18)), we now suppose that the mass is not zero, so that there is an additional, second term in the expansion. Observe that the evaluation of the coefficient $C_{J,2}$ in the first order in g is the same as in the massless case, although the operator J is now mixed with $m^2\mathbf{1}$ under renormalization. The expansion of the triangle diagram contributing to $C_{J,2}$ is again described by Fig. 1.5, where now the lines on the right-hand side are massless with the exception of the two lines in the loop in the second term.

But let us now look at the coefficient functions $C_0(x)$ and $C_{m,2}$. Suppose for a moment that the operator J is generated by the normal-ordered product ϕ^2. In the zero order of perturbation theory, this means that the tadpole diagram which would be generated by fields taken at the same space–time point is now forbidden. Therefore the vacuum expectation value of the operator J vanishes. If we insert the OPE (4.54) between vacuum states, only the contribution from operators proportional to the unity operator will survive. The one-loop diagram contributing to $\langle TJ(x)J(0)\rangle$ in order g^0 is shown on the left-hand side of Fig. 4.3. The UV divergence in the loop is removed by a renormalization constant for the product of the two composite operators. Using the expansion (4.27) at $m_1 = m_2 = m$ in the first two orders in m^2/q^2, we obtain

[6] In [39, 64] it was realized that all the logarithms of the quark masses can be included in the vacuum expectation values of non-trivial composite operators which appear in the OPE of currents in QCD. Later, in [232, 50], it was shown that minimal subtractions within dimensional regularization provide a polynomial dependence of the coefficient functions on the masses in every order of perturbation theory.

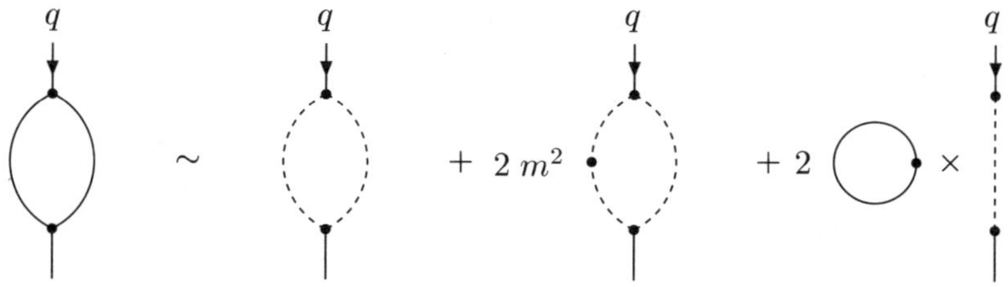

Fig. 4.3. One-loop diagram contributing to the product of two composite operators and its asymptotic expansion when $m \to 0$

$$\langle TJ(x)J(0)\rangle_0 \sim -\frac{\mathrm{i}}{2\cdot 16\pi^2}\left(-\ln\left(-q^2/\mu^2\right)+2+\frac{2m^2}{q^2}\left[\ln\left(-q^2/m^2\right)+1\right]\right)+\ldots . \tag{4.55}$$

A logarithm of m is inevitably present in the coefficient function $C_{m,2}$, so that there is no chance of obtaining a factorization under the assumption of normal order for the composite operators.

To improve the situation, it suffices just to forget the old-fashioned convention about normal order. Moreover, the techniques presented here for expanding Feynman integrals, applied within dimensional regularization, naturally lead to a polynomial dependence of the Wilson coefficients on the masses and therefore to a factorization. To see this in our example, we consider the large-momentum expansion given by (4.24), (4.25) and (4.26) (before taking the limit $\varepsilon \to 0$). For the equal-mass case, the first two orders of the expansion are pictured in Fig. 4.3 The first two terms of the expansion contribute to C_0 and $C_{m,2}$, respectively, while the third term is naturally interpreted as the vacuum expectation value $\langle J\rangle_0$ times a factor $2\mathrm{i}/q^2$, which is the zero-order part of the corresponding coefficient function $C_{J,2}$, in agreement with (1.24). We then consider the renormalized composite operator J. In the massive case in the zero order in g, its renormalization is given by

$$J = J^{\mathrm{B}} + \frac{1}{2\times 16\pi^2}\frac{1}{\varepsilon}m^2\mathbf{1} . \tag{4.56}$$

After we express J^{B} from this equation, we see that the second term gives a contribution to $C_{m,2}$ and subtracts its pole part. We obtain

$$\tilde{C}_{m,2} = -\frac{\mathrm{i}}{16\pi^2}\frac{\ln(-q^2/\mu^2)}{q^2} . \tag{4.57}$$

We see that the logarithm of the mass has been transferred from (4.55) to the vacuum expectation value of J and, now, $\tilde{C}_{m,2}$ does not depend on the mass.

Since the coefficient functions can now depend only polynomially on the masses, it is natural to attribute all the powers of the masses to the composite operators and thereby deal with coefficient functions which do not depend at all on the masses. Thus, any basis of composite operators can now be constructed from monomials of the fields and their derivatives (without normal ordering, of course!) and monomials constructed from the masses.

To write down the OPE of a pair of composite operators in a given theory, one starts with the construction of an appropriate basis. The next problem is the evaluation of the Wilson coefficients. This problem has a very simple solution, given within the so-called method of projectors [128, 127] and corresponding to the explicit formulae for the diagrammatic expansion in the off-shell large-momentum limit (4.34) and (4.36). Let Π_i be a family of projectors[7] corresponding to a chosen basis O_i generated by polynomials o_i. These projectors satisfy $\Pi_i(o_k) = \delta_{ik}$.

For example, in the general monomial basis $J_{\{\lambda\}}(x)$, in the case of an operator F depending on several four-coordinates, we have

$$\Pi_{\{\lambda\}}[F(x_1,\ldots,x_n)] = c_{\{\lambda\}} \langle F(x_1,\ldots,x_n)\tilde{j}_{\{\lambda\}}(0)\rangle^{\mathrm{amp}} , \tag{4.58}$$

where

$$c_{\{\lambda\}} = \mathrm{i}^{-|\{\lambda\}|} / \left(\prod_i l_i!(\lambda_i)! \right) , \quad |\{\lambda\}| = \sum_i |\lambda_i| \equiv \sum_i r_i , \tag{4.59}$$

and

$$\langle F(x_1,\ldots,x_n)\tilde{j}_{\{\lambda\}}(0)\rangle^{\mathrm{amp}} = \left(\frac{\partial}{\partial p_1}\right)^{(\lambda_1)} \cdots \left(\frac{\partial}{\partial p_l}\right)^{(\lambda_l)} \times \langle F(x_1,\ldots,x_n)\tilde{\phi}_1(p_1)\ldots\tilde{\phi}_1(p_l)\rangle^{\mathrm{amp}}\Big|_{p_1=\ldots=p_l=0} . \tag{4.60}$$

The superscript 'amp' denotes the contribution of graphs which are 1PI after the amputation of the external p-lines. In the case $F(x_1,x_2) = J_1(x_1)J_2(x_2)$ (with $x_2 = 0$), these diagrams have to be 1PI after contraction of the external vertices corresponding to the composite fields J_1 and J_2. The presence of the masses is trivially taken into account, by the corresponding parts of the projectors that pick up coefficients of the Taylor expansion in the masses.

Now, the simplest formula for the Wilson coefficients in the OPE (4.53) in a given basis O_i is [128]

$$\tilde{C}_i = \sum_k \Pi_i[\tilde{J}_1(q)J_2(0)] Z^{-1}_{ik} , \tag{4.61}$$

where Z^{-1}_{ik} are the elements of the matrix Z^{-1} which is inverse to the renormalization matrix of the composite operators defined by (2.97).

Let us present a simple 'informal proof' [128] of this very simple formula. Suppose that the OPE (4.52) is true. Let us act with the projector

[7] These are operations that define projections onto the operators of the given basis, rather than real projectors satisfying $\Pi^2 = \Pi$. To obtain projectors with this property, it suffices to perform a trivial redefinition. In this case have $\Pi_i(o_k) = \delta_{ik} o_i$.

Π_i on both sides of this equation. We have $\Pi_i[\tilde{J}_1(q)J_2(0)]$ on the left and $\sum_k C_k(x)\Pi_i(O_k)$ on the right. When acting on diagrams contributing to O_k, the projector Π_i performs Taylor expansion in their masses and external momenta, thereby producing massless Feynman integrals which are zero within dimensional regularization. The only case in which it does not produce zero is the tree contribution, i.e. the contribution without loops, where diagrams have only lines corresponding to external S-matrix vertices. These lines are, however, amputated according to the prescriptions encoded in any projector Π_i. Then $\Pi_i(O_k)$ reduces to $\Pi_i(o_k) = \delta_{ik}$. Observe now that the same result holds for the 'bare' composite operators defined by (2.95), i.e. $\Pi_i(O_k^{\rm B}) = \delta_{ik}$. An important point is that a similar relation does not hold for the renormalized composite operators, which, in contrast to the bare operators, include counterterms for diagrams with a vertex corresponding to the composite operator. Indeed, if such a counterterm for the whole diagram is included, then the projector can act non-trivially on this quantity.

Thus we arrive at the following explicit formula for the OPE:

$$T\tilde{J}_1(q)J_2(0) \sim \sum_i \Pi_i[\tilde{J}_1(q)J_2(0)]O_i^{\rm B}(0) \; . \tag{4.62}$$

Now we turn to the renormalized composite operators by solving the system of linear equations (2.97), to obtain

$$\sum_i \Pi_i[\tilde{J}_1(q)J_2(0)]O_i^{\rm B}(0) = \sum_{ik} Z_{ik}^{-1}\Pi_i[\tilde{J}_1(q)J_2(0)]O_k(0) \tag{4.63}$$

and therefore justify (4.61).

However, we must realize that we have made two (plausible) assumptions: the possibility of changing the order of differentiation with respect to the momenta and masses at zero values and in the summation of the asymptotic series, and the possibility of setting massless vacuum integrals to zero. On the other hand, it is possible to derive (4.61) directly from the diagrammatic expansion (4.36). Consider the expansion of Feynman diagrams with n S-matrix external momenta contributing to $T\tilde{J}_1(q)J_2(0)$. In other words, we consider the large-momentum expansion of the Green function $\langle T\tilde{J}_1(q)J_2(0)\tilde{\phi}(p_1)\tilde{\phi}(p_2)\rangle^{\rm amp}$. These diagrams $F_\Gamma(q, p_1, \ldots, p_n)$ depend on one large and n small external momenta. The expansion (4.36) in the case of one large momentum is of the form

$$RF_\Gamma(q, p_1, \ldots, p_n) \sim \sum_\gamma \bar{R}F_{\Gamma/\gamma}(p_1, \ldots, p_n) \circ \mathcal{M}_\gamma RF_\gamma(q, \ldots) \; , \tag{4.64}$$

where the Taylor operator $\mathcal{M}_\gamma$ performs expansion in the masses of γ and its small external momenta. The first factor, written for the reduced graph Γ/γ, does not depend on q and the second factor depends on the small external momenta of γ, which can be either small external momenta p_i or loop momenta of the reduced graph.

The expansion is shown graphically in Fig. 4.4. Now we turn from summation over Γ and γ to summation over $\gamma' = \Gamma/\gamma$ and γ. We can also distinguish the contribution of graphs γ with a given number of external lines, l, without the large external momentum q. The first factor on the right-hand side of Fig. 4.4 can be recognized as the Green function of the composite operator, while the second factor is the corresponding coefficient function in the OPE (4.62) written in the general monomial basis. The fact that the reduced diagram $F_{\Gamma/\gamma}$ (4.64) is renormalized by the incomplete R-operation, where the counterterm for the reduced vertex is absent, results in the presence of the bare composite operator in (4.62).

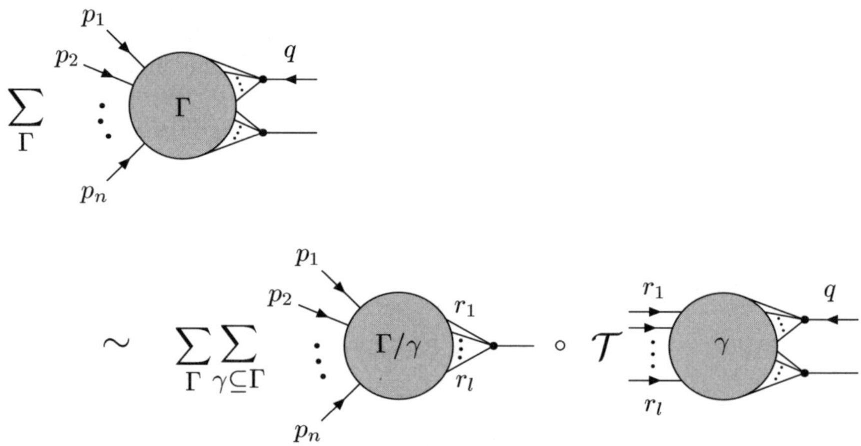

Fig. 4.4. Asymptotic expansion of the sum of the diagrams contributing to the product of two composite operators. The Taylor operator $\mathcal{T}$ expands in $r_1, \ldots, r_{l-1}$ and the masses of γ

4.5 The OPE of Quark Currents and its Applications

A classic example of the application of the OPE is the calculation of the total cross-section for hadron production in e^+e^- annihilation (see [58] for a review), which is described by the ratio

$$R(s) \equiv \frac{\sigma(e^+e^- \to \text{hadrons})}{\sigma_0}\,, \tag{4.65}$$

where $s = q^2$ is the centre-of-mass energy. This ratio is normalized by $\sigma_0 = 4\pi\alpha^2/(3s)$, which is the cross-section for the reaction $e^+e^- \to \mu^+\mu^-$ (given by a tree diagram). In the case of such a process mediated by an intermediate photon,[8] the function $R(s)$ is expressed through the two-point correlator (the vacuum expectation value)

$$\mathrm{i} \int \mathrm{d}x \, \mathrm{e}^{\mathrm{i}q\cdot x} \langle T J_\mu(x) J_\nu(0) \rangle = \left(q_\mu q_\nu - g_{\mu\nu} q^2 \right) \Pi(q^2) \tag{4.66}$$

of the product of two vector currents $J_\mu = \bar{\psi}\gamma_\mu\psi$ in QCD as follows:

$$R(s) = 12\pi \, \mathrm{Im} \, \Pi(s + \mathrm{i}0) \; . \tag{4.67}$$

In the first two orders of perturbative QCD, the function $R(s)$ can be straightforwardly evaluated:

$$\begin{aligned} R(s) &= R^{(0)}(s) + \frac{\alpha_\mathrm{s}}{\pi} R^{(1)}(s) + \left(\frac{\alpha_\mathrm{s}}{\pi}\right)^2 R^{(2)}(s) + \ldots \, , \\ R^{(0)}(s) &= \frac{3}{2}\beta(3-\beta^2) \, , \\ R^{(1)}(s) &= \frac{3}{2}(3-\beta^2) \left\{ (1+\beta^2) \left[2\,\mathrm{Li}_2\left(\frac{1-\beta}{1+\beta}\right) + \mathrm{Li}_2\left(\left(\frac{1-\beta}{1+\beta}\right)^2\right) \right. \right. \\ & \left. + \ln\frac{1-\beta}{1+\beta} \ln\frac{8\beta^2}{(1+\beta)^3} \right] - 2\beta \ln\frac{8\beta^2}{(1+\beta)^3} \\ & \left. + \left(3\beta - \frac{33+22\beta^2-7\beta^4}{8(3-\beta^2)} \right) \ln\frac{1-\beta}{1+\beta} + \frac{3\beta(5-3\beta^2)}{4(3-\beta^2)} \right\} \, , \end{aligned} \tag{4.68}$$

where $\mathrm{Li}_2(z)$ is the dilogarithm (A.57) and

$$\beta = \sqrt{1 - \frac{4m^2}{s}} \; . \tag{4.69}$$

The next-to-next-to-leading order contribution $R^{(2)}(s)$ is not known analytically. However, in specific regimes, the method of asymptotic expansions can be successfully applied. In particular, at large values of s, it is natural to use the OPE, i.e. to expand the product of the currents in the limit of large q. Taking into account the transverse structure, the OPE (4.53) is rewritten as

$$T\tilde{J}_\mu(q) J_\nu(0) \sim \frac{q_\mu q_\nu - g_{\mu\nu} q^2}{(q^2)^2} \sum_i \tilde{C}_i(q) O_i(0) \; . \tag{4.70}$$

Let us illustrate how the formulae of the previous section work. For simplicity, consider the situation with one quark flavour ($n_\mathrm{f} = 1$) with a mass m. As usual, the number of colours N is kept general, and the value $N = 3$ is fixed at the end of the calculation.

This is the list of the corresponding composite operators, up to dimension four, that enter the right-hand side and have non-zero vacuum expectation values:

$$\begin{aligned} & O_0 = \mathbf{1} \, , \quad O_2 = m^2 \mathbf{1} \, , \\ & O_{4,1} = m^4 \mathbf{1} \, , \quad O_{4,2} = G^{a,\mu\nu} G^a_{\mu\nu} \, , \quad O_{4,3} = m\bar{\psi}\psi \; . \end{aligned} \tag{4.71}$$

[8] Processes mediated by a Z boson are also considered.

Three more operators of dimension four [223, 65] are

$$
\begin{aligned}
O_{4,4} &= \bar{\psi}\left(\mathrm{i}\overleftrightarrow{\not{D}}/2 - m\right)\psi\,,\\
O_{4,5} &= A_\nu^a\left(\nabla_\mu^{ab}G^{b,\mu\nu} + g\bar{\psi}t^a\gamma^\nu\psi\right) - \partial_\mu\bar{c}^a\partial^\mu c^a\,,\\
O_{4,6} &= (\nabla_\mu^{ab}\partial^\mu\bar{c}^b)c^a\,,
\end{aligned}
\tag{4.72}
$$

where

$$
\nabla_\mu^{ab} = \delta^{ab}\partial_\mu - gf^{abc}A_\mu^c\,, \tag{4.73}
$$

and t^a is the $\mathrm{SU}(N)$ generator.

The operator $O_{4,4}$ vanishes owing to the equations of motion, and the operators $O_{4,5}$ and $O_{4,6}$ are non-physical because they are not gauge-invariant. They appear because the gauge invariance of the Lagrangian is broken by the gauge fixing. Since matrix elements of these operators vanish for physical states they do not really contribute to the OPE.

These are the corresponding projectors [227] that are applied to the initial product of currents when (4.61) is used:

$$
\Pi_0(F) = \langle F\rangle^{\mathrm{amp}}\big|_{m^2=0}\,,\quad \Pi_2(F) = \frac{\partial}{\partial m^2}\langle F\rangle^{\mathrm{amp}}\Big|_{m^2=0}\,, \tag{4.74}
$$

$$
\Pi_{4,1}(F) = \frac{1}{2}\frac{\partial^2}{\partial(m^2)^2}\langle F\rangle^{\mathrm{amp}}\Big|_{m^2=0}\,, \tag{4.75}
$$

$$
\Pi_{4,2}(F) = \frac{1}{4d(1-d)}\frac{\partial}{\partial p_1}\cdot\frac{\partial}{\partial p_2}\langle F\tilde{A}_\mu^a(p_1)\tilde{A}^{a,\mu}(p_2)\rangle^{\mathrm{amp}}\Big|_{p_1=p_2=0,\,m=0}\,, \tag{4.76}
$$

$$
\Pi_{4,3}(F) = \frac{1}{4N}\,\mathrm{Tr}\left(\frac{\partial}{\partial m} + \frac{1}{d}\gamma_\nu\frac{\partial}{\partial p_\nu}\right)\langle F\tilde{\bar{\psi}}(-p)\tilde{\psi}(p)\rangle^{\mathrm{amp}}\Big|_{p=0,\,m=0}\,, \tag{4.77}
$$

$$
\Pi_{4,5}(F) = \frac{1}{2d}\frac{\partial}{\partial p_1}\cdot\frac{\partial}{\partial p_2}\langle F\tilde{c}^a(p)\tilde{c}^a(-p)\rangle^{\mathrm{amp}}\Big|_{p=0,\,m=0}\,, \tag{4.78}
$$

$$
\Pi_{4,6}(F) = \frac{1}{d}\frac{\mathrm{i}f^{abc}}{gN}\frac{\partial}{\partial p_\mu}\langle F\tilde{A}_\mu^c(0)\tilde{c}^a(p)\tilde{c}^a(-p)\rangle^{\mathrm{amp}}\Big|_{p=0,\,m=0}\,, \tag{4.79}
$$

where no summation over repeated colour indices in (4.76), (4.78) and (4.79) is implied.

The renormalization matrix Z_{ij} of the three physical composite operators of dimension four, $O_{4,1}$, $O_{4,2}$ and $O_{4,3}$, is [223]

$$
\begin{pmatrix}
\frac{1}{1-\beta/\varepsilon} & \frac{-4\gamma_m/\varepsilon}{1-\beta/\varepsilon} & \mu^{-2\varepsilon}4Z_m^{-4}\alpha_s\frac{\partial}{\partial\alpha_s}Z_0\\
0 & 1 & -\mu^{-2\varepsilon}4Z_m^{-4}Z_0\\
0 & 0 & Z_m^{-4}
\end{pmatrix}\,, \tag{4.80}
$$

where β and γ_m are the β-function and the anomalous dimension of the mass, respectively. Z_0 is defined through the equation

$$E_0^{\mathrm{B}} = \mu^{-2\varepsilon} \left[E_0(\mu) - Z_0(\alpha_{\mathrm{s}}) m^4(\mu) \right] , \tag{4.81}$$

and E_0^{B} and E_0 are the bare and renormalized densities, respectively, of the vacuum energy (given by the sum of vacuum diagrams).

The non-physical composite operators do not mix with the physical ones under renormalization, so that the whole renormalization matrix has a block form. However, projectors for the non-physical operators contribute to (4.61) because the inverse matrix is involved there.

So, the coefficient functions of the operators O_0 and O_2 are computed as the first two terms of the Taylor expansion of the polarization function Π in m^2. The corresponding Feynman diagrams can be analytically evaluated by means of an algorithm [161] based on the method of IBP [68]. For the physical operators of dimension four, $O_{4,1}$, $O_{4,2}$ and $O_{4,3}$, the following two-loop results for the coefficient functions C_i in the OPE (4.70) have been obtained [65] by use of (4.61):

$$\tilde{C}_{4,1} = \frac{1}{12} \frac{\alpha_{\mathrm{s}}}{\pi} + \frac{2C_{\mathrm{A}} - C_{\mathrm{F}}}{48} \left(\frac{\alpha_{\mathrm{s}}}{\pi} \right)^2 , \tag{4.82}$$

$$\begin{aligned} \tilde{C}_{4,2} &= 2 + \frac{C_{\mathrm{F}} \alpha_{\mathrm{s}}}{2\pi} \\ &+ \left(\frac{\alpha_{\mathrm{s}}}{12\pi} \right)^2 C_{\mathrm{F}} \left[1161 C_{\mathrm{F}} - 100 C_{\mathrm{A}} - 344 + 432\zeta(3) + 6(16 + 11 C_{\mathrm{A}}) L \right] , \end{aligned} \tag{4.83}$$

$$\begin{aligned} &\tilde{C}_{4,3} \\ &= \frac{C_{\mathrm{A}}}{16\pi^2} \left[-4(1 + 2L) + \frac{\alpha_{\mathrm{s}}}{\pi} C_{\mathrm{F}} \left(-\frac{64}{3} + 64\zeta(3) - 22L - 12L^2 \right) \right] , \end{aligned} \tag{4.84}$$

where the SU(N) colour factors C_{F} and C_{A} are given by (1.29), and $L = \ln(-\mu^2/q^2)$.

Besides the correlator of the vector currents, a lot of other correlators are phenomenologically important. Various currents of the form $\bar{\psi}\Gamma\psi'$ are used, with (generally different) quark fields ψ and ψ' and a matrix Γ which, in particular, defines (pseudo)scalar and (pseudo)vector currents. While the vector and axial currents are relevant to the cross-section $\sigma(e^+e^- \to \text{hadrons})$, the scalar and pseudoscalar currents describe the decay of a scalar or pseudoscalar Higgs boson.

The OPE of the general vector and axial currents,

$$T\tilde{J}_\mu(q) J_\nu(0) \sim \frac{1}{s^2} \sum_i \left[(q_\mu q_\nu - g_{\mu\nu} q^2) \tilde{C}_i(q) + q_\mu q_\nu \tilde{C}_i^L(q) \right] O_i(0) , \tag{4.85}$$

generally contains not only a transverse but also a longitudinal part, which, however, is irrelevant when inserted into the total cross-section.

Consider, for example, the 'non-diagonal' (pseudo)vector currents $\bar{u}\gamma_\mu d$ and $\bar{u}\gamma_\mu\gamma_5 d$, where u and d are quarks of two chosen flavours (not necessarily

u and d quarks), with a general number of flavours n_{f}. In this case, the composite operators in the OPE are a little bit more general. The set of dimension four operators is extended as follows:

$$O_{4,1} \to O^{ij}_{4,1} = m_i^2 m_j^2 \; , \quad O^{i}_{4,1} = m_u m_d m_i^2 \; , \tag{4.86}$$

$$O_{4,3} \to O^{ij}_{4,3} = m_i \bar{q}_j q_j \; , \tag{4.87}$$

$$O_{4,4} \to O^{i}_{4,4} = \bar{q}_i \left(\mathrm{i} \overleftrightarrow{\not{D}} /2 - m_i \right) q_i \; , \tag{4.88}$$

where the q_i are other quark fields. For brevity, we omit the unit operator $\mathbf{1}$ here and in the following formula.

At the two-loop level, the contribution of dimension four physical operators to the transverse part of the OPE of two (pseudo)vector $\bar{u}d$ currents is [65]

$$\begin{aligned}
&\tilde{C}_{4,2} O_{4,2} + \sum_{ij} \tilde{C}^{ij}_{4,3} O^{ij}_{4,3} + \sum_{i} \tilde{C}^{i}_{4,3} O^{i}_{4,3} \\
&= \frac{\alpha_{\mathrm{s}}}{12\pi} \left(1 + \frac{2C_{\mathrm{A}} - C_{\mathrm{F}}}{4} \frac{\alpha_{\mathrm{s}}}{\pi} \right) G^{a,\mu\nu} G^{a}_{\mu\nu} - \frac{3C_{\mathrm{F}}}{4} \frac{\alpha_{\mathrm{s}}}{\pi} \\
&\times \left(1 + \frac{\alpha_{\mathrm{s}}}{24\pi} \left[48C_{\mathrm{A}} - 21C_{\mathrm{F}} - 4n_{\mathrm{f}} + (22C_{\mathrm{A}} - 4n_{\mathrm{f}})L \right] \right) (m_u \bar{u}u + m_d \bar{d}d) \\
&\pm \left\{ 1 + C_{\mathrm{F}} \frac{\alpha_{\mathrm{s}}}{\pi} \left[1 + \frac{\alpha_{\mathrm{s}}}{24\pi} \left(81C_{\mathrm{F}} + \frac{83}{3} C_{\mathrm{A}} - \frac{14}{3} n_{\mathrm{f}} + (22C_{\mathrm{A}} - 4n_{\mathrm{f}})L \right) \right] \right\} \\
&\times (m_u \bar{d}d + m_d \bar{u}u) + C_{\mathrm{F}} \left(\frac{\alpha_{\mathrm{s}}}{\pi} \right)^2 \left(\zeta(3) + \frac{L-3}{4} \right) \sum_i m_i \bar{q}_i q_i \\
&+ \frac{C_{\mathrm{A}}}{16\pi^2} \left[C_{\mathrm{F}} \frac{\alpha_{\mathrm{s}}}{\pi} \left(\frac{38}{3} - 8\zeta(3) \right) m_u^2 m_d^2 - \left(2 + 3C_{\mathrm{F}} \frac{\alpha_{\mathrm{s}}}{\pi} (1+L) \right) (m_u^4 + m_d^4) \right. \\
&\left. \mp \left(4L + C_{\mathrm{F}} \frac{\alpha_{\mathrm{s}}}{\pi} \left[14 - 12\zeta(3) + 8L + 6L^2 \right] \right) (m_u^3 m_d + m_d^3 m_u) \right] \; ,
\end{aligned} \tag{4.89}$$

where the upper and lower signs stand for the vector and pseudovector cases, respectively.

The phenomenological applications of the OPE of two quark currents are numerous. The OPE derived within dimensional renormalization, with operators without normal ordering, provides a factorization of the contributions from small and large distances, as was explained in the previous section. The coefficient functions involve information about short distances and therefore can be computed by perturbation theory – see the above results. The vacuum expectation values (condensates) of the composite operators present in the OPE can be used effectively to *parameterize* non-perturbative contributions. A classic example is provided by the QCD sum rule method [205]. In particular, the OPE of two quark currents has been used for testing QCD in hadronic tau decay and in the determination of the masses of the light quarks (u, d, s) (for a recent review see [70]).

However, there is no obligation to treat the matrix elements of the composite operators as non-perturbative objects. Another important application of OPE is the analysis of quark mass effects in e^+e^- annihilation into hadrons, performed by evaluating not only the coefficient functions but also the vacuum expectation values of the composite operators by perturbation theory. For example, in [56] the OPE of the two vector and axial quark currents discussed in this section has been used to obtain the terms of order $\alpha_s m^4/s^2$ and $\alpha_s^2 m^4/s^2$. One could proceed by computing the polarization of the vacuum $\Pi(q^2)$ using the off-shell large-momentum expansion, i.e. by expanding it in the ratio m^2/q^2 using the prescriptions of Sect. 4.2, where the expansion of a given diagram is written as a sum over AI subgraphs according to (4.34). A complication that arises when this is done is the problem of handling numerous contributions[9] from subgraphs of three-loop graphs. It has turned out to be more effective [56] to use the OPE of the (pseudo)vector non-diagonal currents and evaluate, in addition to the coefficient functions given by (4.82) and (4.89), the vacuum expectation values of the physical composite operators $O_{4,2}$ and $O_{4,3}^{ij}$ defined by (4.71) and (4.87), respectively. The relevant one- and two-loop diagrams are shown in Fig. 4.5.

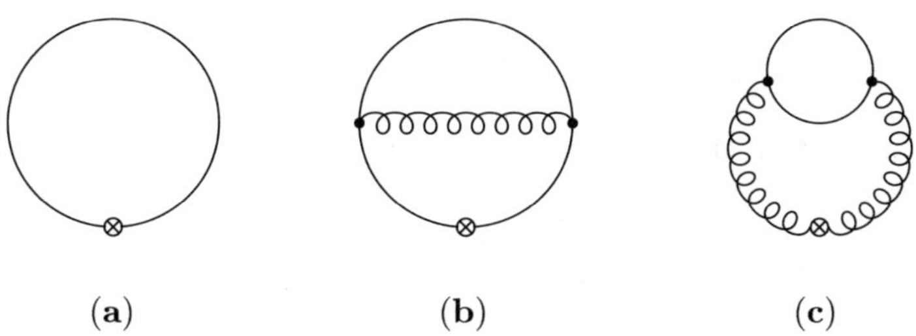

Fig. 4.5a,b. One- and two-loop diagrams contributing to the vacuum expectation values of the composite operator $O_{4,2}$. (**c**) Two-loop diagram contributing to the vacuum expectation value of the composite operator $O_{4,3}^{ij}$

The corresponding results are [39, 64]

$$\langle O_{4,3}^{ij} \rangle = \frac{3 m_i m_j^3}{4\pi^2} \left[1 + \ln \frac{\mu^2}{m_j^2} + 2\frac{\alpha_s}{\pi} \left(\ln^2 \frac{\mu^2}{m_j^2} + \frac{5}{3} \ln \frac{\mu^2}{m_j^2} + \frac{5}{3} \right) \right] \qquad (4.90)$$

and [34]

$$\langle O_{4,2} \rangle = -\frac{\alpha_s}{2\pi^3} \sum_i \left(9 + 8 \ln \frac{\mu^2}{m_i^2} + 3 \ln^2 \frac{\mu^2}{m_i^2} \right) m_i^4 \, . \qquad (4.91)$$

[9] Nevertheless, this problem can be solved with the help of a computer. Probably the first non-trivial example of a realization of the procedure of asymptotic expansion in a situation where there are contributions from several subgraphs can be found in [51].

Since the polynomial part in q^2 does not contribute to the imaginary part, one needs only terms involving logarithms $\ln(-q^2 - \mathrm{i}0) = \ln|q^2| - \mathrm{i}\pi\theta(q^2)$. It turns out, however, that to find the correction of order $\alpha_s^2 m^4/s^2$ to $R(s)$ there is no need to compute three-loop contributions of order α_s^2 to these vacuum expectation values, and that the two-loop results for the coefficient functions and vacuum expectation values of the composite operators are sufficient. To do this, one applies [56] renormalization group equations for the coefficient functions.

The terms of order m^2 can be found in [57] and are given simply by Taylor expansion in m^2. The final result for the cross-section $R(s)$, with contributions up to order $\alpha_s^2 m^4/s^2$, in the case of the vector current and $n_f - 1$ massless flavours, is [56]

$$
\begin{aligned}
R(s) = 3 \Bigg(& 1 - 6\frac{\bar{m}^4}{s^2} + \frac{\alpha}{\pi}\left(1 + 12\frac{\bar{m}^2}{s} - 22\frac{\bar{m}^4}{s^2}\right) \\
& + \left(\frac{\alpha}{\pi}\right)^2 \Bigg\{ n_f\left(-\frac{11}{12} + \frac{2}{3}\zeta(3)\right) + \frac{365}{24} - 11\zeta(3) + \frac{\bar{m}^2}{s}\left(\frac{253}{2} - \frac{13}{3}n_f\right) \\
& \quad + \frac{\bar{m}^4}{s^2}\Bigg[n_f\left(\frac{1}{3}\ln\frac{\bar{m}^2}{s} - \frac{2}{3}\pi^2 - \frac{8}{3}\zeta(3) + \frac{143}{18}\right) \\
& \qquad - \frac{13}{2}\ln\frac{\bar{m}^2}{s} + 27\pi^2 + 108\zeta(3) - \frac{2977}{12}\Bigg]\Bigg\}\Bigg) ,
\end{aligned}
\tag{4.92}
$$

This result can be written in terms of the $\overline{\mathrm{MS}}$ mass given in the one-loop approximation by $\bar{m}^2 = m^2/\left\{1 + (\alpha_s/\pi)[-2\ln(m^2/s) + 8/3]\right\})$ at the scale $\mu = s$. (With this choice, all the logarithms of s disappear in the terms of order m^0 and m^2.) The number of colours has been here set to $N = 3$.

The $\alpha_s m^4/s^2$ and $\alpha_s^2 m^4/s^2$ corrections are essential for calculating the electron–positron annihilation cross-sections at LEP and at lower energies, in particular, in the region between the charm and the bottom thresholds and at energies of several GeV above the $b\bar{b}$ threshold – see [56, 58] for a discussion.

5 Large-Mass Expansion

At the diagrammatic level, the large-mass expansion, i.e. the expansion in the large-mass limit, has almost the same structure as the off-shell large-momentum expansion, so that we apply the method of expansion by subgraphs again here. We start with a one-loop example and then formulate general prescriptions. These prescriptions are given by the same formula (4.34) as for the off-shell large-momentum limit, with a suitable change of the class of subgraphs involved. This general formula is illustrated through two-loop examples. The large-mass expansion at the operator level is then described. The asymptotic expansion of Green functions of light fields is governed by an effective Lagrangian with coefficients containing information about the full initial theory. The chapter concludes with examples of effective Lagrangians that arise in the large-mass limit and are connected with typical physical situations.

5.1 One-Loop Example

Let us consider the following example.

Example 5.1. The one-loop diagram of Fig. 5.1 with a large mass M, a small mass m and a small external momentum q.

Fig. 5.1. One-loop self-energy diagram. The thick internal line denotes the propagator with the heavy mass

The Feynman integral can be written as

$$F_{5.1}(M^2, m^2, q^2; d) = \int \frac{\mathrm{d}^d k}{(k^2 - m^2)^2[(q-k)^2 - M^2]} \; . \tag{5.1}$$

Although this integral is convergent we have introduced dimensional regularization in advance, for the same reason as in the previous chapter.

An obvious (but too straightforward) way to construct a remainder in the limit of large M is to include, in addition to the operator $\mathcal{T}_{q,m}$ corresponding to the whole graph, the operator $\mathcal{T}_k$ acting on the propagator with the large mass, $1/[(q-k)^2 - M^2]$. This operator could then be used to subtract the IR divergences caused by Taylor expansion in m in another propagator. However, we would not provide homogeneity of the terms of the corresponding expansion. It is more reasonable to follow our previous strategy of performing subtractions in *all* small parameters of subgraphs. So let us expand this factor $1/[(q-k)^2 - M^2]$ in all small parameters, namely, q (which is small in the given limit) and k (which is considered small, within our conventions, like any other loop momentum). Thus we construct the remainder as follows:

$$\mathcal{R}^n F_{5.1}(M^2, m^2, q^2, d) = \int \mathrm{d}^d k \left(1 - \mathcal{T}^n_{m,q}\right) \frac{1}{(k^2 - m^2)^2} \times \left(\left(1 - \mathcal{T}^n_{q,k}\right) \frac{1}{(q-k)^2 - M^2} \right) . \tag{5.2}$$

Note that the operator $\mathcal{T}^n_{m,q}$ acts on the whole expression, rather than only on the first factor.

An asymptotic behaviour of order $n+1$ up to logarithms will be achieved because of the operator $\left(1 - \mathcal{T}^n_{m,q}\right)$ if the remainder is UV and IR finite. To see the UV convergence, let us investigate the integration at large k. Only terms in $\mathcal{T}^n_{q,k}$ are dangerous here, because they generate positive powers of k. Let us consider the contribution of order n. Using again the properties of Taylor operators (4.16) and (4.18), we obtain, from

$$\mathcal{T}^{(n)}_{q,k} \frac{1}{(q-k)^2 - M^2} = \sum_{j=0}^{n} p_{1,j}(k) p_{2,n-j}(q) , \tag{5.3}$$

where $p_{1,j}$ and $p_{2,j}$ are monomials of degree j, a sum of terms

$$\left[\left(1 - \mathcal{T}^n_{m,q}\right) \frac{1}{(k^2 - m^2)^2} \right] \sum_{j=0}^{n} p_{1,j}(k) p_{2,n-j}(q) = \sum_{j=0}^{n} p_{1,j}(k) p_{2,n-j}(q) \left[\left(1 - \mathcal{T}^j_{m,q}\right) \frac{1}{(k^2 - m^2)^2} \right] . \tag{5.4}$$

Counting powers of k gives k^j from $p_{1,j}(k)$ and $1/k^{5+j}$ from the expressions in square brackets. This provides convergence at large k.

Consider now the region of small k. In this case it is sufficient to investigate the convergence of the terms originating from $\mathcal{T}^n_{m,q}$:

$$\mathcal{T}^n_{m,q} \frac{1}{(k^2 - m^2)^2} \left[\left(1 - \mathcal{T}^n_{q,k}\right) \frac{1}{(q-k)^2 - M^2} \right] . \tag{5.5}$$

The expression in the square brackets can be represented in a form similar to (4.21),

$$\sum_{i=0}^{n+1} p_{1,j}(k) p_{2,n+1-j}(q) f_j(k,q) , \tag{5.6}$$

where $p_{1,j}$ and $p_{2,j}$ are monomials of degree j, and $f_j(k,q)$ is finite in the vicinity of the point $k = 0$, $q = 0$. Using the commutation relations (4.16), we obtain

$$\mathcal{T}^n_{m,q} \frac{1}{(k^2-m^2)^2} p_{1,j}(k) p_{2,n+1-j}(q) f_j(k,q) \\ = p_{1,j}(k) p_{2,n+1-j}(q) \mathcal{T}^{j-1}_{m,q} \frac{1}{(k^2-m^2)^2} f_j(k,q) . \tag{5.7}$$

Power counting gives k^j from $p_{1,j}(k)$ and, at worst, $1/k^{j+3}$ from the last factors, so that the integral is convergent at small k.

Using the same manipulations as in our previous examples and for the general diagram in the off-shell large-momentum limit, we straightforwardly arrive at the asymptotic expansion starting from the remainder (5.2):

$$F_{5.1} \sim \mathcal{M}_\Gamma F_{5.1} + \mathcal{M}_\gamma F_{5.1} \\ = \int \mathrm{d}^d k \left(\frac{1}{(k^2)^2} + \frac{2m^2}{(k^2)^3} + \ldots \right) \left(\frac{1}{k^2-M^2} + \frac{2q\cdot k - k^2}{(k^2-M^2)^2} + \ldots \right) \\ + \int \frac{\mathrm{d}^d k}{(k^2-m^2)^2} \left(-\frac{1}{M^2} - \frac{(q-k)^2}{M^4} + \ldots \right) . \tag{5.8}$$

The integrals in the first and the second contributions can be computed by means of (A.5) and (A.2), respectively. The poles in ε in the two contributions are cancelled, and we obtain the following result:

$$F_{5.1} \sim \mathrm{i}\pi^{d/2} \left(-\frac{\Gamma(\varepsilon-1)}{(M^2)^{1+\varepsilon}} + \frac{2\Gamma(\varepsilon-2)}{(M^2)^{2+\varepsilon}} \left[(\varepsilon-2)m^2 - (1+\varepsilon)q^2 \right] + \ldots \\ - \frac{\Gamma(\varepsilon)}{M^2(m^2)^\varepsilon} + \frac{\Gamma(\varepsilon-1)}{(M^2)^2(m^2)^\varepsilon} \left[(\varepsilon-2)m^2 - (1-\varepsilon)q^2 \right] + \ldots \right) \\ = \mathrm{i}\pi^2 \left[\frac{1-\ln(M^2/m^2)}{M^2} + \frac{2m^2+5q^2+2(2m^2+q^2)\ln(M^2/m^2)}{2M^4} \right. \\ + \frac{3m^4+24m^2q^2+10(q^2)^2-3[3m^4+6m^2q^2+(q^2)^2]\ln(M^2/m^2)}{3M^6} \\ \left. + O\left(\varepsilon, \frac{\ln M}{M^8}\right) \right] . \tag{5.9}$$

5.2 General Prescriptions

For a general Feynman diagram F_Γ which depends on the large masses $M_1, \ldots$, small masses $m_1, \ldots$ and small external momenta $q_1, \ldots$, the derivation of the general prescriptions is almost identical to the case of the off-shell large-momentum limit. We start from the remainder in the form (4.28), with another class of subgraphs involved in the forest formula. The choice of these subgraphs is explained in a way similar to the previous case.

When the integrand is expanded into a Taylor series in the small parameters of the problem, the IR divergences can appear only in the lines with the small masses. Thus subtractions in the remainder should be performed only in AI subgraphs and, this time, we call a subgraph γ AI if it

(*i*) contains all the heavy lines,
(*ii*) is 1PI with respect to the light lines.

We call a line *heavy* if the corresponding mass is large, and *light* otherwise. The second requirement is of the same origin as a similar requirement for the off-shell large-momentum limit. Note that the AI subgraphs can be disconnected and consist of several connectivity components, i.e., generally, $\gamma = \cup_i \gamma_i$, where every component is 1PI with respect to the light lines.

When defining the subtraction operators that enter the forest formula for the remainder of the large-mass expansion, we follow the same rule as in the previous case: they are Taylor expansion operators (4.30) with respect to the small masses and small external momenta of AI subgraphs, with the subtraction degrees chosen in the same way (4.31). The difference is that F_Γ and Π_γ depend now on the large masses M_i, rather than on the large momenta Q_i. (The factor $\Pi_{\Gamma/\gamma}$ is still independent of the large parameters.)

In this case the remainder

$$\mathcal{R}^n F_\Gamma \left(M_1/\varrho, \ldots, M_i/\varrho, \ldots, q_1, \ldots, q_n, m_1, \ldots\right)$$

has order $\varrho^{\bar{a}+1}$ modulo logarithms when $\varrho \to 0$. A proof of this estimate can be obtained from a similar proof (see Appendix B.2) for the off-shell large-momentum limit by trivial modifications.

Explicit prescriptions for the large-mass limit can be derived from the remainder in the same way as in the previous case. The final formula is

$$F_\Gamma \sim \sum_\gamma F_{\Gamma/\{\gamma_1 \cup \ldots \cup \gamma_i\}} \circ \prod_i \mathcal{M}_{\gamma_i} F_{\gamma_i} \,, \tag{5.10}$$

where the sum runs over all AI subgraphs of Γ and we have explicitly indicated that the AI subgraphs generally consist of several connectivity components γ_i.

For the large-mass expansion, the strategy of expansion by regions also leads to the same prescription (5.10). There are two scales in the problem: the large masses are of the same order, $M_i \sim M$, and the momenta and the small masses are much smaller, $q_i \sim m_l \sim m$, with $m \ll M$. The regions for a given loop momentum are defined as follows:

$$\text{large,} \quad k \sim M \,, \tag{5.11a}$$

$$\text{small,} \quad k \sim m \,. \tag{5.11b}$$

The set of regions labelled by 1PI subgraphs of the given graph is defined as follows:

$$k_i \sim M, \text{ if } k_i \text{ is a loop momentum of } \gamma \,;$$
$$k_i \sim m, \text{ if } k_i \text{ is not a loop momentum of } \gamma \,. \tag{5.12}$$

The same arguments as for the off-shell large-momentum limit then show that these regions reproduce all the contributions in (5.10).

5.3 Two-Loop Examples

In the case of the large-momentum expansion (4.34), the contribution from the whole graph has IR divergences starting from some minimal order and there is at least one additional (to $\gamma = \Gamma$) contribution due to some subgraph $\gamma \neq \Gamma$. In contrast, the large-mass expansion can be combinatorially trivial, i.e. involve only a contribution from $\gamma \equiv \Gamma$ (the contribution from the region of the small loop momenta):

$$F_\Gamma \sim \mathcal{T}_{m_1,\ldots,q_1,\ldots} F_\Gamma \,. \tag{5.13}$$

This happens when the given graph does not have cuts composed of light lines. Observe that although in this situation the Taylor expansion of the integrand formally breaks down for large loop momenta, i.e. when the loop momenta are of the order of the large mass, the above contribution gives the *whole* expansion in the large-mass limit. Moreover, in the case of one external momentum, this Taylor expansion has a radius of convergence equal to the value of the first threshold. This fact follows from the well-known analytic properties of the Feynman amplitudes [98].

Consider, for example, the same diagram that we dealt with in Sect. 4.3 but in the 'opposite' limit:

Example 5.2. The master two-loop diagram (see Fig. 2.5) with non-zero masses, which are all considered large with respect to the external momentum.

The Feynman integral is (4.41). The large-mass expansion includes only the contribution from the whole graph and is given by (5.13). Since all the masses are large and this is a scalar diagram, the expansion reduces to a power series in the external momentum squared [95]:

$$J_m(a_1,\ldots,a_5,m_1,\ldots,m_5,q) \sim \sum_{n=0}^{\infty} \frac{(q^2)^n}{4^n n!(d/2)_n} \left(\Box_q^n J\right)\Big|_{q=0} \,, \tag{5.14}$$

where $(x)_n$ is the Pochhammer symbol (A.53).

In order to recursively evaluate coefficients in this power series one can apply the following relations [94], which are obtained by differentiation of the integrand with the flow of the external momentum through lines 1 and 3:

$$\Box = 4\left[(a_1+a_2+1-d/2)(a_1\mathbf{1}^+ + a_3\mathbf{3}^+) + a_1(a_1+1)m_1^2\mathbf{1}^{++} \right. \\ \left. + a_3(a_3+1)m_3^2\mathbf{3}^{++} + a_1a_3(m_1^2+m_3^2-m_5^2-\mathbf{5}^-)\mathbf{1}^+\mathbf{3}^+\right] \,. \tag{5.15}$$

(The notation for the increasing and lowering operators is explained after (2.66).) Then, after using partial fractions, all the integrals involved are expressed through vacuum integrals of the form (4.47).

Let us now consider a combinatorially non-trivial two-loop diagram.

Example 5.3. The master two-loop diagram of Fig. 2.5 with the large masses $m_1 = m_2 = m_5 = m$, two other zero masses and a small external momentum – see Fig. 5.2.

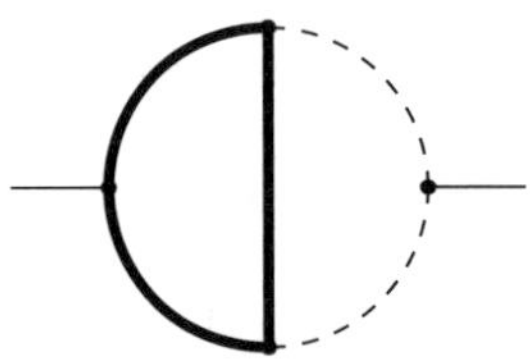

Fig. 5.2. Two-loop master diagram. The *dashed* and *thick internal lines* denote propagators with zero and heavy masses, respectively.

There are contributions from two AI subgraphs: Γ and $\gamma = \{1, 2, 5\}$. Any term of the first contribution can be evaluated by means of (5.14) and (5.15) and the tabulated two-loop formulae (A.38). (Non-trivial vacuum integrals of the form (4.47) with three non-zero masses do not arise here.) We obtain the following series:

$$-\pi^d \frac{\Gamma(1+\varepsilon)^2}{\varepsilon(1-\varepsilon)(m^2)^{1+2\varepsilon}} \left(\frac{1}{2(1+2\varepsilon)} - \frac{(3-\varepsilon)(1+\varepsilon)}{3(3+2\varepsilon)} \frac{q^2}{m^2} \right) + \ldots . \tag{5.16}$$

The contribution of γ can be evaluated by means of (A.12) and (A.3):

$$\int \frac{\mathrm{d}^d l}{l^2 (q-l)^2} \int \frac{\mathrm{d}^d k}{k^2 - m^2} \mathcal{T}_l \frac{1}{(k-l)^2 - m^2} \mathcal{T}_q \frac{1}{(q-k)^2 - m^2}$$
$$= -\pi^d \frac{\Gamma(1+\varepsilon)^2 \Gamma(1-\varepsilon)^2}{2\varepsilon \Gamma(2-2\varepsilon)(-q^2)^{\varepsilon} (m^2)^{1+\varepsilon}} + \ldots . \tag{5.17}$$

It is possible to derive an explicit formula for a general term of this contribution, written in terms of a three-fold finite sum.

Summing the two contributions, we observe that the UV and IR poles are cancelled, and we obtain the following result:

$$F_{5.3} \sim -\frac{\pi^4}{m^2} \left[\frac{1}{2} \ln\left(-q^2/m^2\right) - \frac{3}{2} + \left(\frac{1}{24} \ln\left(-q^2/m^2\right) - \frac{1}{16} \right) \frac{q^2}{m^2} \right] + \ldots \tag{5.18}$$

in agreement with an explicit result obtained in [38]:

$$F_{5.3} = -\pi^4 \frac{\sqrt{\pi}}{4m^2} \sum_{n=0}^{\infty} \frac{n!}{\Gamma(n+3/2)(n+1)} \left(\ln\left(-q^2/m^2\right) - \frac{3}{n+1} \right) \left(\frac{q^2}{4m^2} \right)^n . \tag{5.19}$$

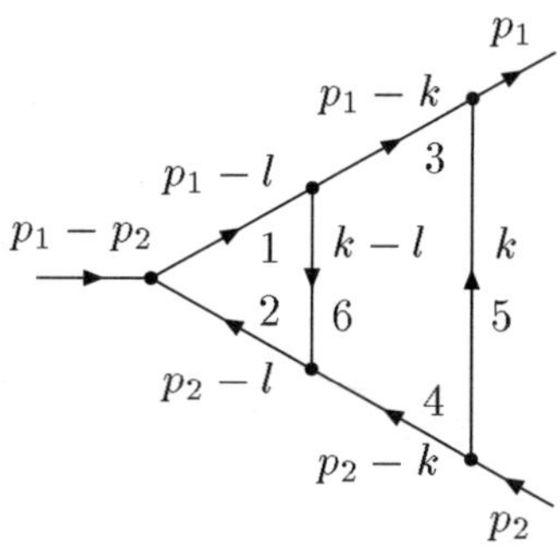

Fig. 5.3. Two-loop vertex diagram

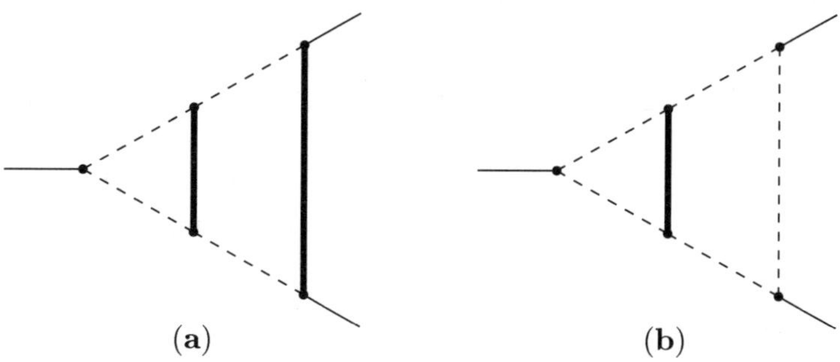

Fig. 5.4a,b. The two diagrams considered in Examples 5.4 and 5.5. The *dashed* and *thick internal lines* denote propagators with zero and heavy masses, respectively

Other examples of expansions of two-loop self-energy diagrams with zero thresholds in the large-mass limit can be found in [19].

Now we turn to examples of two-loop vertex diagrams in the large-mass limit. We shall consider the planar two-loop diagram of Fig. 5.3 with two external momenta on the massless mass shell, $p_1^2 = p_2^2 = 0$, and two different distributions of masses shown as in Figs. 5.4a,b. Choosing the loop momenta as in Fig. 5.3, we obtain the following expression for the corresponding Feynman integral:

$$V(m_1, \ldots, m_6, q^2) = \int \frac{\mathrm{d}^d l}{(l^2 - 2p_1 \cdot l - m_1^2)(l^2 - 2p_2 \cdot l - m_2^2)}$$

$$\times \int \frac{\mathrm{d}^d k}{(k^2 - 2p_1 \cdot k - m_3^2)(k^2 - 2p_2 \cdot k - m_4^2)(k^2 - m_5^2)[(k-l)^2 - m_6^2]} \,. \tag{5.20}$$

For any choice of the masses, all the terms in the contribution

$$\mathcal{T}_q V(m_1, \ldots, m_6, q^2) = \sum_{n=0}^{\infty} A_n(m_1, \ldots, m_6)(q^2)^n \tag{5.21}$$

corresponding to $\gamma = \Gamma$ in (5.10) can be computed analytically using an algorithm developed in [115, 116], based on relations following from differentiation of the integrand with the given choice of the flow of the external momenta. We obtain the following expression for the resulting two-loop vacuum massive integrals [115, 116]:

$$A_n(m_1,\ldots,m_6) = \int\int \frac{\mathrm{d}^d k\,\mathrm{d}^d l}{(l^2-m_1^2)(l^2-m_2^2)} \times \frac{F_n(m_1,\ldots,m_6,k,l)}{(k^2-m_3^2)(k^2-m_4^2)(k^2-m_5^2)[(k-l)^2-m_6^2]}\,, \tag{5.22}$$

where

$$F_n = \frac{1}{n+1}\sum_{j=0}^{n}\sum_{j'=0}^{n}\sum_{i=\max\{0,j+j'-n\}}^{\min\{n,[(j+j')/2]\}} a_{jj'}^{ni} \times \frac{(k^2)^i(l^2)^{n-j-j'+i}(k\cdot l)^{j+j'-2i}}{(k^2-m_3^2)^j(k^2-m_4^2)^{j'}(l^2-m_1^2)^{n-j}(l^2-m_2^2)^{n-j'}}\,. \tag{5.23}$$

The coefficients $a_{jj'}^{ni}$ have been evaluated up to $n=30$ for $\varepsilon=0$ in [115] and in an expansion in ε up to ε^2 in [116]. Here are a few of the initial values:

$$\begin{aligned} a_{11}^{20} &= \frac{1}{3}+\frac{\varepsilon}{6}+\frac{\varepsilon^2}{108}\,,\\ a_{11}^{30} &= \frac{4}{9}+\frac{\varepsilon}{3}+\frac{31\varepsilon^2}{324}\,,\quad a_{03}^{30}=2+\frac{17\varepsilon}{6}+\frac{199\varepsilon^2}{72}\,,\quad a_{12}^{30}=\frac{2}{9}+\frac{\varepsilon}{6}+\frac{31\varepsilon^2}{648}\,,\\ a_{11}^{40} &= \frac{1}{2}+\frac{19\varepsilon}{40}+\frac{1459\varepsilon^2}{7200}\,,\quad a_{03}^{40}=2+\frac{97\varepsilon}{30}+\frac{6139\varepsilon^2}{1800}\,,\\ a_{12}^{40} &= \frac{1}{3}+\frac{19\varepsilon}{60}+\frac{1459\varepsilon^2}{10800}\,,\ldots\,. \end{aligned} \tag{5.24}$$

Our next two-loop diagram is the following.

Example 5.4. The vertex two-loop diagram of Fig. 5.4a with large masses $m_5=m_6=m$, four zero masses and a small external momentum q.

The set of subgraphs contributing to (5.10) consists of $\gamma_1=\Gamma$, $\gamma_2=\{3456\}$, $\gamma_3=\{1256\}$ and $\gamma_4=\{56\}$. Note that γ_3 and γ_4 are disconnected. To compute the contribution of γ_1 we apply (5.21), (5.22) and (5.23) and observe that the factor $(k\cdot l)^{j+j'-2i}$ in (5.23) can be replaced by $\left[(k^2+l^2-m^2)/2\right]^{j+j'-2i}$ because cancelling the denominator $(k-l)^2-m^2$ produces one-loop massless tadpoles, which are equal to zero. The product $(k^2)^i(k^2+l^2-m^2)^{j+j'-2i}$ is then rewritten in terms of a combination of products of powers of k^2 and l^2, and, by means of partial fractions (2.60), the resulting expressions containing powers of k^2 and k^2-m^2 in the denominator are expressed in terms of ratios containing either k^2 or k^2-m^2. The result is (5.21), where

$$A_n = \frac{\left(i\pi^{d/2}\right)^2}{(n+1)(m^2)^{2+2\varepsilon+n}}\sum_{j=0}^{n}\sum_{j'=0}^{n}\sum_{i=\max\{0,j+j'-n\}}^{\min\{n,[(j+j')/2]\}}\frac{(j+j'-2i)!\,a_{jj'}^{ni}}{2^{j+j'-2i}}$$

$$\times \sum_{j_1=0}^{j+j'-2i} \sum_{j_2=0}^{j_1} \frac{(-1)^{n+i+j+j'+j_2}}{(j+j'-2i-j_1)!j_2!} \left[V_{mm0}(1, n+j_1+i-j-j'+2, 1) \right.$$
$$\left. - \sum_{r=0}^{j+j'-i-j_2+1} (-1)^r V_{00m}(n+j_1+i-j-j'+2, 1+r, 1) \right] . \quad (5.25)$$

Here V_{mm0} and V_{00m} are the ratios of gamma functions on the right-hand sides of (A.38) and (A.39), respectively; they correspond to vacuum scalar integrals with three lines.

We have the following contributions from the remaining three subgraphs: for γ_2,

$$\int \frac{\mathrm{d}^d l}{l^2(q-l)^2} \int \frac{\mathrm{d}^d k}{k^2} \mathcal{T}_{p_1,q,l} \frac{1}{[(p_1-k)^2-m^2][(l-k)^2-m^2](q-k)^2} ; \quad (5.26)$$

for γ_3,

$$\int \frac{\mathrm{d}^d k}{k^2(q-k)^2} \mathcal{T}_{p_1,k} \frac{1}{(p_1-k)^2-m^2} \int \frac{\mathrm{d}^d l}{l^2} \mathcal{T}_{q,k} \frac{1}{[(l-k)^2-m^2](q-l)^2} ; \quad (5.27)$$

and for γ_4,

$$\int \frac{\mathrm{d}^d k}{k^2(q-k)^2} \mathcal{T}_{p_1,k} \frac{1}{(p_1-k)^2-m^2} \int \frac{\mathrm{d}^d l}{l^2(q-l)^2} \mathcal{T}_{l,k} \frac{1}{(l-k)^2-m^2} . \quad (5.28)$$

Each term in these contributions can be analytically evaluated by recursive integration over the two loop momenta. The contributions of the subgraphs γ_2 and γ_3 take the form

$$\frac{\left(\mathrm{i}\pi^{d/2}\right)^2}{(m^2)^{2+\varepsilon}(-q^2)^\varepsilon} \sum_{n=0}^{\infty} c_n^{(2,3)}(\varepsilon)\, (q^2/m^2)^n . \quad (5.29)$$

Using (A.5) and (A.10) and the summation formulae (A.50), we obtain [113]

$$c_n^{(2)}(\varepsilon) = \sum_{\substack{i_1,i_2,n_3\geq 0,\ i_1+i_2+n_3 \text{ even} \\ i_1+i_2+n_3\leq 2n}} \sum_{\substack{j_3\geq 0, \\ j_3+n_3 \text{even}}}^{n_3} (-1)^{(i_1+i_2+n_3)/2}$$
$$\times \frac{\Gamma(\varepsilon)\Gamma(1-\varepsilon)\Gamma((i_1+i_2-j_3)/2+1-\varepsilon)(n-(i_1+i_2-n_3)/2)!}{\Gamma((i_1+i_2-j_3)/2+2-2\varepsilon)(n-(i_1+i_2+n_3)/2)!((n_3-j_3)/2)!}$$
$$\times \frac{i_1!i_2!\theta(i_1+i_2-j_3)\theta(i_1-i_2+j_3)\theta(-i_1+i_2+j_3)}{((i_1+i_2-j_3)/2)!((i_1-i_2+j_3)/2)!((-i_1+i_2+j_3)/2)!}$$
$$\times C(2+i_1+i_2, 2+n-(i_1+i_2-n_3)/2; (i_1+i_2+n_3)/2) , \quad (5.30)$$

where $C(\lambda_1, \lambda_2; n)$ is the ratio of gamma functions on the right-hand side of (A.5), and $\theta(n) = 1$ for $n \geq 0$ and $\theta(n) = 0$ otherwise.

Using (A.5), (A.10) and (A.47), we also obtain

$$c_n^{(3)}(\varepsilon) = (-1)^n \Gamma(\varepsilon)\Gamma(1-\varepsilon) \sum_{\substack{i_1 \geq 0 \\ i_1-(n_1-j_1)/2 \leq n}}^{i_1} \sum_{n_1=0}^{n_1} \sum_{\substack{j_1 \geq 0, \\ j_1+n_1 \text{even}}} (-1)^{i_1+j_1}$$
$$\times \frac{i_1!}{(i_1-n_1)!((n_1-j_1)/2)!} \frac{\Gamma(n-i_1+(n_1-j_1)/2+1-\varepsilon)}{\Gamma(n-i_1+(n_1-j_1)/2+2-2\varepsilon)}$$
$$\times C(j_1+1, i_1+2; (n_1+j_1)/2) , \tag{5.31}$$

with the same function $C(\lambda_1, \lambda_2; n)$.

Finally, for γ_4, we obtain

$$\frac{\left(\mathrm{i}\pi^{d/2}\right)^2}{(m^2)^2(-q^2)^{2\varepsilon}} \sum_{n=0}^{\infty} c_n^{(4)}(\varepsilon)\,(q^2/m^2)^n , \tag{5.32}$$

where

$$c_n^{(4)}(\varepsilon) = (-1)^n [\Gamma(\varepsilon)\Gamma(1-\varepsilon)]^2 \sum_{j=0}^{n} \frac{\Gamma(j+1-\varepsilon)\Gamma(n-j+1-\varepsilon)}{\Gamma(j+2-2\varepsilon)\Gamma(n-j+2-2\varepsilon)} . \tag{5.33}$$

Summing the four contributions, we see that the poles in ε cancel and we obtain, at $\varepsilon = 0$, the following series [113]:

$$F_{5.4} \sim -\frac{\pi^4}{m^4} \left[L^2 - L + \frac{\pi^2}{2} - 4 - \frac{q^2}{m^2} \left(L^2 - \frac{5}{6} L + \frac{\pi^2}{2} - \frac{341}{72} \right) \right.$$
$$+ \left(\frac{q^2}{m^2}\right)^2 \left(\frac{11}{12} L^2 - \frac{79}{120} L + \frac{11\pi^2}{24} - \frac{2617}{600} \right)$$
$$\left. - \left(\frac{q^2}{m^2}\right)^3 \left(\frac{5}{6} L^2 - \frac{89}{168} L + \frac{5\pi^2}{12} - \frac{63051}{15680} \right) + \ldots \right] , \tag{5.34}$$

where $L = \ln\left(-q^2/m^2\right)$.

Note that there is again an interplay of UV and IR poles in the individual contributions, which are IR in γ_1, UV and IR in $\gamma_{2,3}$ and IR in γ_4. Now we turn to the next example, where a new kind of divergence participates in the game:

Example 5.5. The vertex two-loop diagram of Fig. 5.4b with a large mass $m_6 = m$, five zero masses and a small external momentum q.

This time, we have another set of AI subgraphs in (5.10): $\gamma_1 \equiv \Gamma, \gamma_2 = \{3456\}, \gamma_3 = \{126\}$ and $\gamma_4 = \{6\}$.

The contribution of γ_1 is evaluated as in the previous case. We obtain (5.21), where

$$
\begin{aligned}
A_n &= \frac{\left(\mathrm{i}\pi^{d/2}\right)^2}{(n+1)(m^2)^{2+2\varepsilon+n}} \sum_{j=0}^{n} \sum_{j'=0}^{n} \sum_{i=\max\{0,j+j'-n\}}^{\min\{n,[(j+j')/2]\}} \frac{(j+j'-2i)!a_{jj'}^{ni}}{2^{j+j'-2i}} \\
&\times \sum_{j_1=0}^{j+j'-2i} \sum_{j_2=0}^{j_1} \frac{(-1)^{n+i+j+j'+j_2}}{(j+j'-2i-j_1)!j_2!} \\
&\times V_{00m}(n+j_1+i-j-j'+2, j+j'-i-j_2+3, 1)\,, \qquad (5.35)
\end{aligned}
$$

where V_{00m} again denotes the ratio of gamma functions on the right-hand side of (A.39).

The contribution of γ_2 is computed as in the previous example, with a similar cumbersome result, given by (5.29) and

$$
\begin{aligned}
c_n^{(2)}(\varepsilon) = &\sum_{\substack{i_1,i_2,n_3\geq 0,\ i_1+i_2+n_3 \text{ even} \\ i_1+i_2+n_3 \leq 2n}} \sum_{\substack{j_3\geq 0, \\ j_3+n_3 \text{even}}}^{n_3} (-1)^{(i_1+i_2+n_3)/2} \\
&\times \frac{\Gamma(1-\varepsilon)\Gamma(1+\varepsilon)\Gamma((i_1+i_2-j_3)/2+1-\varepsilon)(n-(i_1+i_2-n_3)/2)!}{\Gamma((i_1+i_2-j_3)/2+2-2\varepsilon)(n-(i_1+i_2+n_3)/2)!} \\
&\times \frac{i_1!i_2!\theta(i_1+i_2-j_3)\theta(i_1-i_2+j_3)\theta(-i_1+i_2+j_3)}{((i_1+i_2-j_3)/2)!((i_1-i_2+j_3)/2)!((-i_1+i_2+j_3)/2)!} \\
&\times C(1+i_2, 3+n+(i_1-i_2+n_3)/2; (i_1+i_2+n_3)/2)\,. \qquad (5.36)
\end{aligned}
$$

The third and the fourth contributions can be described graphically as Taylor expansions of the triangle γ_3 and the isolated line γ_4, respectively, in their external momenta, inserted into the corresponding reduced diagrams, which are the light triangle consisting of lines $\{3,4,5\}$ and the same triangle connected with a one-loop propagator diagram with lines $\{1,2\}$, respectively. In any case, the insertion factors are scalar functions of the three invariants, $(p_1+k)^2, (p_2+k)^2$ and q^2. When these Taylor expansions are performed, the factor $(p_1+k)^2$ leads to a cancellation of one of the propagators in the light triangle. This produces a one-loop massless diagram with its external momentum on the light cone. This diagram is zero within dimensional regularization – see Chap. 2. Thus we come to the conclusion that only terms without such factors survive, and one can set k_2 to zero in the last two contributions. So, the sum of the two terms for the subgraphs γ_3 and γ_4 happens to be just a product of two factors: the light triangle with zero masses given by (2.79), and the expansion in the large-mass limit of the lower triangle γ_3 with p_1, p_2, p_1-p_2 as external momenta. The last expansion can be performed by our general technique and leads to the following result for the sum of the contributions of γ_3 and γ_4:

$$
-\frac{\left(\mathrm{i}\pi^{d/2}\right)^2}{(M^2)^{1+\varepsilon}(-q^2)^{1+\varepsilon}} \frac{1}{\varepsilon^3} \frac{\Gamma(1-\varepsilon)^3}{\Gamma(1-2\varepsilon)} \left(\sum_{n=0}^{\infty} \frac{n!}{\Gamma(n+2-\varepsilon)} (-q^2/M^2)^n \right.
$$

$$
-(-q^2/M^2)^{-\varepsilon} \sum_{n=0}^{\infty} \frac{\Gamma(n+1-\varepsilon)}{\Gamma(n+2-2\varepsilon)} (-q^2/M^2)^n \Bigg) . \tag{5.37}
$$

Combining all four contributions, we arrive at the following result [109], written as an expansion in ε up to the finite part:

$$
\begin{aligned}
F_{5.5} \sim & -\frac{\left(\mathrm{i}\pi^{d/2}\mathrm{e}^{-\gamma_{\mathrm{E}}\varepsilon}\right)^2}{m^2 q^2} \Bigg\{ (-L+1)\frac{1}{\varepsilon^2} + \left(\frac{3}{2}L^2 - 3L - \frac{\pi^2}{6} + 3\right)\frac{1}{\varepsilon} \\
& -\frac{7}{6}L^3 + \frac{7}{2}L^2 + \left(\frac{\pi^2}{3} - 7\right)L - 2\zeta(3) - \frac{\pi^2}{3} + 7 \\
& -\frac{q^2}{m^2}\left[\left(-\frac{1}{2}L + \frac{1}{4}\right)\frac{1}{\varepsilon^2} + \left(\frac{3}{4}L^2 - \frac{9}{4}L - \frac{\pi^2}{12} + \frac{5}{8}\right)\frac{1}{\varepsilon}\right. \\
& \left. -\frac{7}{12}L^3 + \frac{17}{8}L^2 + \left(\frac{\pi^2}{6} - \frac{49}{8}\right)L - \zeta(3) - \frac{2\pi^2}{3} + \frac{89}{16}\right] \\
+ & \left(\frac{q^2}{m^2}\right)^2 \left[\left(-\frac{1}{3}L + \frac{1}{9}\right)\frac{1}{\varepsilon^2} + \left(\frac{1}{2}L^2 - \frac{11}{6}L - \frac{\pi^2}{18} + \frac{31}{36}\right)\frac{1}{\varepsilon}\right. \\
- & \left.\frac{7}{18}L^3 + \frac{59}{36}L^2 + \left(\frac{\pi^2}{9} - \frac{160}{27}\right)L - \frac{2\zeta(3)}{3} - \frac{67\pi^2}{108} + \frac{4709}{648}\right] + \ldots \Bigg\} .
\end{aligned} \tag{5.38}
$$

The dimensional regularization parameter μ (which is always implied) is set to m here (so that all the logarithms $\ln(m^2/\mu^2)$ drop out).

To see that the poles in ε in the sum of all four contributions are the same as in the initial Feynman integral, let us apply the following formula for the pole part in ε of the product of three propagators, namely $1/k^2, 1/(k^2-2p_1k)$ and $1/(k^2-2p_2k)$, formally considered as a generalized function of k:

$$
\begin{aligned}
& \frac{\mathrm{i}\pi^{d/2}\mathrm{e}^{-\gamma_{\mathrm{E}}\varepsilon}}{(-q^2)^{1+\varepsilon}} \Bigg(-\frac{1}{\varepsilon^2}\delta^{(d)}(k) \\
& \qquad + \frac{1}{\varepsilon}\int_0^1 \frac{\mathrm{d}z}{z}[\delta^{(d)}(k-zp_1) + \delta^{(d)}(k-zp_2) - 2\delta^{(d)}(k)] \Bigg) .
\end{aligned} \tag{5.39}
$$

(This formula can be obtained by transforming to coordinate space and calculating the pole part using alpha parameters.) We observe that the term containing $\delta^{(d)}(k)$ corresponds to the pole part of the sum of the contributions from γ_3 and γ_4 given by (5.37). Thus, the term containing the integral over z should give the pole part of the sum of the contributions from γ_1 and γ_2. So, we perform integration over z of the one-loop triangle diagram with one non-zero and two zero masses. We then apply the general formulae for the large-mass expansion to this triangle and conclude that the pole part in ε of the sum of the contributions from γ_1 and γ_2 should be equal to the pole part of the following quantity:

$$
-\pi^4 \left[\frac{2\Gamma(\varepsilon)}{(m^2)^{1+\varepsilon}q^2} \sum_{n=1}^{\infty} n! \left(\sum_{j=1}^{n} \frac{C(j+1, 2n-j+2; n)}{j} \right) (q^2/m^2)^n \right.
$$

$$
\left. + \frac{2\Gamma(\varepsilon)^2\Gamma(1-\varepsilon)}{m^2(-q^2)^{1+\varepsilon}} \sum_{n=1}^{\infty} \frac{s_n \Gamma(n+1-\varepsilon)}{\Gamma(n+2-2\varepsilon)} \left(-q^2/m^2\right)^n \right] , \tag{5.40}
$$

where $s_n = \sum_{j=1}^{n} j^{-1}$. Note that the double poles in (5.40) cancel so that there must also be a cancellation of the double poles in the sum of the contributions from γ_1 and γ_2. This is indeed the case. One also can easily check that the single-pole part agrees with that obtained by direct calculation of the sum of the contributions from γ_1 and γ_2.

The double pole in ε originates only from the sum of the contributions from γ_3 and γ_4. From (5.37), we obtain the following formula for its coefficient:

$$
\frac{-\pi^4}{(q^2)^2} \left[\ln(-q^2/m^2) \ln(1+q^2/m^2) + \mathrm{Li}_2(-q^2/m^2) \right] . \tag{5.41}
$$

Although the given diagram can be expanded by formulae relevant to limits typical of Euclidean space, where we usually have an interplay between UV and IR divergences, there is now an interplay between UV, IR and collinear divergences. We have IR divergences in γ_1, UV and IR divergences in γ_2, UV divergences in γ_4, and collinear and IR divergences in γ_3. It turns out that a part of the original collinear divergence is transformed into divergences of other types because the collinear divergences in γ_3 correspond only to the double pole arising from the 'product' of the original IR and collinear divergences.

5.4 Decoupling and the Effective Lagrangian

At a fixed energy scale, the particles of a given theory are naturally subdivided into heavy and light particles, with masses M_i and m_j. Let us assume that the theory is exactly renormalizable, i.e. all the terms in the corresponding interaction Lagrangian have dimension four, in the mass units. A term of the Lagrangian is called 'heavy' if it involves at least one heavy field (i.e. a field with a heavy mass), and 'light' otherwise. So the Lagrangian can be represented as $\mathcal{L} = \mathcal{L}_{\text{heavy}} + \mathcal{L}_{\text{light}}$. Let $\left\{ T \exp \left[\mathrm{i} \int \mathrm{d}^4 x \, \mathcal{L}_{\mathrm{I}}(x) \right] \right\}_{\text{light}}$ be the part of the renormalized S-matrix composed of diagrams which have only light external lines. As before, the index 'I' denotes the sum of the interaction Lagrangian, with a renormalized value of the coupling, and counterterms corresponding to the free part of the Lagrangian. Equivalently, we can consider the family of Green functions of the light fields. The so-called decoupling theorem [5] says that in the leading power of the asymptotic behaviour in the limit $m_j/M_i \to 0$, these Green functions are determined by the light part, $\mathcal{L}_{\text{light}}$, of the Lagrangian. Formally,

$$
\left[T \exp \left(\mathrm{i} \int \mathrm{d}^4 x \, \mathcal{L}_{\mathrm{I}}(x) \right) \right]_{\text{light}} \sim T \exp \left(\mathrm{i} \int \mathrm{d}^4 x \, \mathcal{L}_{\mathrm{I,light}}(x) \right) . \tag{5.42}
$$

This theorem in fact applies only for special renormalization schemes, for example for the BPHZ renormalization [26, 137, 247] (valid only for theories without massless particles). For example, for minimal subtractions within dimensional regularization, the decoupling theorem is not true, in its naive sense.[1] Nevertheless, it is true in a generalized form,

$$\left[T \exp \left(\mathrm{i} \int \mathrm{d}^4 x \, \mathcal{L}_{\mathrm{I}}(x) \right) \right]_{\mathrm{light}} \sim T \exp \left(\mathrm{i} \int \mathrm{d}^4 x \, \mathcal{L}_{\mathrm{eff}}(x) \right) , \tag{5.43}$$

where the *effective* Lagrangian on the right-hand side is obtained from $\mathcal{L}_{\mathrm{I,light}}$ by adjusting *finite* counterterms. In the general order of the expansion in the inverse large masses, the left-hand side of (5.43) is again given by the right-hand side, where the effective Lagrangian is written as a series of light composite operators. In particular, in the case of one heavy mass, M, the effective Lagrangian is

$$\mathcal{L}_{\mathrm{eff}}(x) = \sum_{\delta_i \geq 4} \frac{C_i(M)}{M^{\delta_i - 4}} O_i(x) , \tag{5.44}$$

where δ_i is the mass dimension of the operator O_i. The first terms with dimension four form the light Lagrangian, with finite changes in the corresponding coupling, mass and wave-function renormalizations as compared with the corresponding values of the pure light theory. After the power dependence on the heavy mass has been explicitly taken out, the coefficient functions $C_i(M)$ are polynomials in $\ln M^2$. The next terms in the effective Lagrangian typically involve operators of dimension six, so that they are power suppressed by a factor $1/M^2$. These terms are already described by non-renormalizable interactions.

To perform the large-mass operator expansion one can apply the general strategy of constructing a low-energy effective theory, by assuming that the latter is described by an effective Lagrangian of the form (5.44), with unknown coefficients C_i for all possible operators O_i. These coefficients are then found, order by order in perturbation theory, from a system of (matching) equations that express the fact that the same results in the limit $M \to \infty$ are obtained by starting from the full and effective theories. We shall shortly describe this procedure and related procedures used to construct effective Lagrangians.

One well-known example of an effective Lagrangian associated with the large-mass limit describes the weak four-fermion interaction between quarks, u and d, a charged lepton, l, and neutrino, ν, due to the exchange of a W boson. This effective Lagrangian is proportional to the square of the weak interaction coupling g and is given by

$$\frac{g^2}{2m_W^2} \bar{l} \gamma_\mu \frac{1-\gamma_5}{2} \nu \bar{u} \gamma^\mu \frac{1-\gamma_5}{2} d , \tag{5.45}$$

[1] This observation was first made in [238, 186].

plus conjugate terms. The first factor is written in terms of the Fermi constant $G_F = g^2/(4\sqrt{2}m_W^2)$, which determines the strength of the weak interaction at energies much smaller than the mass of the W boson, m_W. This Lagrangian is obtained by taking into account only tree exchange diagrams, where the propagator of the W boson is replaced by its asymptotic behaviour in the limit of large m_W, and the factor $1/m_W^2$ arises from this.

At the tree level, the construction of effective Lagrangians is rather transparent. Examples where one-loop information is encoded are also well known. One of them, the Euler–Heisenberg Lagrangian [136, 202] (see also [149]), is related to one of the earliest quantum-field theoretical calculations and describes one-loop quantum corrections, due to creation of electron–positron pairs, to the Lagrangian $-(1/4)F_{\mu\nu}^2$ in the case of a constant classical electromagnetic field. This Lagrangian, evaluated within the Schwinger proper-time method [202], has the form

$$\mathcal{L}_{\rm EH} = \frac{1}{8\pi^2}\int_0^\infty \frac{{\rm d}T}{T}{\rm e}^{-{\rm i}m^2T}\left(\frac{e^2ab}{\tan(eaT)\tanh(ebT)} - \frac{1}{T^2}\right) , \tag{5.46}$$

where a and b are related to the constant electric and magnetic fields by $a^2 - b^2 = \mathbf{E}^2 - \mathbf{B}^2$, $ab = \mathbf{E}\cdot\mathbf{B}$. Expanding this exact result in e^2, one obtains a representation of this effective Lagrangian as a series in inverse powers of the electron mass. The first power-suppressed term contains operators of dimension eight:

$$\begin{aligned}\mathcal{L}_{\rm EH}^{(8)} &= \frac{2\alpha^2}{45m^4}\left[(\mathbf{E}^2-\mathbf{B}^2)^2 + 7(\mathbf{E}\cdot\mathbf{B})^2\right] \\ &\equiv \frac{\alpha^2}{90m^4}\left[(F_{\mu\nu}F^{\mu\nu})^2 + \frac{7}{4}(F_{\mu\nu}\tilde{F}^{\mu\nu})^2\right] ,\end{aligned} \tag{5.47}$$

where $\alpha = e^2/(4\pi)$ is the fine-structure constant and $\tilde{F}^{\mu\nu} = (1/2)\varepsilon^{\mu\nu\rho\sigma}F_{\rho\sigma}$ is the dual electromagnetic tensor.

Let us now describe in more detail how the coefficient functions in the effective Lagrangian are calculated. As in the previous chapter, in the case of the off-shell large-momentum limit, let us start from the diagrammatic expansion. Using the expansion of unrenormalized Feynman integrals (5.10) and applying combinatorial arguments related to exponentiation, similar to those applied in the derivation of (2.94) in Chap. 2, we arrive at the corresponding operator expansion (first in the monomial basis and then rewritten in a general basis) which has the form

$$\left[T\exp\left({\rm i}\int {\rm d}^4x\,\mathcal{L}_{\rm I}^{\rm B}(x)\right)\right]_{\rm light} \sim T\exp\left({\rm i}\int {\rm d}^4x\,\mathcal{L}_{\rm eff}^{\rm B}(x)\right) , \tag{5.48}$$

where

$$\mathcal{L}_{\rm eff}^{\rm B}(x) = \mathcal{L}_{\rm light}^{\rm B} - {\rm i}\sum_i \Pi_i\left[\exp\left({\rm i}\int {\rm d}^4x\mathcal{L}_{\rm I}^{\rm B}(x)\right) - 1\right] o_i(x) , \tag{5.49}$$

and, in contrast to (5.42), the quantities are not renormalized, i.e. the perturbative expansions are performed with the bare couplings, masses and fields.

Here the o_i are composite operators from some basis and the Π_i are the corresponding projectors. Any projector involves Taylor expansion in the external momenta and light masses, so that only heavy Feynman diagrams (including at least one heavy line) which are 1PI with respect to the light lines give non-zero contributions.

If we switch on the renormalization of the initial full theory, we arrive at an operator expansion which differs from (5.48) and (5.49) by using the renormalized Lagrangians $\mathcal{L}_{\rm I}$ and $\mathcal{L}_{\rm I,light}$, i.e. we obtain (5.43), where

$$\mathcal{L}_{\rm eff}(x) = \mathcal{L}_{\rm light} - {\rm i} \sum_i \Pi_i \left[\exp\left({\rm i} \int {\rm d}^4 x \, \mathcal{L}_{\rm I}(x) \right) - 1 \right] o_i(x) \, . \tag{5.50}$$

This operator expansion is similar to the OPE in the form (4.62) and involves an interplay between UV and IR divergences characteristic of any expansion in momenta in masses. If a Feynman diagram generated by the right-hand side of (5.43) involves a vertex associated with the second term in (5.50), with some monomial $o_i(x)$, the corresponding overall UV divergence is not removed. On the other hand, the projectors Π_i generate IR divergences so that, in every perturbation order, the total right-hand side of (5.43) is finite.[2] Anyway, the expansion (5.50), which is a consequence of a diagrammatic expansion valid for general diagrams, provides a kind of existence theorem for the operator large-mass expansion governed by the effective Lagrangian.

Observe that the terms of dimension four in the second part of the right-hand side of (5.50) are similar to those of the first part, so that it is not natural to consider the same monomials on a different footing when generating diagrams. If we confine ourselves to only one insertion of the effective Lagrangian, i.e. take only the first order of expansion of the exponent, we obtain, from the second term, a bare composite operator which can be rewritten in terms of renormalized operators of the same dimension by means of the inverse matrix of the composite operators defined by (2.97). But, in any case, these two parts of the effective Lagrangian are naturally combined, when one is analysing the leading-power large-mass behaviour, in the framework of a strategy based on direct matching of the coefficients in (5.44). Equivalently, one can introduce, in addition to the renormalization constants of the given theory, so-called decoupling constants. We shall shortly see an example of such a procedure.

It can happen that, in a given order of perturbative expansion in the given coupling constants, the projector Π_i does not generate IR divergences. In particular, this is the case when no massless cuts arise in Feynman diagrams of the given perturbation order. In this case one can straightforwardly apply (5.50) for the evaluation of the coefficients of the effective Lagrangian.

---

[2] It is possible to write down a manifestly finite form of the large-mass operator expansion, at the cost of introducing the R^*-operation [62] – see [208].

For example, starting from QED in the limit of a large mass of the electron, the power-suppressed contributions to the effective Lagrangian consisting of the electromagnetic field start from operators of dimension eight, namely $(F_{\mu\nu}F^{\mu\nu})^2$ and $(F_{\mu\nu}\tilde{F}^{\mu\nu})^2$. At the one-loop level, only the diagram shown in Fig. 5.5, generated by the first term in the expansion of $\exp\left[\mathrm{i}\int \mathrm{d}x\, \mathcal{L}_\mathrm{I}(x)\right]$ in (5.50), is needed for the evaluation of the corresponding coefficients. Eventually, one reproduces (5.47).

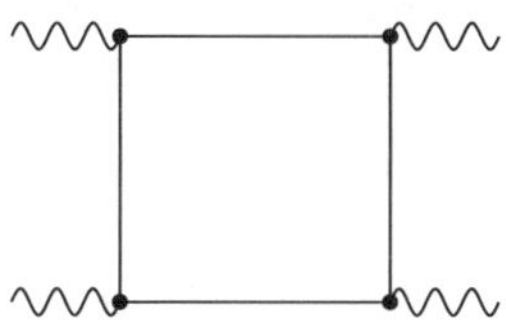

Fig. 5.5. One-loop diagram corresponding to dimension-eight operators in the Euler–Heisenberg effective Lagrangian

This is an example of a situation where an old-fashioned calculation turns out to be more effective and provides, in the compact explicit form (5.46), one-loop results for *all* operators in the effective Lagrangian. However, such calculations, for a constant classical electromagnetic field, can hardly be relevant, in higher loops, to effective Lagrangians corresponding to the quantum electromagnetic potential, so that it is reasonable to turn our attention to the evaluation of coefficients of individual operators in the effective Lagrangian by the techniques presented earlier.

5.5 An Example: Decoupling of a Heavy Quark

Let us consider QCD with one heavy quark and $n_\mathrm{l} = n_\mathrm{f} - 1$ light quark flavours, where the heavy part of the Lagrangian is $\bar{\psi}_\mathrm{h}(\mathrm{i}\not{D} - m_\mathrm{h})\psi_\mathrm{h}$. The resulting effective Lagrangian describes a low-energy theory with a decoupled ('integrated out') heavy quark. In the leading order of the limit $m_\mathrm{h} \to \infty$, the construction of the effective Lagrangian at the three-loop level has been performed in [55] as follows. The renormalization of the full QCD is given by (2.91), and the renormalization of the resulting QCD with light flavours is given by similar relations, where all the quantities related to the effective theory will be indicated by primes. To describe the transition to the effective theory, it is convenient to introduce decoupling (matching) constants at the bare level, where they determine relations between bare quantities:

$$g'^\mathrm{B} = \zeta_g^\mathrm{B} g^\mathrm{B}\,, \quad m_i'^\mathrm{B} = \zeta_m^\mathrm{B} m_i^\mathrm{B}\,, \quad \xi'^\mathrm{B} - 1 = \zeta_3^\mathrm{B}(\xi^\mathrm{B} - 1)\,,$$
$$\psi_i'^\mathrm{B} = \sqrt{\zeta_2^\mathrm{B}}\psi_i^\mathrm{B}\,, \quad A_\nu'^{\mathrm{B},a} = \sqrt{\zeta_3^\mathrm{B}} A_\nu^{\mathrm{B},a}\,, \quad c'^{\mathrm{B},a} = \sqrt{\tilde{\zeta}_3^\mathrm{B}} c_a^\mathrm{B}\,, \tag{5.51}$$

where i denotes only the light flavours.

With these definitions, the couplings and the masses of the two theories are connected by

$$\alpha_s' = \left(\frac{Z_g}{Z_g'}\zeta_g^{\rm B}\right)^2 \alpha_s \equiv \zeta_g^2 \alpha_s \,, \tag{5.52}$$

$$m_i' = \frac{Z_m}{Z_m'}\zeta_m^{\rm B} m_i \equiv \zeta_m m_i \,. \tag{5.53}$$

Note that the bare decoupling constants are singular in ε, in contrast to the renormalized decoupling constants ζ_g and ζ_m.

The bare decoupling constants are determined from the decoupling (matching) relations between the 1PI Green functions of the full and effective theories. Let $\Sigma(p) = \not{p}\Sigma_V(p^2) + m\Sigma_S(p^2)$ be the light-quark propagator (a 1PI Green function of two light-quark fields). The decoupling relation for the massless propagator then takes the form

$$\frac{\rm i}{\not{p}[1+\Sigma_V^{\rm B}(p^2)]} = \frac{1}{\zeta_2^{\rm B}} \frac{\rm i}{\not{p}[1+\Sigma'{\rm B}_V(p^2)]} \,. \tag{5.54}$$

In the limit $m_{\rm h} \to \infty$, this relation is written at the point $p = 0$, where $\Sigma_V'^{\rm B}(p^2)$ vanishes because it involves massless tadpole integrals. This procedure, of course, reproduces the action of the projector operators present in (5.49) and (5.50). Therefore,

$$\zeta_2^{\rm B} = 1 + \Sigma_V^{\rm B}(0) \,. \tag{5.55}$$

In a similar way, other expressions for the decoupling constants can be obtained [55]:

$$\zeta_m^{\rm B} = \frac{1-\Sigma_S^{\rm B}(0)}{1+\Sigma_V^{\rm B}(0)} \,, \quad \zeta_3^{\rm B} = 1 + \Pi_G^{\rm B}(0) \,, \tag{5.56}$$

$$\tilde{\zeta}_3^{\rm B} = 1 + \Pi_c^{\rm B}(0) \,, \quad \zeta_g^{\rm B} = \frac{1+\Gamma_{G\bar{c}c}^{\rm B}(0,0)}{\tilde{\zeta}_3^{\rm B}\sqrt{\zeta_3^{\rm B}}} \,, \tag{5.57}$$

where $\Pi_G(p^2)$ and $\Pi_c(p^2)$ are the gluon and ghost vacuum polarizations, respectively, and $\Gamma_{G\bar{c}c}^{\rm B}(p,k)$ is defined through the 1PI part of the amputated gluon–ghost–ghost Green function. Observe that only heavy diagrams, i.e. those involving a heavy quark loop, contribute to these relations.

Thus the calculation, at the two- and three-loop levels, reduces to the evaluation of two- and three-loop vacuum diagrams. For example, the only two-loop diagram contributing to ζ_m is shown in Fig. 5.6. The three-loop bubbles can be evaluated by the program MATAD [224], which is based on an algorithm described in [37].

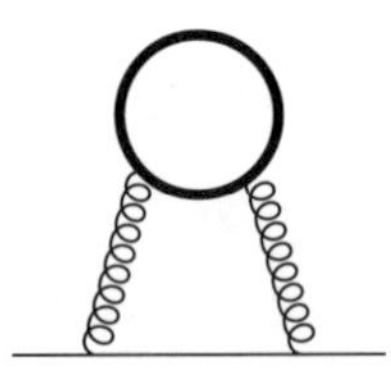

Fig. 5.6. Two-loop diagram contributing to ζ_m

After taking into account the QCD renormalization constants Z_m and Z_g (which are known up to the three-loop level [197, 49, 236]), the following results for the three-loop decoupling constants ζ_g and ζ_m have been obtained [55]:

$$\zeta_m = 1 + \left(\frac{\alpha_s}{\pi}\right)^2 \left(\frac{89}{432} - \frac{5}{36}\ln\frac{\mu^2}{m_h^2} + \frac{1}{12}\ln^2\frac{\mu^2}{m_h^2}\right)$$
$$+ \left(\frac{\alpha_s}{\pi}\right)^3 \left[\frac{2951}{2916} - \frac{407}{864}\zeta(3) + \frac{5}{4}\zeta(4) - \frac{1}{36}B_4\right.$$
$$+ \left(-\frac{311}{2592} - \frac{5}{6}\zeta(3)\right)\ln\frac{\mu^2}{m_h^2} + \frac{175}{432}\ln^2\frac{\mu^2}{m_h^2} + \frac{29}{216}\ln^3\frac{\mu^2}{m_h^2}$$
$$\left. + n_l\left(\frac{1327}{11664} - \frac{2}{27}\zeta(3) - \frac{53}{432}\ln\frac{\mu^2}{m_h^2} - \frac{1}{108}\ln^3\frac{\mu^2}{m_h^2}\right)\right], \qquad (5.58)$$

where $B_4 = 16\,\mathrm{Li}_4(1/2) - (13/2)\zeta(4) - 4\zeta(2)\ln^2 2 + (2/3)\ln^4 2$, and α_s is the coupling of the full theory with n_f flavours at the scale μ. The two-loop part of this result has been obtained in [239, 20].

Furthermore [55, 54],

$$\zeta_g^2 = 1 + \frac{\alpha_s}{\pi}\left(-\frac{1}{6}\ln\frac{\mu^2}{m_h^2}\right) + \left(\frac{\alpha_s}{\pi}\right)^2\left(\frac{11}{72} - \frac{11}{24}\ln\frac{\mu^2}{m_h^2} + \frac{1}{36}\ln^2\frac{\mu^2}{m_h^2}\right)$$
$$+ \left(\frac{\alpha_s}{\pi}\right)^3\left[\frac{564731}{124416} - \frac{82043}{27648}\zeta(3) - \frac{955}{576}\ln\frac{\mu^2}{m_h^2} + \frac{53}{576}\ln^2\frac{\mu^2}{m_h^2}\right.$$
$$\left. - \frac{1}{216}\ln^3\frac{\mu^2}{m_h^2} + n_l\left(-\frac{2633}{31104} + \frac{67}{576}\ln\frac{\mu^2}{m_h^2} - \frac{1}{36}\ln^2\frac{\mu^2}{m_h^2}\right)\right]. \qquad (5.59)$$

The two-loop part of this result agrees with [162].

These results for the decoupling constants have been used to derive low-energy theorems for the ggH and $\bar{q}qH$ interactions [55], by expressing effective couplings of the Higgs boson to gluons and light quarks through the decoupling constants ζ_g and ζ_m.

An example of an effective Lagrangian with power suppressed terms is given by the same situation with the heavy quark decoupled. In the one-loop approximation, one can use (5.43) and (5.50). The dimension-six correction to the effective light QCD Lagrangian comes from the diagram of Fig. 5.7 and is related to the operator $\nabla_\nu^{ab} G^{a\mu\nu} \nabla_{\nu'}^{ab'} G_\mu^{b'\nu'}$, where the covariant derivative ∇_ν^{ab} is given by (4.73). The corresponding projector can be chosen (at least in the given one-loop approximation) to be proportional to $g_{\mu\nu}\Box_q$, where q is the

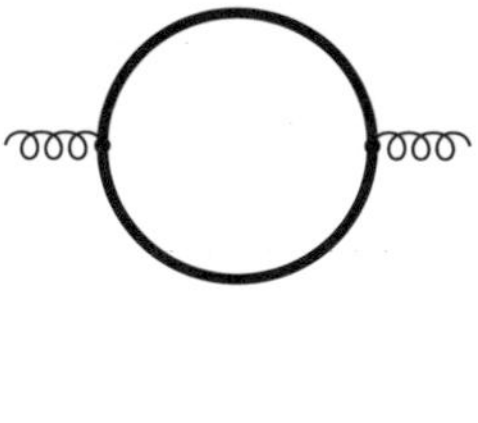

Fig. 5.7. One-loop diagram contributing to the dimension-six term in the effective QCD Lagrangian

external momentum of Fig. 5.7 and μ and ν are external vector indices. After the calculation of the coefficient of this operator, one can use the equation of motion, $\nabla_\nu^{ab} G^{b,\mu\nu} = -g\bar{\psi}\gamma^\mu t^a \psi$, to transform to four-fermion operators. The well-known result is (see, e.g., [184])

$$-\frac{\alpha_s^2}{15m_h^2}\bar{\psi}\gamma^\mu t^a\psi\,\bar{\psi}\gamma_\mu t^a\psi\,. \tag{5.60}$$

The operator expansion in inverse powers of the heavy masses can be performed not only for the S-matrix elements with light external lines but also for light composite fields and their products. For example, the contribution of operators of dimension four and six to the heavy-mass expansion of the light vector current takes the form [48]

$$\bar{\psi}\gamma_\mu\psi \sim \bar{\psi}'\gamma_\mu\psi' + \left(\frac{\alpha_s}{\pi}\right)^2\left(\frac{1}{135}\ln\frac{\mu^2}{m_h^2} - \frac{56}{2025}\right)\frac{\Box(\bar{\psi}'\gamma_\mu\psi')}{m_h^2} + \ldots\,. \tag{5.61}$$

Note that the contribution of the dimension six operators to the right-hand side of (5.61) is in fact renormalization group invariant because the presence of the extra term (5.60) in the effective QCD Lagrangian forces the vector current to have a non-zero anomalous dimension:

$$\mu^2\frac{\mathrm{d}}{\mathrm{d}\mu^2}\bar{\psi}\gamma_\mu\psi = -\frac{1}{135}\left(\frac{\alpha_s}{\pi}\right)^2\frac{\Box(\bar{\psi}\gamma_\mu\psi)}{m_h^2}\,. \tag{5.62}$$

6 Threshold Expansion. One Heavy Mass in the Threshold

In this and the next chapter, we investigate asymptotic expansions of Feynman integrals near threshold, i.e. when an external momentum squared tends to a threshold value, and their operator analogues. This chapter deals with the case where only one particle with a heavy mass contributes to the threshold. As is well known, thresholds are located at $q^2 = \left(\sum m_i\right)^2$, where q is the sum of the external momenta flowing through a cut in a diagram, and the m_i are the masses of the lines in the cut. We shall consider two kinds of such threshold limits:

(A) The external momentum is on the mass shell corresponding to the heavy mass M, and there is another mass m (or several masses) which is (are) small. The limit is $m/M \to 0$.
(B) The masses in the diagram are either heavy or zero. The threshold is composed of one heavy mass and some zero masses. The external momentum squared tends to its threshold value.

Limit (B) can be also considered as the on-shell limit, because the mass shell and the threshold coincide. We prefer to use the variant given above because of a similarity to Limit (A). For Limit (B), we shall keep in mind the situation where one of the quarks is much heavier than all the other quarks. In Chap. 7, we shall then consider the threshold expansion for thresholds generated by quarks with non-zero masses.

We follow the strategy of expansion by regions. After studying typical one-loop examples that show what regions are relevant to the limits under consideration, we formulate general prescriptions in Sect. 6.2, where we also use the language of subgraphs for Limit (A). Typical two-loop examples for both threshold limits are then presented in Sect. 6.3. A typical phenomenological application of the expansion of diagrams in Limit (A), connected with the calculation of the muon anomalous magnetic moment in QED, is presented in Sect. 6.4. Finally, in Sect. 6.5, we describe how Limit (B) leads, at the operator level, to the Heavy Quark Effective Theory (HQET) [183, 174, 132].

6.1 One-Loop Examples

Let us consider first an example of a one-loop diagram with two masses, m and M, in the threshold, with the external momentum on the mass shell $q^2 = M^2$.

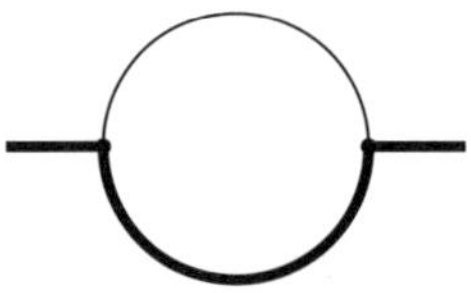

Fig. 6.1. One-loop propagator on-shell diagram with a small mass m and a large mass M

Example 6.1. The on-shell diagram of Fig. 6.1 with $q^2 = M^2$, in Limit (A).

The position of the threshold, $(M+m)^2$, and the external momentum squared, M^2, tend to each other in the limit. It is reasonable to let the external momentum flow through the heavy line when the Feynman integral takes the form

$$F_{6.1}(M^2, m^2; d) = \int \frac{\mathrm{d}^d k}{[(k-q)^2 - M^2](k^2 - m^2)} \equiv \int \frac{\mathrm{d}^d k}{(k^2 - 2q \cdot k)(k^2 - m^2)} . \quad (6.1)$$

Before expanding this integral in the limit $m^2 \ll q^2 = M^2$, let us observe that in the general case, where $m^2 \ll q^2 \sim M^2$ and $m^2 \ll |q^2 - M^2|$, we could safely apply (5.10). The expansion would consist of contributions from two subgraphs. The graph itself would generate the Taylor expansion of $1/(k^2 - m^2)$ in m^2, and the subgraph consisting of the line with mass M would generate a Taylor expansion of its propagator in the loop momentum k of the whole graph, which is, by definition, considered small. The leading contribution from the latter subgraph is

$$\frac{1}{q^2 - M^2} \int \frac{\mathrm{d}^d k}{k^2 - m^2} .$$

In our on-shell example, we can again write down a similar contribution from the whole graph. However, we cannot write down the contribution of the above subgraph because, now, $q^2 - M^2 = 0$ so that the propagator $1/(k^2 - 2qk)$ cannot be expanded into a Taylor series in k.

However, another formulation of the large-mass expansion, in the language of regions, admits a straightforward generalization to the on-shell case. The contribution from the whole subgraph is now interpreted as the contribution from the region of large loop momenta. Let us use the word 'hard' instead of 'large' from now on. So, the hard contribution to the expansion is obtained as a Taylor expansion of the propagator with mass m at $m = 0$:

$$F_{6.1}^{(\mathrm{h})} = \sum_{n=0}^{\infty} (m^2)^n \int \frac{\mathrm{d}^d k}{(k^2 - 2q\cdot k)(k^2)^{n+1}} . \tag{6.2}$$

Evaluating this integral by means of (A.13), we obtain

$$F_{6.1}^{(\mathrm{h})} = \mathrm{i}\pi^{d/2} (M^2)^{-\varepsilon} \sum_{n=0}^{\infty} \frac{\Gamma(n+\varepsilon)\Gamma(-2n-2\varepsilon+1)}{\Gamma(-n-2\varepsilon+2)} \left(-\frac{m^2}{M^2}\right)^n . \tag{6.3}$$

In the off-shell case, the contribution from the subgraph with the heavy line is connected to the region of small loop momenta, $k \sim m$. From now on we shall use the word 'soft' instead of 'small'. It turns out that in the on-shell case, we again have to consider the region of soft loop momenta. In this region, the propagator with mass m is not expanded, while the terms $k^2 \sim m^2$ and $2q\cdot k \sim \sqrt{q^2}m$ in the second propagator are of different order, so that the corresponding contribution generates a Taylor expansion of $1/(k^2 - 2q\cdot k)$ in k^2:

$$F_{6.1}^{(\mathrm{s})} = \sum_{n=0}^{\infty} (-1)^n \int \mathrm{d}^d k \frac{(k^2)^n}{(-2q\cdot k)^{n+1}(k^2 - m^2)} . \tag{6.4}$$

We can again label this contribution in graph-theoretical language and attribute it to the same subgraph consisting of the heavy line, as before. The only difference is that instead of Taylor expanding the corresponding propagator, we have to prescribe that we expand it into a Taylor series with respect to k^2.

In the above integral, $(k^2)^n$ can be substituted by $(m^2)^n$ because the factor $k^2 - m^2$ cancels the massive denominator and produces massless tadpole integrals without scale. Calculating the above one-loop integral by means of (A.3) gives

$$F_{6.1}^{(\mathrm{s})} = \mathrm{i}\pi^{d/2} \frac{\sqrt{\pi}}{2(m^2)^{\varepsilon}} \sum_{n=0}^{\infty} \frac{\Gamma((n-1)/2+\varepsilon)}{2^n \Gamma(n/2+1)} \left(\frac{m^2}{M^2}\right)^{(n+1)/2} . \tag{6.5}$$

In the sum of the two contributions, the artificial IR and UV poles are cancelled, with the following result:

$$\begin{aligned} F_{6.1}(M^2, m^2; d) \sim \mathrm{i}\pi^{d/2} \mathrm{e}^{-\gamma_{\mathrm{E}}\varepsilon} \Bigg[& \frac{1}{\varepsilon} + 2 - \ln M^2 + \frac{1}{2}\frac{m^2}{M^2}\left(\ln \frac{M^2}{m^2} + 2\right) \\ & + \frac{1}{2}\frac{m}{M} \sum_{n=0}^{\infty} \frac{\Gamma(n+1/2)\Gamma(n-1/2)}{(2n)!} (m^2/M^2)^n \\ & - \frac{1}{2}\frac{m^4}{M^4} \sum_{n=0}^{\infty} \frac{(n+1)!n!}{(2n+3)!} (m^2/M^2)^n \Bigg] . \end{aligned} \tag{6.6}$$

The pole in ε, which is present from the beginning, is of UV nature. This expansion reproduces a known analytic result:

$$F_{6.1}(M^2, m^2; d)$$
$$= \mathrm{i}\pi^{d/2} \mathrm{e}^{-\gamma_{\mathrm{E}}\varepsilon} \left(\frac{1}{\varepsilon} + 2 - \ln M^2 - x^2 \ln x - 2x\sqrt{4 - x^2} \arctan \sqrt{\frac{2-x}{2+x}} \right) , \tag{6.7}$$

where $x = m/M$.

Now we turn to the second variant of tending to the threshold, when there is one zero mass in the cut and the external momentum squared tends to its threshold value, $q^2 \to m^2$. The corresponding one-loop example is once again Fig. 3.1 but in another limit:

Example 6.2. The Feynman diagram of Fig. 3.1 in Limit (B), i.e. $q^2 \to m^2$.

It is again reasonable to use the previous routing of the external momentum which is different from the choice used in Chap. 3 in the case of the large-momentum limit. In particular, this makes explicit the dependence of the integrand on the expansion parameter, $y = m^2 - q^2$, which goes to zero at the threshold. In this case our Feynman integral takes the form

$$F_{6.2}(q^2, y; d) = \int \frac{\mathrm{d}^d k}{k^2[(q-k)^2 - m^2]^2} \equiv \int \frac{\mathrm{d}^d k}{k^2(k^2 - 2q\cdot k - y)^2} . \tag{6.8}$$

As in the previous chapters, we have introduced, in advance, dimensional regularization for an initially finite quantity.

The situation is similar to the large-momentum or large-mass expansion: if we expand the integrand into a Taylor series in y and do not do anything else we obtain a wrong result, and the appearance of an IR divergence in the zero-order term tells us about that. This is a typically pseudo-Euclidean limit, so we shall use the strategy of expansion by regions. Thus, the hard contribution is obtained by expanding the propagator $1/(k^2 - 2q\cdot k - y)^2$ in y. The resulting terms can be evaluated by means of (A.13) in terms of gamma functions and give

$$F_{6.2}^{(\mathrm{h})} = \int \frac{\mathrm{d}^d k}{k^2} \mathcal{T}_y \frac{1}{(k^2 - 2q\cdot k - y)^2} = \int \frac{\mathrm{d}^d k}{k^2(k^2 - 2q\cdot k)^2} + \ldots$$
$$= \mathrm{i}\pi^{d/2} \frac{1}{(q^2)^{1+\varepsilon}} \sum_{n=0}^{\infty} \frac{(-1)^n \Gamma(n+1+\varepsilon)}{n!(n+2\varepsilon)} \left(\frac{y}{q^2} \right)^n . \tag{6.9}$$

Let us now try the region of soft loop momentum, $k \sim \sqrt{y}$ (i.e. where k is of order one, in mass units). In this region, k^2 is not expanded, while $k^2 - y$ is of a higher order in y as compared with $-2q\cdot k$ in $k^2 - 2q\cdot k - y$. However, we obtain an integral without scale, which we set to zero according to one of the prescriptions of the strategy of expansion by regions:

$$F_{6.2}^{(\mathrm{s})} = \int \frac{\mathrm{d}^d k}{k^2(-2q\cdot k)^2} + \ldots = 0 . \tag{6.10}$$

The fact that this is an integral without scale can be seen immediately, because it is an integral of a homogeneous function, so that one can introduce spherical coordinates and the result is an integral of a pure power of the radial variable.

It turns out that there is a missing contribution that is non-zero, unlike the contribution from the region of soft k. We shall call the corresponding region *ultrasoft* (us). This region is defined by the relation $k \sim y/\sqrt{q^2}$. In this region, k^2 is of order y^2 while $-2q{\cdot}k - y$ is of order y, so that one has to expand in k^2. Starting from this region, we obtain

$$F_{6.2}^{(\mathrm{us})} = \int \frac{\mathrm{d}^d k}{k^2} \mathcal{T}_{k^2} \frac{1}{(k^2 - 2q{\cdot}k - y)^2} = \int \frac{\mathrm{d}^d k}{k^2(-2q{\cdot}k - y)^2} + \ldots$$
$$= -\mathrm{i}\pi^{d/2} \frac{\Gamma(1-\varepsilon)\Gamma(2\varepsilon)}{(q^2)^{1-\varepsilon} y^{2\varepsilon}} \,. \tag{6.11}$$

Only the leading term survives, because, in the next terms, the factor k^2 resulting from expansion cancels the massless propagator so that scaleless integrals appear.

Observe that the remainder of the expansion obtained is of the form

$$\int \frac{\mathrm{d}^d k}{k^2} \left(1 - \mathcal{T}_y^N\right) \left(1 - \mathcal{T}_{k^2}^N\right) \frac{1}{(k^2 - 2q{\cdot}k - y)^2} \,. \tag{6.12}$$

To show its convergence and the desired asymptotic behaviour, one can apply arguments similar to those used for the one-loop examples in Chaps. 3–5.

Note that the first term in (6.9) has an IR pole in ε, while (6.11) has a UV pole. These poles are cancelled, and the resulting series reproduces the known explicit result (2.58) at $\varepsilon = 0$:

$$\mathrm{i}\pi^{d/2} \frac{1}{q^2} \left[\ln \frac{y}{q^2} + \sum_{n=1}^{\infty} \frac{(-1)^n}{n} \left(\frac{y}{q^2} \right)^n \right] = \mathrm{i}\pi^2 \frac{\ln\left(1 - q^2/m^2\right)}{q^2} \,. \tag{6.13}$$

We have worked here with the variables (q^2, y). By a trivial process of expanding powers of q^2 in powers of m^2 and y, all the contributions and the final result can be written in terms of the variables (m^2, y).

The next example is the same diagram without the dot:

Example 6.2a. The Feynman diagram of Fig. 3.1 with the first power of the massive propagator, in Limit (B).

The Feynman integral is

$$\int \frac{\mathrm{d}^d k}{k^2(k^2 - 2q{\cdot}k - y)} \,. \tag{6.14}$$

It is expanded in a similar way. The hard contribution is

$$\mathrm{i}\pi^{d/2} \frac{1}{(q^2)^{\varepsilon}} \sum_{n=0}^{\infty} \frac{(-1)^n \Gamma(n+\varepsilon)}{n!(1-n-2\varepsilon)} \left(\frac{y}{q^2} \right)^n \,. \tag{6.15}$$

The (us) contribution is

$$i\pi^{d/2} \frac{\Gamma(1-\varepsilon)\Gamma(2\varepsilon-1)}{(q^2)^{1-\varepsilon} y^{2\varepsilon-1}} . \tag{6.16}$$

The pole in the term with $n = 0$ in (6.15) is nothing but the UV pole that is present from the beginning. The IR pole for $n = 1$ in (6.15) is cancelled by the UV pole in (6.16).

6.2 General Prescriptions

Let us consider a general propagator-type diagram in one of the threshold limits with one heavy line in the threshold. We suppose that there is a path consisting of heavy lines connecting the two chosen external vertices. In Limit (A), the two end points are on the mass shell, with $q^2 = M^2$, and all the other masses m_i (generally different) are of the same order and much smaller than M. In Limit (B), the external momentum squared, $q^2 = m^2 - y$, tends to the threshold value $q^2 = m^2$. Let us suppose that all the other masses are zero. In both cases, it is natural to choose the loop momenta in such a way that the external momentum q flows through the path consisting of the heavy lines.

It turns out that we have already met all the kinds of regions relevant to threshold regimes with one heavy mass in the threshold. So, to expand a given diagram in these two limits, we have to consider the following regions in Limit (A):

$$\text{(h)}, \quad k \sim M\,,$$
$$\text{(s)}, \quad k \sim m\,;$$

and the following regions in Limit (B):

$$\text{(h)}, \quad k \sim \sqrt{q^2}\,,$$
$$\text{(us)}, \quad k \sim y/\sqrt{q^2}\,.$$

We then follow the other two prescriptions for expansion by regions.

The rules for Limit B have been formulated in the variables (q^2, y). In order to switch to the variables (m^2, y) it is sufficient to expand powers of $q^2 = m^2 - y$ in y. In the leading order, this just means the replacement $\sqrt{q^2} \to m$.

In the case of diagrams with more than two endpoints, it is necessary to characterize the external momenta squared. In particular, when the other external momenta are off-shell, i.e. $\left(\sum_{i\in I} Q_i\right)^2 \neq M_j^2$ for any subset of indices I different from the two vertices with momentum q, and hard, i.e. of order M, the above prescriptions can be extended straightforwardly. Observe also that in this case, the prescriptions for the expansion in Limit (A) can be described in graph-theoretical language [87]. These prescriptions resemble very much the prescriptions for the off-shell large-mass limit based on (5.10). The class of relevant (AI) subgraphs is the same:

(a) in γ, there is a path between any pair of external vertices associated with large external momenta Q_i;
(b) γ contains all the lines with large masses;
(c) every connectivity component γ_j of the graph $\hat{\gamma}$ obtained from γ by collapsing all the external vertices with large external momenta to a point is 1PI with respect to the lines with small masses.

Note that, generally, γ can be disconnected. One can distinguish the connectivity component γ_0 that contains external vertices with large momenta. The corresponding subtraction operator $\mathcal{M}_\gamma$ turns out to be a product $\prod_i \mathcal{M}_{\gamma_i}$ of operators of Taylor expansion in certain momenta and masses. For connectivity components γ_i other than γ_0, the corresponding operator performs Taylor expansion of the Feynman integral F_{γ_i} in its small masses and external momenta. Now consider $\mathcal{M}_{\gamma_0}$. The component γ_0 can be naturally represented as the union of its 1PI components and cut lines (after a cut line is removed, the subgraph becomes disconnected; here these cut lines are of course lines with large masses). By definition, $\mathcal{M}_{\gamma_0}$ is again factorized, and the Taylor expansion of the 1PI components of γ_0 is performed as in the case of the c-components γ_i, $i = 1, 2, \ldots$.

Finally, the action of the operator $\mathcal{M}$ on the cut lines with propagators $1/(k^2 - 2q{\cdot}k)$, where k is some loop momentum, is defined as Taylor expansion in k^2. If other external momenta flow through such a line, then this is just a Taylor expansion in k.

Note that in all cases apart from the cut lines with propagators of the type $1/(k^2 - 2q \cdot k)$, the action of the corresponding operator $\mathcal{M}$ can be described graphically (as for the off-shell limit) by contraction of the corresponding subgraph to a point and insertion of the resulting polynomial into the reduced vertex of the reduced graph.

In another typical case of Limit (A), where some external momentum tends to the light cone, the situation becomes more complicated. The corresponding rules for the expansion turn out to be of a mixed type, with hard and soft regions typical of the threshold limit and with collinear regions that are typical of the Sudakov limits and related limits studied in Chap. 8.

Limit (B) is very similar to Limit (A): now the ultrasoft regions play the role of the soft regions. This similarity will be traced in the next section, where we study two-loop examples. The corresponding prescriptions can also be described similarly in graph-theoretical language. But let us now change this language, using another labelling by subgraphs. In the language of regions, the expansion of a given diagram is given in both limits by a sum of contributions labelled by regions, i.e. in the present case, by decompositions of the loop momenta into two subsets: hard and (ultra)soft. In other words, these decompositions can be labelled by hard parts (with the rest of the loop momenta being (ultra)soft), which are subgraphs consisting of 1PI connectivity components. To illustrate the difference between these graphical languages consider Examples 6.1 and 6.2. The hard contributions to the expansion are naturally associated with the whole graph. In the first language, the contribution from the expansion of the factor $1/(k^2 - 2q\cdot k)$ (in Example 6.1) and from the factor $1/(k^2 - 2q\cdot k - y)$ (in Example 6.2) in k^2 is labelled by the subgraph consisting of the corresponding line. In the language that we are now going to use, these expansions in k^2 correspond to the empty subgraph when the loop momentum is not hard, i.e. when it is soft (in Example 6.1) or ultrasoft (in Example 6.2).

So, the asymptotic expansion in these limits can be written in the following form:

$$F_\Gamma \sim \sum_\gamma \mathcal{T}_{k^2} F_{\Gamma/\gamma} \circ \mathcal{M}_\gamma F_\gamma \,, \tag{6.17}$$

where the operator $\mathcal{M}_\gamma$ performs Taylor expansion of the integrand of the Feynman integral F_γ in its small external momenta, which are either loop momenta of the whole graph or small momenta of the graph, i.e. all the external momenta of Γ apart from the given large momentum q. By definition, this operator reduces to a product of Taylor operators when the subgraph γ is disconnected. Taylor expansion is also performed in the small masses and the parameter y in Limits (A) and (B), respectively. The first Taylor operator $\mathcal{T}_{k^2}$ expands propagators with a dependence of the type $1/(k^2-2q{\cdot}k)$ (in the first case) or $1/(k^2-2q{\cdot}k-y)$ (in the second case) in k^2. Equation (6.17) resembles very much the prescriptions (5.10) for the large-mass limit. The only new feature is the expansion in k^2, which is non-local, from the momentum space point of view, in contrast to the Taylor expansions performed by the operators $\mathcal{M}_\gamma$.

We must realize, however, that (6.17) has not been completely written in graph-theoretical language, because the definition of the operator $\mathcal{T}_{k^2}$, in contrast to $\mathcal{M}_\gamma$, is connected with some specific choice of the loop momenta. A natural way to make the definition unrelated to the loop momenta is to use the alpha representation. See Sect. 9.2 and Sect. 9.3 for some steps in this direction.

6.3 Two-Loop Examples

Let us follow the strategy of expansion by regions, adjusted to the limits under consideration. The next two examples are the master self-energy diagrams of Fig. 6.2 with masses m and M in the limit $m/M \to 0$.

Example 6.3. The on-shell ($q^2 = M^2$) master diagram of Fig. 6.2a with two large masses M, two small masses m and a central line with zero mass, in Limit (A).

Choosing the canonical routing of the external momentum through the diagram, we have

$$F_{6.3}(M^2, m^2; d)$$

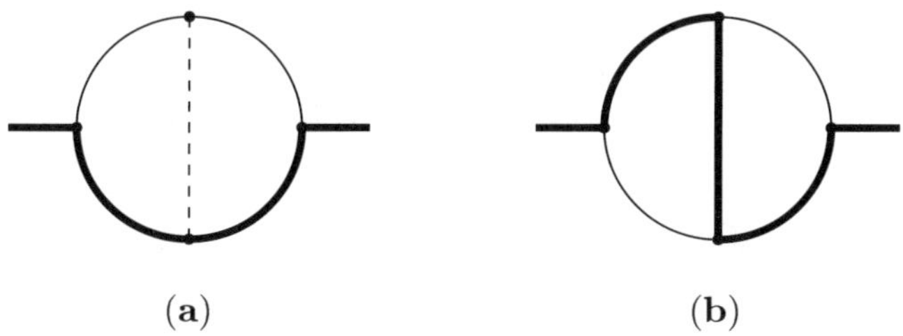

Fig. 6.2a,b. Two-loop on-shell master diagrams in the threshold limit $m/M \to 0$

$$= \int\int \frac{\mathrm{d}^d k\, \mathrm{d}^d l}{(k^2-m^2)(l^2-m^2)(k^2-2q\cdot k)(l^2-2q\cdot l)(k-l)^2} \,. \tag{6.18}$$

The list of regions that contribute to the expansion of (6.18) consists of (h–h), (h–s), (s–h) and (s–s). Any term of the hard contribution reduces to one of the following family of on-shell integrals:

$$J_{\mathrm{on},1}(a_1,\ldots,a_5) = \int\int \frac{\mathrm{d}^d k\, \mathrm{d}^d l}{(k^2)^{a_1}(k^2-2q\cdot k)^{a_2}(l^2-2q\cdot l)^{a_3}(l^2)^{a_4}[(k-l)^2]^{a_5}} \,. \tag{6.19}$$

An algorithm for the analytic evaluation of general integrals with integer powers of the propagators, based on IBP, has been developed in [129, 40, 37, 114]. The leading (h–h) contribution is given by [37]

$$J_{\mathrm{on},1}(1,\ldots,1) = \frac{\left(\mathrm{i}\pi^{d/2}\mathrm{e}^{-\gamma_{\mathrm{E}}\varepsilon}\right)^2}{3q^2}\left[\pi^2\left(\frac{1}{2\varepsilon}+1-\ln q^2\right)+6\zeta(3)\right] \,. \tag{6.20}$$

where the result is written up to the finite part in ε.

The two equal (h–s) and (s–h) contributions can be recursively evaluated by means of (A.18) and (A.23), in every order of expansion, in terms of gamma functions for general ε. The leading term in this series starts from the order m/M^3:

$$F_{6.3}^{(\mathrm{s-h})}(M^2,m^2;d) = -\frac{\left(\mathrm{i}\pi^{d/2}\right)^2\sqrt{\pi}\Gamma(\varepsilon-1/2)\Gamma(\varepsilon)m^{1-2\varepsilon}}{4(1+2\varepsilon)M^{3+2\varepsilon}} + \ldots \,. \tag{6.21}$$

The evaluation of the (s–s) contribution, with Taylor expansion of the heavy propagators in k^2 and l^2, requires a knowledge of the following family of integrals:

$$\int\int \frac{\mathrm{d}^d k\, \mathrm{d}^d l}{(-2q\cdot k)^{a_1}(k^2-m^2)^{a_2}(-2q\cdot l)^{a_3}(l^2-m^2)^{a_4}[(k-l)^2]^{a_5}} \,. \tag{6.22}$$

These integrals can be evaluated using IBP, which provides the following recurrence relation:

$$d - a_1 - a_2 - 2a_5 + a_1\mathbf{1}^+\mathbf{3}^- + a_2\mathbf{2}^+(\mathbf{4}^- - \mathbf{5}^-) = 0 \tag{6.23}$$

and the corresponding symmetrical relation. The boundary integrals, with a_2, a_4 or a_5 equal to zero, turn out to be recursively one-loop and can be easily evaluated in terms of gamma functions by means of (A.22) and (A.25). The boundary two-loop integral with $a_1 = a_3 = 0$ can be evaluated by means of (A.38).

Combining the leading means of (h–h) and (s–s) terms, we see that the poles are cancelled, and we obtain the leading order of expansion. At $\varepsilon = 0$, this is

$$F_{6.3}(M^2, m^2; 4) \sim \frac{\left(\mathrm{i}\pi^2\right)^2}{M^2} \left[\pi^2 \left(\frac{1}{3} \ln \frac{m^2}{M^2} + \ln 2 \right) + \frac{3}{2}\zeta(3) \right] . \tag{6.24}$$

The next diagram possesses not only two-particle cuts, $q^2 = (M+m)^2$, but also a three-particle cut at $q^2 = (M + 2m)^2$. Nevertheless, no complications arise and this diagram can be expanded in the limit $m/M \to 0$ by the same method [87].

Example 6.4. The on-shell ($q^2 = M^2$) master diagram of Fig. 6.2b with three large masses M and two small masses m, in Limit (A).

The Feynman integral is written as

$$F_{6.4}(M^2, m^2; d) = \int\int \frac{\mathrm{d}^d k \, \mathrm{d}^d l}{(k^2 - m^2)(l^2 - m^2)(k^2 - 2q\cdot k)} \times \frac{1}{(l^2 - 2q\cdot l)[(k+l)^2 - 2(q\cdot k + q\cdot l)]} . \tag{6.25}$$

We have the same list of regions as in the previous example: (h–h), (h–s)=(s–h) and (s–s). The evaluation of the hard contribution reduces to the evaluation of the following on-shell integrals:

$$J_{\mathrm{on},2}(a_1, \ldots, a_5) = \int\int \frac{\mathrm{d}^d k \, \mathrm{d}^d l}{(k^2)^{a_1}(k^2 - 2q\cdot k)^{a_2}} \times \frac{1}{(l^2 - 2q\cdot l)^{a_3}(l^2)^{a_4}[(k+l)^2 - 2(q\cdot k + q\cdot l)]^{a_5}} . \tag{6.26}$$

For these integrals, an analytic algorithm based on IBP has been developed in [129, 40, 37, 114].

The leading term of the whole expansion reduces [36] to

$$J_{\mathrm{on},2}(1, \ldots, 1) = \frac{\left(\mathrm{i}\pi^2\right)^2}{q^2} \left(\frac{3}{2}\zeta(3) - \pi^2 \ln 2 \right) + O(\varepsilon) . \tag{6.27}$$

The (h–s) contribution equals

$$\int \frac{\mathrm{d}^d l}{l^2 - m^2} \mathcal{T}_{l^2} \frac{1}{l^2 - 2q\cdot l} \int \frac{\mathrm{d}^d k}{k^2 - 2q\cdot k} \mathcal{T}_{m^2} \frac{1}{k^2 - m^2} \times \mathcal{T}_l \frac{1}{(k+l)^2 + 2q\cdot(k+l)}$$

$$= \sum_{j_1,j_2,j_3=0}^{\infty} (-1)^{j_1} m^{2(j_1+j_2)} \sum_{n=0}^{j_3} \int \mathrm{d}^d l \frac{(m^2 - 2q\cdot l)^{j_3-n}}{(l^2 - m^2)(-2q\cdot l)^{j_1+1}} \binom{j_3}{n}$$
$$\times \int \mathrm{d}^d k \frac{(2k\cdot l)^n}{(k^2)^{j_2+1}(k^2 - 2q\cdot k)^{j_3+2}} . \qquad (6.28)$$

The factors l^2 in the numerator have been substituted by m^2 because, whenever $l^2 - m^2$ appears, it cancels the corresponding factor in the denominator and we obtain a massless tadpole integral, which vanishes. The calculation of the two one-loop integrals in (6.28) can be performed straightforwardly by means of (A.23) and (A.18). The leading contribution from the (h–s) and (s–h) regions is

$$\left(\mathrm{i}\pi^{d/2}\right)^2 \sqrt{\pi}\Gamma(\varepsilon - 1/2)\Gamma(\varepsilon)\frac{m^{1-2\varepsilon}}{2M^{3+2\varepsilon}} . \qquad (6.29)$$

The (s–s) contribution is

$$\int\int \frac{\mathrm{d}^d k\, \mathrm{d}^d l}{(k^2 - m^2)(l^2 - m^2)}$$
$$\times \mathcal{T}_\kappa \frac{1}{(\kappa k^2 - 2q\cdot k)[\kappa(k+l)^2 - 2q\cdot(k+l)](\kappa l^2 - 2q\cdot l)}\Bigg|_{\kappa=1}$$
$$\equiv \sum_{j_1,j_2,j_3=0}^{\infty} \int\int \frac{\mathrm{d}^d k\, \mathrm{d}^d l}{(k^2 - m^2)(l^2 - m^2)}$$
$$\times \frac{(-m^2)^{j_1+j_2}\left[-(k+l)^2\right]^{j_3}}{(-2q\cdot k)^{j_1+1}[-2q\cdot(k+l)]^{j_3+1}(-2q\cdot l)^{j_2+1}} . \qquad (6.30)$$

The problem therefore reduces to the evaluation of the following family of integrals:

$$J_{\mathrm{ss}}(a_1, a_2, a_3, a_4, a_5)$$
$$= \int\int \frac{\mathrm{d}^d k\, \mathrm{d}^d l\, (k\cdot l)^{a_5}}{(k^2 - m^2)^{a_1}(l^2 - m^2)^{a_2}(2q\cdot k + i0)^{a_3}(2q\cdot k + 2q\cdot l + i0)^{a_4}} . \qquad (6.31)$$

An algorithm for evaluating these integrals based on IBP has been presented in [87]. After reducing the indices by use of recurrence relations, an explicit result valid for general ε in the following specific case was used there:

$$J_{\mathrm{ss}}(1,1,0,a,0) = -\left(\frac{2m^{2-2\varepsilon}\pi^{5/2-\varepsilon}}{\Gamma(3/2-\varepsilon)}\right)^2 \frac{2^{1-2\varepsilon}}{(-4Mm)^a} \frac{\cos(\pi\varepsilon + \pi a/2)}{\cos(\pi\varepsilon)}$$
$$\times B\left(\frac{3}{2} - \varepsilon, \frac{a}{2} - 1 + \varepsilon\right) B\left(\frac{a}{2} - 2 + 2\varepsilon, -\frac{a}{2} + \frac{3}{2} - \varepsilon\right) , \qquad (6.32)$$

where $B(a,b) = \Gamma(a)\Gamma(b)/\Gamma(a+b)$ is the second Euler integral.

The leading (s–s) contribution is

$$- \left(\mathrm{i}\pi^{d/2}\right)^2 \sqrt{\pi}\Gamma(\varepsilon - 1/2)\Gamma(\varepsilon)\frac{m^{1-4\varepsilon}}{2M^3} \,. \tag{6.33}$$

The first two leading terms of the expansion when $m/M \to 0$ are therefore

$$F_{6.4}(M^2, m^2; 4) \sim \frac{\left(\mathrm{i}\pi^2\right)^2}{M^2}\left(\frac{3}{2}\zeta(3) - \pi^2 \ln 2 - \frac{\pi m}{M}\ln\frac{m^2}{M^2}\right) + \ldots \,. \tag{6.34}$$

As was demonstrated in [87] by evaluating the first 18 terms, the expansion converges very well, at least up to $m/M = 0.5$.

We now turn to the second type of threshold regime, with one non-zero mass in the threshold.

Example 6.5. The Feynman diagram of Fig. 6.3a in Limit (B).

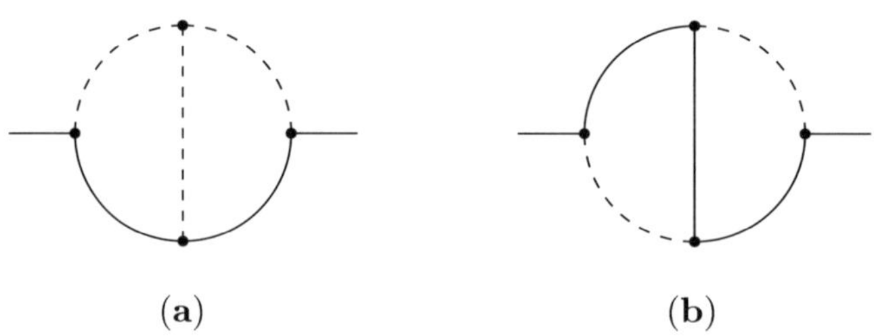

Fig. 6.3a,b. Two-loop master diagrams in the threshold limit $q^2 \to m^2$

With the canonical routing of the external momentum through heavy lines, we have

$$F_{6.5}(q^2, y; d) = \int\int \frac{\mathrm{d}^d k\, \mathrm{d}^d l}{k^2(k^2 - 2q\cdot k - y)l^2(l^2 - 2q\cdot l - y)(k-l)^2} \,, \tag{6.35}$$

where again $y = m^2 - q^2$.

The non-zero contributions are generated by the (h–h), (h–us), (us–h) and (us–us) regions. Any term of the hard contribution can be expressed in terms of the on-shell integrals (6.19), with the leading term given again by (6.20).

The equal (us–h) and (h–us) contributions can easily be evaluated in every order of expansion by use of (A.16) and (A.25). The leading term starts from the NLO of the whole expansion:

$$F_{6.5}^{(\mathrm{us-h})}(q^2, y; d) = -\frac{\left(\mathrm{i}\pi^{d/2}\right)^2 \Gamma(1-\varepsilon)\Gamma(\varepsilon)\Gamma(2\varepsilon - 1)y^{1-2\varepsilon}}{(1+2\varepsilon)(q^2)^2} \,. \tag{6.36}$$

The evaluation of the (us–us) contribution generated by Taylor expansion of the heavy propagators in k^2 and l^2 requires a knowledge of the following family of integrals:

$$\int\int \frac{\mathrm{d}^d k\, \mathrm{d}^d l}{(-2q\cdot k - y)^{a_1}(k^2)^{a_2}(-2q\cdot l - y)^{a_3}(l^2)^{a_4}[(k-l)^2]^{a_5}} \,. \tag{6.37}$$

These integrals can be evaluated[1] [41] using IBP, which provides exactly the same relation (6.23) as in the case of the integrals (6.22). The boundary integrals, with a_3, a_4 or a_5 equal to zero, turn out to be recursively one-loop and can easily be evaluated in terms of gamma functions by means of (A.7), (A.16) and (A.25).

Combining the (h–h) and (us–us) terms, we see that the poles are cancelled, and we obtain the leading order of expansion as follows:

$$F_{6.5}(q^2, y; d) \sim \frac{\left(\mathrm{i}\pi^2\right)^2}{q^2}\left(\frac{2}{3}\pi^2 \ln\frac{y}{q^2} + 6\zeta(3)\right) + \ldots \,. \tag{6.38}$$

Finally, we consider one more example.

Example 6.6. The Feynman diagram of Fig. 6.3b in Limit (B).

With the canonical routing of the external momentum, the Feynman integral takes the form

$$F_{6.6}(q^2, y; d) = \int\int \frac{\mathrm{d}^d k\, \mathrm{d}^d l}{k^2(k^2 - 2q\cdot k - y)l^2(l^2 - 2q\cdot l - y)} \times \frac{1}{(k+l)^2 - 2(q\cdot k + q\cdot l) - y} \,. \tag{6.39}$$

We are again left with the (h–h), (h–us), (us–h) and (us–us) contributions. The hard contribution can be expressed in terms of the integrals (6.26), with the leading term again given by (6.27). The equal (us–h) and (h–us) contributions can be evaluated by means of (A.13) and (A.25). The leading term is

$$F_{6.6}^{(\text{us}-\text{h})}(q^2, y; d) = \left(\mathrm{i}\pi^{d/2}\right)^2 \Gamma(1-\varepsilon)\Gamma(\varepsilon)\Gamma(2\varepsilon - 1)\frac{y^{1-2\varepsilon}}{(q^2)^2} \,. \tag{6.40}$$

The (us–us) contribution is obtained by expanding the three heavy propagators in k^2, l^2 and $(k+l)^2$. To evaluate the corresponding integrals which are typical of HQET, we first get rid of one of the three linear propagators using the identity

$$1 = \frac{1}{y}\left[(-2q\cdot k - 2q\cdot l - y) - (-2q\cdot k - y) - (-2q\cdot l - y)\right] \,.$$

Then (A.26) is applied, with the results in terms of gamma functions.

Combining the first two leading (h–h) terms and the leading terms for the other contributions, we arrive at the following first two orders of expansion:

[1] Even in the case of three loops, similar integrals, which are typical of the HQET, have been evaluated by means of IBP recurrence relations – see [131].

$$F_{6.6}(q^2, y; d) \sim \frac{\left(\mathrm{i}\pi^2\right)^2}{q^2} \left[\frac{3}{2}\zeta(3) - \pi^2 \ln 2 \right.$$
$$\left. + \frac{y}{q^2}\left(\ln^2\frac{y}{q^2} - 3\ln\frac{y}{q^2} + \frac{\pi^2}{8} + 3\right)\right] + \ldots . \tag{6.41}$$

6.4 An Application: the Muon Anomalous Magnetic Moment in QED

As an example of the application[2] of the general prescriptions in the case of Limit (A), let us consider the evaluation of the muon anomalous magnetic moment at the three-loop level. The electron and muon anomalous magnetic moments, $a_l = (g_l - 2)/2$, $l = e, \mu$, have traditionally provided precision tests of the Standard Model and stringent constraints on effects potentially connected with 'new physics'. From the point of view of multiloop calculations, the muon anomalous magnetic moment is more challenging, because the existence of a lighter lepton (electron) enhances some diagrams by a logarithmic factor of $\ln(m_\mu/m_e) \simeq 5.33$. As a result, diagrams with up to five photonic and fermionic loops have to be included to match the present experimental accuracy.[3] The resulting QED prediction is expressed as a truncated series in the fine-structure constant,

$$a_\mu^{\mathrm{QED}} = \sum_{n=1}^{5} C_n \left(\frac{\alpha}{\pi}\right)^n .$$

Let us consider $C_3^{\mathrm{vac.pol.}}(e,\tau)$, which is a part of the coefficient C_3 in this expansion. This part arises from the diagram of Fig. 6.4c, with electron and tau lepton loops inserted into the photon propagator. Historically, this was the last three-loop piece that was evaluated analytically [86]. Let us describe how this was done.

Consider first the two two-loop diagrams shown in Figs. 6.4a,b, with $p_{1,2}^2 = m_\mu^2$. In fact the quantity that is needed is the first derivative with respect to q at $q = 0$, when $p_1 = p + q/2$ and $p_1 = p - q/2$, so that, effectively, we are dealing with propagator-type diagrams.

In the case of the tau lepton loop, the mass of this loop is much larger than the muon mass; $m_\mu \equiv m \ll M \equiv m_\tau$. This is the large-mass limit typical of Euclidean space. In the language of regions, to expand the corresponding Feynman integral in this limit it is sufficient to consider contributions from the (h–h) and (s–h) regions, where the loop momentum of the loop is indicated by the second symbol, and the word 'soft' means $k \sim m = m_\mu$. The (h–h) contribution is obtained by Taylor expanding the integrand in the external momenta and the muon mass, and reduces to vacuum diagrams. The

[2] See [82, 6] for some other applications.

[3] Four loops are sufficient for a_e: for a recent review see [81].

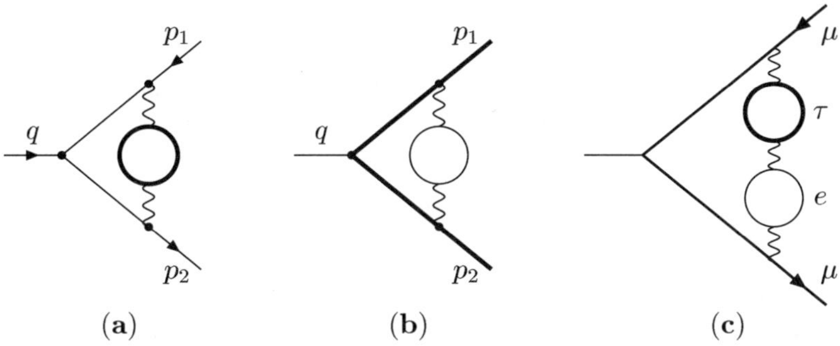

Fig. 6.4a–c. Two- and three-loop graphs contributing to the muon anomalous magnetic moment: (**a**) with a tau lepton loop, (**b**) with an electron loop, (**c**) with a tau lepton loop and an electron loop

(s–h) contribution naturally factorizes and is given by the Taylor expansion of the Feynman integral of the tau loop in its external momentum, inserted into the photon line.

In the case of the graph of Fig. 6.4b with the electron loop, the mass of this loop is much smaller than the muon mass; $m_e \equiv m \ll M \equiv m_\mu$. This is nothing but our threshold Limit (A) (typical of pseudo-Euclidean space). Now the muon mass is considered large rather than small, and a loop momentum is called 'soft' if $k \sim m = m_e$ The regions that are relevant here are (h–h), (h–s) and (s–s), where the loop momentum of the loop is again indicated by the second symbol. The (h–h) contribution is now given by Taylor expanding the integrand in the electron mass. The (h–s) contribution takes a factorized form and reduces to electron vacuum bubbles. Finally, the (s–s) contribution is typically Minkowskian and reduces to the expansion of the muon propagators in k^2, where k is the loop momentum flowing through these lines. Namely, one needs [86] to calculate the following family of integrals with integer indices:

$$J(a_1, a_2) = \int \frac{\mathrm{d}^d k \, \mathrm{d}^d l}{(k^2)^{a_1}(-2p\cdot k)^{a_2}(l^2 - m^2)[(k-l)^2 - m^2]} \, . \tag{6.42}$$

Using Feynman parameters, these integrals can be evaluated in terms of gamma functions for general ε.

These expansions of the two diagrams in m_μ/m_τ and m_e/m_μ, respectively, successfully reproduce the expansion of the known analytic result [102] (see also a simpler form in [90]).

Analytic results for the three-loop diagram of Fig. 6.4c are not (yet?) known. Within the semi-analytic approach based on expansion by regions, this diagram has been expanded [86] up to the first three terms in m_μ/m_τ and the first two terms in m_e/m_μ. This is a problem with three scales, and it is natural to consider the highest scale, m_τ, as hard. Let us keep the word 'soft' for the middle scale, m_μ, and characterize loop momenta of the order of the lowest scale, $k \sim m_e$, as 'supersoft' (ss). (We keep the word 'ultrasoft' for the scale which is the square of the soft scale.) Let us now list the loop momenta in the following order: the loop momenta of the τ lepton insertion, the loop momentum flowing through the muon lines, and the loop momenta of the electron insertion. The list of regions that generate non-zero contributions in the given limit is

$$(\text{h–h–h}),\ (\text{h–h–ss}),\ (\text{h–s–s}),\ (\text{h–s–ss}),\ (\text{h–ss–ss}) \ .$$

The evaluation of the corresponding contributions is similar to the two-loop case. In renormalized form, the contribution of Fig. 6.4c to the muon anomalous magnetic moment, evaluated in an expansion in m_μ/m_τ and m_e/m_μ, is [86]

$$\begin{aligned} C_3^{\text{vac.pol.}}(e,\tau) = {} & \frac{m_\mu^2}{m_\tau^2}\left(\frac{4}{135}\ln\frac{m_\mu}{m_e} - \frac{1}{135}\right) + \frac{m_\mu^4}{m_\tau^4}\left(-\frac{229213}{12348000} + \frac{\pi^2}{630}\right. \\ & \left. - \frac{37}{11025}\ln\frac{m_\tau}{m_\mu} - \frac{1}{105}\ln\frac{m_\tau}{m_\mu}\ln\frac{m_\tau m_\mu}{m_e^2} + \frac{3}{4900}\ln\frac{m_\mu}{m_e}\right) \\ & + \frac{m_\mu^6}{m_\tau^6}\left(-\frac{1102961}{75014100} + \frac{4\pi^2}{2835} - \frac{398}{297675}\ln\frac{m_\tau}{m_\mu} - \frac{8}{945}\ln\frac{m_\tau}{m_\mu}\ln\frac{m_\tau m_\mu}{m_e^2}\right. \\ & \left. - \frac{524}{297675}\ln\frac{m_\mu}{m_e}\right) + \frac{2}{15}\frac{m_e^2}{m_\tau^2} - \frac{4\pi^2}{45}\frac{m_e^3}{m_\tau^2 m_\mu} \ . \end{aligned} \tag{6.43}$$

This result agrees with the numerical calculation of [160].

Another recent example of the use of the strategy of expansion by regions for a limit related to those considered in this chapter can be found in [85], where an expansion in the ratio of a small and a large mass was applied in the framework of Non-Relativistic QCD (NRQCD). Spin-dependent and spin-independent contributions to energy levels in simple atoms, such as hydrogen or muonium, have been evaluated as expansions in the ratio of the mass of the electron to that of the nucleus, m/M. A prescription for such expansions has been given to all orders in m/M. The relevant regions turned out again to be 'hard' and 'soft'.

6.5 Threshold Expansion and HQET

If we are interested in processes with no heavy particles in the initial and final states, we can use the large-mass off-shell limit and the corresponding expansion, which leads, at the operator level, to a description of the theory by an effective Lagrangian composed of the light fields, as explained in Chap. 5. But if we want to study processes with external heavy particles, we need another large-mass limit. It turns out that the threshold (on-shell) Limit (B) leads straightforwardly, at the operator level, to the Heavy Quark Effective Theory (HQET) [183, 174, 132], which is an effective theory designed to reproduce QCD processes where the heavy quarks in the initial and final states are at energy scales much smaller than the mass of the heavy quark. So it is supposed that one quark, with mass m, is much heavier than the quarks of other flavours. For a heavy quark that stays almost at rest, the four-momentum $p = (p_0, \mathbf{p})$ is such that $|\mathbf{p}|, |p_0 - m| \ll m$ and the gluons and the light quarks have smaller momenta, i.e. $|\mathbf{k}_i|, |k_{0i}| \ll m$.

For the heavy-quark propagator in QCD, this means that

$$
\begin{aligned}
S(p) = \mathrm{i}\frac{\not{p} + m}{p^2 - m^2 + \mathrm{i}0} &= \mathrm{i}\frac{m(1+\gamma^0) + (p_0 - m)\gamma^0 - \mathbf{p}\cdot\gamma}{2m(p_0 - m) + (p_0 - m)^2 - \mathbf{p}^2 + \mathrm{i}0} \\
&= \mathrm{i}\frac{1+\gamma^0}{2}\frac{1}{p_0 - m + \mathrm{i}0} + O\left(\frac{1}{m}\right) . \qquad (6.44)
\end{aligned}
$$

The leading term in the last line of (6.44) defines the HQET quark propagator, which is used in the corresponding Feynman rules in the leading order in $1/m$. The residual energy is denoted by $\omega = p_0 - m$.

So, the limit corresponding to HQET is nothing but the threshold Limit (B). The parameter $y \equiv m^2 - p^2$ used for that limit is related to ω as follows:

$$
y = -2m\omega - \omega^2 , \quad \omega = \sqrt{m^2 - y} - m = -\frac{y}{2m} + O(y) .
$$

In a manifestly covariant formulation, the four-momentum of the heavy quark is written as $p = mv + \tilde{p}$, where v^μ is the quark velocity, with $v^2 = 1$. The components of $\tilde{p}$ are considered small with respect to m. The covariant HQET quark propagator generalizes the term in the last line in (6.44) and takes the form

$$
\tilde{S}(p) = \mathrm{i}\frac{1+\not{v}}{2}\frac{1}{\tilde{p}\cdot v} . \qquad (6.45)
$$

In order to see how the limit of HQET is performed in Feynman integrals, consider now a quark propagator inserted into a diagram. Let us choose the canonical routing of the external momentum of the heavy quark, p, through a given heavy line, and let k be the sum of the loop momenta and other (small) external momenta flowing through this line. We easily observe that the quark QCD propagator reduces to the HQET quark propagator in the leading order of expansion in y/m^2,

$$
\begin{aligned}
S(p-k) = \mathrm{i}\frac{\not{p} - \not{k} + m}{(p-k)^2 - m^2} &\equiv \mathrm{i}\frac{\not{p} - \not{k} + m}{k^2 - 2k\cdot p - y} \\
\sim \mathrm{i}\frac{\not{p} + m}{-2k\cdot p - y} &\sim \mathrm{i}\frac{1+\not{v}}{2}\frac{1}{(\tilde{p} - k)\cdot v} = \tilde{S}(p-k) , \qquad (6.46)
\end{aligned}
$$

if we expand it while thinking of the momentum k as *ultrasoft*. This means that, in order to arrive at HQET diagrams starting from QCD, we have to think about the region of ultrasoft loop momenta and consider, at the same time, the other masses and external momenta to be of order $y/m \sim \omega$.

The gluon propagator stays the same, and the QCD vertex factor $\mathrm{i}g\gamma^\mu t^a$ between the two factors $(1+\not{v})/2$ corresponding to heavy-quark lines is replaced by $\mathrm{i}gv^\mu t^a$. Thus the Feynman rules of the HQET are reproduced by the pure ultrasoft contributions within the threshold (on-shell) Limit (B).

At the Lagrangian level, the transition to the HQET means that the quark part of the QCD Lagrangian (2.15),

$$\sum \bar{q}_i(\mathrm{i}\not{D} - m_i)q_i = \bar{Q}(\mathrm{i}\not{D} - m)Q + \sum_{\text{light flavours}} \bar{q}_i(\mathrm{i}\not{D} - m_i)q_i \tag{6.47}$$

is replaced according to

$$\bar{Q}(\mathrm{i}\not{D} - m)Q \to \overline{\tilde{Q}}_v \mathrm{i}v\cdot D\tilde{Q}_v + \cdots , \tag{6.48}$$

where the static quark field $\tilde{Q}_v$ is a four-component spinor field satisfying the relation $\not{v}\tilde{Q}_v = \tilde{Q}_v$. This first term in the effective Lagrangian generates the HQET quark propagator (6.45). To construct the HQET in the next orders in the expansion in ω/m [183, 174, 132] one uses the standard strategy for constructing an effective theory and writes down an effective Lagrangian as a series in $1/m$, by including in it all operators having the necessary symmetries, with arbitrary (matching) coefficients, which are computed by solving a system of equations that express the fact that the same results are obtained for the HQET amplitudes and the corresponding QCD amplitudes expanded at threshold.

The following are the operators of dimension four and five [99, 107] in the HQET effective Lagrangian:

$$\tilde{Q}^+ \mathrm{i}D_0\tilde{Q} + \frac{C_k}{2m}\tilde{Q}^+\mathbf{D}^2\tilde{Q} - \frac{C_m}{2m}\tilde{Q}^+\mathbf{B}\cdot\sigma\tilde{Q} , \tag{6.49}$$

where $\mathbf{B}$ is the magnetic field and the σ_i are the Pauli matrices.

For simplicity, the non-covariant notation is used here. The two terms of dimension five are the kinetic energy and the chromomagnetic interaction, respectively. When dealing with the HQET, one uses the reparameterization invariance [171], which is based on the fact that the velocity of the heavy quark can be changed a little without changing the physical content of the HQET. This property relates terms of different order in $1/m$ and, in particular, leads to the equation $C_k = 1$ to all orders of perturbation theory. The matching coefficient C_m and the anomalous dimension of the chromomagnetic operator have been calculated to two loops in [1, 79]. To do this, matching conditions for on-shell scattering amplitudes in QCD (in an expansion in ω) and in the HQET have been used in [79].

To arrive at the HQET, one can start from the general form (6.17) of the asymptotic expansion in the threshold Limit (B). As in the case of the off-shell large-momentum or large-mass limit, the factor $\mathcal{M}_\gamma F_\gamma$ originating from a hard contribution is a polynomial in the external momenta of γ inserted into the reduced diagram $F_{\Gamma/\gamma}$. This polynomial is a Taylor series in $y \sim -2m\omega$ and other small parameters (light masses and small external momenta), i.e.

it is given by formal Taylor expansion of the diagram γ at threshold (on-shell). Therefore the hard contributions are exactly those on-shell diagrams that participate in the matching calculations for the coefficients of the HQET effective Lagrangian. On the other hand, the first factor on the right-hand side of (6.17) determines a Feynman integral for $F_{\Gamma/\gamma}$ with the heavy-quark propagators expanded in k^2, i.e. written within the Feynman rules of the HQET.

Thus, for the HQET, one needs either calculations of HQET diagrams or calculations of on-shell QCD diagrams. We have already seen some techniques for calculating these diagrams in Sects. 6.1 and 6.3: in the first case, the (us–us–...–us) contributions are calculated and, in the second case, (h–h–...–h) contributions in Limit (B).

The HQET is a well-established effective theory that has been successfully applied in practice – see, for example, [183, 175, 132, 174]. We should realize that it has been developed by the standard 'matching' strategy for constructing effective theories, without reference to expanding Feynman diagrams at threshold (on-shell). The transition to the HQET was performed by a straightforward generalization of the construction of an effective theory related to the off-shell large-momentum limit. Although the limit of the HQET is specifically Minkowskian, the combinatorial structure is almost the same as in the off-shell case. In particular, this is also a problem with two scales. In the next chapter we shall turn our attention to the case of a threshold expansion with at least two non-zero masses in the threshold, where the structure of the expansion becomes much more complicated because this is essentially a problem with three scales. We shall see that the diagrammatic expansions help us very much to understand what happens at the operator level.

7 Threshold Expansion. Two Masses in the Threshold

We continue to investigate expansion near threshold and now turn our attention to thresholds composed of at least two particles with non-zero masses. We shall usually consider the case of two equal non-zero masses in the threshold, so that we shall typically deal with expansion at $q^2 \to 4m^2$. Guided by expansion by regions, we start with one-loop examples and a two-loop example, which will help us to characterize all the relevant regions, i.e. hard, soft, potential and ultrasoft, for the threshold expansion. Then general prescriptions are formulated and subtle points about their application are explained. We continue, in Sect. 7.3, with examples of two-loop diagrams for which analytic results are unknown.

We then turn to an operator implementation of the threshold expansion. The particles move slowly close to the threshold, so that a non-relativistic description is naturally connected with this regime. For QCD, this means a transition to Non-Relativistic QCD (NRQCD) [231, 164, 25] and, further, to potential NRQCD [193, 17]. The complexity of the threshold expansion can be seen from the presence of four characteristic regions and also of three different scales in the problem: the mass of the heavy quark m, a typical relative momentum mv and a typical kinetic energy mv^2.

Finally, typical applications of the threshold expansion are briefly described in Sects. 7.5 and 7.6: the two-loop correction to the leptonic decay of quarkonium, with two-loop matching coefficients of the vector currents in QCD and NRQCD used as an input, and the next-to-next-to-leading order (NNLO) total cross-section for production of the top quark near threshold.

7.1 One-Loop Examples and a Two-Loop Example

Our first example in this chapter is the following.

Example 7.1. The Feynman diagram of Fig. 2.1 with equal non-zero masses in the limit $q^2 \to 4m^2$.

The Feynman integral is

$$F_{7.1}(q^2, y; d) = \int \frac{\mathrm{d}^d k}{k^2[(q-k)^2 - m^2]}$$

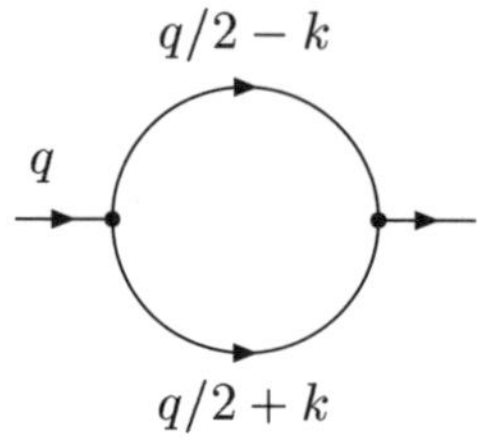

Fig. 7.1. One-loop propagator diagram with two non-zero masses in the threshold

$$\equiv \int \frac{\mathrm{d}^d k}{(k^2 + q\cdot k - y)(k^2 - q\cdot k - y)}\,, \tag{7.1}$$

where the loop momentum is chosen in another way (see Fig. 7.1) in order to make explicit the dependence of the propagators on the expansion parameter, $y = m^2 - q^2/4$, of the problem. This will be our usual convention: in the case of two equal non-zero masses in the threshold, to let half of the external momentum flow through one of the massive lines and the other half of it through the other massive line. So we have transformed to the new variables $(q^2, m^2) \to (q^2, y)$, which are more natural than the variables (m^2, y).

Let us perform expansion by regions. The primary task is to identify all relevant regions in the problem. The hard region always contributes to the expansion in any limit. In this example, the hard region generates a 'naive' Taylor expansion of the integrand in y, where the integrals can be evaluated by means of partial fractions and the tabulated formula (A.13):

$$\begin{aligned} F_{7.1}^{(\mathrm{h})} &= \int \mathrm{d}^d k\, \mathcal{T}_y \frac{1}{(k^2 + q\cdot k - y)(k^2 - q\cdot k - y)} \\ &= \int \frac{\mathrm{d}^d k}{(k^2 + q\cdot k)(k^2 - q\cdot k)} + \ldots \\ &= \mathrm{i}\pi^{d/2} \left(\frac{4}{q^2}\right)^{\varepsilon} \sum_{n=0}^{\infty} \frac{\Gamma(n+\varepsilon)}{n!(1-2\varepsilon-2n)} \left(\frac{-4y}{q^2}\right)^n . \end{aligned} \tag{7.2}$$

Observe now that, in contrast to the expansion considered in the previous chapters, we have not generated new poles by this procedure. The pole in ε is present only in the leading term on the right-hand side and is nothing but the UV pole present from the beginning. Nevertheless, (7.2) is not the whole result because this expansion procedure is really valid only in the hard region, i.e. that of large k.

Let us look for other regions. The soft region region, $k \sim \sqrt{y}$, gives

$$\begin{aligned} F_{7.1}^{(\mathrm{s})} &= \int \frac{\mathrm{d}^d k}{(q\cdot k + \mathrm{i}0)(-q\cdot k + \mathrm{i}0)} + \ldots \\ &= -\int \frac{\mathrm{d}^d k}{(q\cdot k + \mathrm{i}0)(q\cdot k - \mathrm{i}0)} + \ldots = 0\,, \end{aligned} \tag{7.3}$$

because these are integrals without scale. Although the product in the integrand is, generally speaking, ill-defined because of pinching singularities, the presence of a scaleless integral in the components of k additional to the linear combination $q \cdot k$ shows that the integral is zero, according to our prescriptions.

The ultrasoft region, $k \sim y/\sqrt{q^2}$, also generates (for the same reason) a zero contribution:

$$\int \frac{\mathrm{d}^d k}{(q\cdot k - y + i0)(q\cdot k + y - i0)} + \ldots = 0 . \tag{7.4}$$

In order to find the missing contribution, let us choose the frame $q = \{q_0, \mathbf{0}\}$ (keeping in mind the non-relativistic flavour of the problem). Although the problem of the expansion has been stated in a completely Lorentz-invariant way, its solution is greatly simplified by breaking down the Lorentz invariance. We have

$$F_{7.1} = \int \frac{\mathrm{d}k_0 \, \mathrm{d}^{d-1}\mathbf{k}}{(\mathbf{k}^2 - k_0^2 + q_0 k_0 + y - \mathrm{i}0)(\mathbf{k}^2 - k_0^2 - q_0 k_0 + y - \mathrm{i}0)} . \tag{7.5}$$

In any region other than the hard one, we have to suppose that some component of k is small. It is easy to observe that we have no chance of arriving at a non-zero contribution if we do not suppose that k_0 is small. When k_0 is small, i.e. at least $|k_0| \le \sqrt{y}$, we can neglect k_0^2 in comparison with $q_0 k_0$. Thus both propagators are expanded in k_0^2, and we obtain

$$\int \frac{\mathrm{d}k_0 \, \mathrm{d}^{d-1}\mathbf{k}}{(\mathbf{k}^2 + q_0 k_0 + y - \mathrm{i}0)(\mathbf{k}^2 - q_0 k_0 + y - \mathrm{i}0)} + \ldots . \tag{7.6}$$

It turns out that this series is already composed of quantities that are homogeneous with respect to the expansion parameter, y. Each term can be evaluated by, first, integrating over k_0 using Cauchy's theorem. To be consistent we have to decide that, for any term arising from the Taylor expansion in k_0^2, we shall close the integration contour in the same half-plane. Let this be the upper half-plane, for definiteness. Observe that in this example and other examples in this subsection, the results for the contributions of this type do not depend on this choice. Note also that, starting from some order of the expansion, the integrand does not vanish when $k_0 \to \infty$. Nevertheless, we do not pay attention to this fact and all the resulting integrals are by definition obtained by use of Cauchy's theorem.

It turns out, however, that in this example, only the leading term of the contribution (7.6) survives because, for any of the subsequent terms, the resulting integrals in the vector component, $\mathbf{k}$, are integrals without scale. The leading term gives

$$\frac{\pi \mathrm{i}}{\sqrt{q^2}} \int \frac{\mathrm{d}^{d-1}\mathbf{k}}{\mathbf{k}^2 + y} = \mathrm{i}\pi^{d/2} \Gamma(\varepsilon - 1/2) \sqrt{\frac{\pi y}{q^2}} y^{-\varepsilon} , \tag{7.7}$$

where the spatial integral has been evaluated by means of the $(d-1)$-dimensional variant (with the replacement $\varepsilon \to \varepsilon + 1/2$) of (A.1).

Let us now come back to (7.6) and remember that we supposed that we had started from the region with small k_0. We have to say something about $\mathbf{k}$. We do not want the combination $\mathbf{k}^2 + q_0 k_0 + y$ to be expanded further because, otherwise, we shall arrive at zero scaleless integrals. This requirement fixes absolutely the order of all the quantities involved, and we arrive at the following characterization of this new region [18]:

$$\textit{potential (p)}, \quad k_0 \sim y/\sqrt{q^2}, \quad \mathbf{k} \sim \sqrt{y}\,. \tag{7.8}$$

We call it 'potential' because it is connected with the Coulomb potential (see Sect. 7.6).

Thus we have contributions from two regions, and the whole expansion of the given diagram near threshold consists of (7.2) and

$$F_{7.1}^{(\mathrm{p})} = \mathrm{i}\pi^{d/2}\Gamma(\varepsilon - 1/2)\sqrt{\frac{\pi y}{q^2}}\,y^{-\varepsilon}\,. \tag{7.9}$$

The sum of these two contributions successfully reproduces the known analytic result for the given diagram,

$$F_{7.1}(q^2, y; d) = \mathrm{i}\pi^{d/2}\Gamma(\varepsilon) y^{-\varepsilon}\;{}_2F_1\left(\frac{1}{2}, \varepsilon; \frac{3}{2}; -\frac{q^2}{4y}\right)\,, \tag{7.10}$$

which can be obtained by means of Feynman parameters. By use of (A.54), the right-hand side of (7.10) can be rewritten as a sum of two terms, exactly corresponding to the hard and potential contributions (7.2) and (7.9).

The next example is the following.

Example 7.2. The Feynman diagram of Fig. 7.2 with two legs on the mass shell, $p_{1,2}^2 = m^2$, and equal non-zero masses in the limit $q^2 \to 4m^2$.

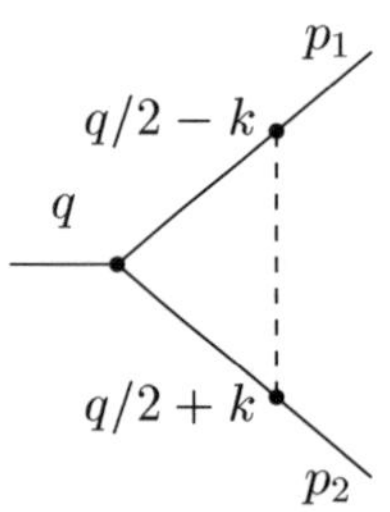

Fig. 7.2. Triangle diagram with two non-zero masses in the threshold

The Feynman integral is

$$F_{7.2}(q^2, y; d) = \int \frac{\mathrm{d}^d k}{(k^2 + q\cdot k - y)(k^2 - q\cdot k - y)(k-p)^2}\,, \tag{7.11}$$

where $q = p_1 + p_2$, $p = (p_1 - p_2)/2$. We again have $y = m^2 - q^2/4 \to 0$. We choose a frame where $q = (q_0, \mathbf{0})$ and $p_{1,2} = (p_0, \pm\mathbf{p})$, $p^2 = -\mathbf{p}^2 = y$.

The situation is quite similar to the previous diagram. The soft and ultrasoft regions generate zero contributions because of the appearance of scaleless integrals. There are two non-zero contributions, generated by the hard and potential regions. Each term of the (h) contribution obtained by Taylor expansion of the integrand in y and p,

$$\begin{aligned} F_{7.2}^{(\mathrm{h})} &= \int \frac{\mathrm{d}^d k}{(k^2 + q\cdot k)(k^2 - q\cdot k)k^2} \\ &= -\mathrm{i}\pi^{d/2} \left(\frac{4}{q^2}\right)^{1+\varepsilon} \frac{\Gamma(\varepsilon)}{2(1+2\varepsilon)} + \ldots , \end{aligned} \tag{7.12}$$

can be evaluated by means of partial fractions and (A.13).

Each term of the (p) contribution,

$$\begin{aligned} F_{7.2}^{(\mathrm{p})} &= -\int \frac{\mathrm{d}k_0 \, \mathrm{d}^{d-1}\mathbf{k}}{(\mathbf{k}^2 - q_0 k_0 + y - i0)(\mathbf{k}^2 + q_0 k_0 + y - i0)(\mathbf{k} - \mathbf{p})^2} + \ldots \\ &= \mathrm{i}\pi^{d/2} \frac{y^{-\varepsilon}}{\sqrt{q^2 y}} \frac{\sqrt{\pi}\,\Gamma(\varepsilon + 1/2)}{2\varepsilon} , \end{aligned} \tag{7.13}$$

can be evaluated by closing the integration contour over k_0 in the upper half-plane (as we have previously specified) and then evaluating $(d-1)$-dimensional integrals by means of Feynman parameters. As in the previous example, only the leading term in the potential contribution is non-zero because, in the subsequent terms, all the resulting $(d-1)$-dimensional integrals are scaleless.

One can check that the sum of the single potential term (7.13) and the series (7.12) originating from the hard contribution equals the whole analytic result for the given diagram, which can be evaluated by means of Feynman parameters, for example:

$$\begin{aligned} F_{7.2}(q^2, y; d) &= \mathrm{i}\pi^{d/2} \frac{\Gamma(\varepsilon)}{2y^{1+\varepsilon}} \, {}_2F_1\left(\frac{1}{2}, 1+\varepsilon; \frac{3}{2}; -\frac{q^2}{4y}\right) \\ &= \mathrm{i}\pi^{d/2} \left(\frac{4}{q^2}\right)^{1+\varepsilon} \left[\sqrt{\frac{\pi q^2}{y}} \left(\frac{q^2}{y}\right)^{\varepsilon} \frac{\Gamma(\epsilon + 1/2)}{2^{3+2\varepsilon}\varepsilon} \right. \\ & \left. -\frac{\Gamma(\varepsilon)}{2(1+2\varepsilon)} \sum_{n=0}^{\infty} \frac{\Gamma(1+\epsilon+n)}{\Gamma(1+\varepsilon)} \frac{1+2\varepsilon}{(1+2\varepsilon+2n)n!} \left(\frac{-4y}{q^2}\right)^n \right] . \end{aligned} \tag{7.14}$$

We now turn to the scalar scattering box diagram in the limit of small relative momentum, where we encounter a non-zero contribution from the soft region, $k_0 \sim \sqrt{y}$, $\mathbf{k} \sim \sqrt{y}$.

Example 7.3. The master box diagram of Fig. 7.3 with all four legs on the mass shell, $p_i^2 = m^2$, $i = 1, 2, 3, 4$, and equal non-zero masses in the limit $t \sim y = m^2 - q^2/4 \to 0$.

Here $t = (p_1 - p_3)^2$ is the Mandelstam variable. The Feynman integral is written as

$$F_{7.3}(q^2, y, t; d) = \int \frac{\mathrm{d}^d k}{[(k+p)^2 + q\cdot k - y][(k+p)^2 - q\cdot k - y]k^2(k-r)}$$

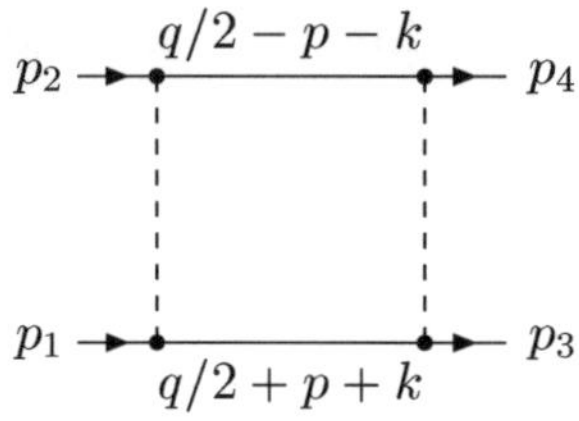

Fig. 7.3. Box diagram with two non-zero masses in the threshold

$$\equiv \int \frac{\mathrm{d}^d k}{[k_0^2 - (\mathbf{k}+\mathbf{p})^2 + q_0 k_0 - y][k_0^2 - (\mathbf{k}+\mathbf{p})^2 - q_0 k_0 - y]} \times \frac{1}{(k_0^2 - \mathbf{k}^2)[k_0^2 - (\mathbf{k}-\mathbf{r})^2]} , \tag{7.15}$$

where $r = p' - p$, $q = p_1 + p_2 = p_3 + p_4$, $p = (p_1 - p_2)/2$ and $p' = (p_3 - p_4)/2$. We choose a frame where $q = (q_0, \mathbf{0})$, $p = (p_0, \mathbf{p})$ and $p' = (p_0, \mathbf{p}')$. We have $p^2 = (p')^2 = y$, $p{\cdot}p' = y - t/2$ and $r^2 = t$.

Let us confine ourselves to terms up to order y^0, and let us consider contributions generated by the regions characterized above. The hard region generates an expansion of the integrand in (7.15) in p, p' and y:

$$F_{7.3}^{(\mathrm{h})} = \int \frac{\mathrm{d}^d k}{(k^2 + q{\cdot}k)(k^2 - q{\cdot}k)(k^2)^2} + \ldots = \frac{1}{(q^2)^2}\left(-\frac{8}{3}\right) + \ldots . \tag{7.16}$$

Every integral in the hard contribution can be evaluated by means of partial fractions and also by (A.13) and (A.14). We have, however, kept only the finite part in ε.

The potential contribution is generated by Taylor expansion of all the propagators in k_0^2. This contribution is evaluated by closing the integration contour over k_0 in the upper half-plane and then evaluating $(d-1)$-dimensional integrals by means of Feynman parameters, as an expansion in ε:

$$F_{7.3}^{(\mathrm{p})} = \int \frac{\mathrm{d}k_0\, \mathrm{d}^{d-1}\mathbf{k}}{(\mathbf{k}^2 - q_0 k_0 + y - i0)(\mathbf{k}^2 + q_0 k_0 + y - i0)\mathbf{k}^2(\mathbf{k}-\mathbf{p})^2}$$
$$= \frac{\mathrm{i}\pi^{d/2+1}}{(q^2)^{1/2+\varepsilon} t\sqrt{y}}\left(\frac{1}{\varepsilon} - \ln(-t/q^2)\right) + O(\sqrt{y}, \sqrt{t}) . \tag{7.17}$$

Now not only the leading term gives a non-zero result. Nevertheless, at $\varepsilon = 0$, the subleading terms vanish. Observe that starting from some order, the potential contribution involves terms that do not decrease as $k \to \pm\infty$ and are just polynomials in k_0. According to our convention, we set such integrals to zero. Let us stress that these integrals are not regularized at all by dimensional regularization, which in fact regularizes only integration over the vector components.

The soft contribution

$$F_{7.3}^{(\mathrm{s})} = \int \frac{\mathrm{d}^d k}{(q_0 k_0 + \mathrm{i}0)(-q_0 k_0 + \mathrm{i}0)k^2(k-r)^2} + \ldots$$
$$= -\frac{1}{q^2} \int \frac{\mathrm{d}^d k}{(k_0 + \mathrm{i}0)(k_0 - \mathrm{i}0)k^2(k-r)^2} + \ldots \tag{7.18}$$

contains pinching singularities. In the language of distribution theory [124], products of the functionals $(x+\mathrm{i}0)^{\lambda_1}$ and $(x-\mathrm{i}0)^{\lambda_2}$ are generally ill-defined. However, the product involved here can be defined as

$$\frac{1}{(k_0 + \mathrm{i}0)(k_0 - \mathrm{i}0)} \to \frac{1}{(k_0 \pm \mathrm{i}0)^2} ,$$

or, according to the principal-value prescription. These three variants differ by terms proportional to $\delta'(k_0)$. Since the rest of the integrand is an even function of k_0, no ambiguity arises. For definiteness, we can suppose that we have defined this product with the $+\mathrm{i}0$ prescription.

Each integral resulting from (7.18) can be evaluated by means of (A.27). We obtain

$$F_{7.3}^{(\mathrm{s})} = \frac{4\mathrm{i}\pi^{d/2}\mathrm{e}^{-\gamma_{\mathrm{E}}\varepsilon}}{(q^2)^{2+\varepsilon}} \left[\left(-\frac{1}{\varepsilon} + \ln(-t/q^2) \right) \left(\frac{q^2}{t} - \frac{4y}{3t} \right) + \frac{2}{3} + O(y,t) \right] . \tag{7.19}$$

Finally, the ultrasoft contribution

$$F_{7.3}^{(\mathrm{us})} = -\frac{1}{t} \int \frac{\mathrm{d}^d k}{(q_0 k_0 + \mathrm{i}0)(q_0 k_0 - \mathrm{i}0)k^2} + \ldots = 0 \tag{7.20}$$

is zero because it consists of scaleless integrals.

The sum of the (h), (s) and (p) contributions reproduces a very simple exact result (up to a finite part in ε)

$$F_{7.3} = \frac{2\mathrm{i}\pi^{d/2}\mathrm{e}^{-\gamma_{\mathrm{E}}\varepsilon}}{(q^2)^{1/2+\varepsilon} t\sqrt{y}} \left(\frac{\pi}{2} - \arctan\sqrt{4y/q^2} \right) \left(\frac{1}{\epsilon} - \ln(-t/q^2) \right) . \tag{7.21}$$

This result can be derived, for example, using the following two-parameter Mellin–Barnes representation, obtained by use of Feynman parameters:

$$F_{7.3}(q^2, y; d) = \frac{\mathrm{i}\pi^{d/2}}{\sqrt{\pi}\Gamma(-2\varepsilon)(q^2)^{2+\varepsilon}} \frac{1}{(2\pi\mathrm{i})^2} \int_{-\mathrm{i}\infty}^{+\mathrm{i}\infty} \int_{-\mathrm{i}\infty}^{+\mathrm{i}\infty} \mathrm{d}z\, \mathrm{d}z' \left(\frac{4y}{q^2} \right)^z$$
$$\times \left(\frac{-t}{4y} \right)^{z'} \frac{\Gamma(2+\varepsilon+z)}{3/2+\varepsilon+z} \Gamma(1+z')^2 \Gamma(-1/2-\varepsilon-z')$$
$$\times \Gamma(-1-\varepsilon-z')\Gamma(-z+z')\Gamma(-z') . \tag{7.22}$$

Since we are interested only in terms up to the finite part in ε, and the overall factor $1/\Gamma(-2\varepsilon)$ contains a power of ε, we need to pick up only the singular terms of the integral, and the only way to generate them is to take the (negative) residue at $z' = -1-\varepsilon$. After that, the resulting integral over z

can be evaluated by expanding in ε and taking the sum of the residues which are to the right of the integration contour, and (7.21) is reproduced.[1]

The representation (7.22) can be also used for picking up individual contributions to the expansion. The terms of the asymptotic behaviour when $t \sim y \to 0$ are obtained by shifting the integration contour over z to the right and taking residues at the poles of the gamma functions with a $(-z)$ dependence. (The pole $1/(3/2+\varepsilon+z)$ is of this type because it has been obtained from such a gamma function.) The pole at $z = -3/2 - \varepsilon$ corresponds to the leading term of the potential contribution. All other poles in z appear because of integration over z'. The hard contribution is generated by taking residues of $\Gamma(-z')$ and then of the resulting gamma functions $\Gamma(-z)$, etc. The pole at $z = -1/2 - \varepsilon$ appears after taking the residue at $z' = -1/2 - \varepsilon$. After taking the residue at this pole, we obtain the sum of the leading soft and subleading potential contributions.

Here we would like again to compare expansion by (local) methods such as Mellin–Barnes integrals and the (global) method of expansion by regions: such a Mellin–Barnes representation must be individually derived for each specific diagram, while the rules for expanding by regions for the given limit have been formulated; the application of these rules requires one only to consider various regions and to perform simple power counting, and analytic manipulations and arguments are not used.

Let us now modify the previous diagram a little.

Example 7.3a. The box diagram of Fig. 7.3 as in Example 7.3 but with the second power of the propagator of the lower line:

$$F_{7.3a}(q^2, y; d) = \int \frac{\mathrm{d}^d k}{[(k+p)^2 + q\cdot k - y]^2[(k+p)^2 - q\cdot k - y]k^2(k-r)^2}$$
$$\equiv \int \frac{\mathrm{d}^d k}{[k_0^2 - (\mathbf{k}+\mathbf{p})^2 + q_0 k_0 - y]^2[k_0^2 - (\mathbf{k}+\mathbf{p})^2 - q_0 k_0 - y]}$$
$$\times \frac{1}{(k_0^2 - \mathbf{k}^2)(k_0^2 - (\mathbf{k}-\mathbf{r})^2)} \,. \qquad (7.23)$$

By considering only the first singular terms of the expansion in y and t, we intend to illustrate some subtle points in the definition of the soft and potential contributions. Therefore we ignore the hard contribution, which starts from order y^0 as usual. The ultrasoft contribution is again zero here.

The potential contribution, obtained by expanding all the propagators in k_0^2,

$$F_{7.3a}^{(\mathrm{p})} = -\int \frac{\mathrm{d}^d k}{[(\mathbf{k}+\mathbf{p})^2 - q_0 k_0 + y - \mathrm{i}0]^2[(\mathbf{k}+\mathbf{p})^2 + q_0 k_0 + y - \mathrm{i}0]}$$

[1] This result obtained by expansion in ε also follows from a similar result [14] obtained by introducing a non-zero mass M into the massless lines, using a 'folklore' correspondence rule $\ln M^2 \to 1/[\varepsilon\Gamma(1-\varepsilon)]$, which, presumably works in one loop.

$$\times \frac{1}{\mathbf{k}^2(\mathbf{k}-\mathbf{r})^2} + \dots \,, \tag{7.24}$$

is dominant and starts from order $-5/2$, measured in powers of the small expansion parameters, y and t, of the problem. When evaluating this contribution by means of integration over k_0 using Cauchy's theorem, we encounter an obstacle because, in contrast to the previous example with powers of propagators equal to one, the result appears ambiguous.

Nevertheless, the leading potential contribution is unambiguous and can be evaluated as an expansion in ε, up to the finite part:

$$F_{7.3a}^{(\mathrm{p}),\,\mathrm{LO}} = -\mathrm{i}\pi^{d/2}\mathrm{e}^{-\gamma_{\mathrm{E}}\varepsilon}\frac{\sqrt{\pi}}{4\sqrt{q^2}ty^{3/2}}\left(\frac{1}{\varepsilon}+2-\ln(-t)\right) . \tag{7.25}$$

However, in the calculation of the subleading potential contribution, terms that are logarithmically divergent at $k_0 \to \pm\infty$, such as $1/[(\mathbf{k}+\mathbf{p})^2 - q_0k_0 + y - \mathrm{i}0]$, appear. If these terms are treated straightforwardly using Cauchy's theorem (i.e. without paying attention to unsafe behaviour at large values of k_0), we obtain different results when the integration contour over k_0 is closed in the upper or the lower half-plane.

Further, the soft contribution

$$\begin{aligned} F_{7.3a}^{(\mathrm{s})} &= \int \frac{\mathrm{d}^dk}{(q_0k_0+\mathrm{i}0)^2(-q_0k_0+\mathrm{i}0)k^2(k-r)^2} + \dots \\ &= -\frac{1}{(q^2)^{3/2}}\int \frac{\mathrm{d}^dk}{(k_0+\mathrm{i}0)^2(k_0-\mathrm{i}0)k^2(k-r)^2} + \dots \end{aligned} \tag{7.26}$$

turns out also to be ill-defined. In the present case, the following two variants

$$\frac{1}{(k_0+\mathrm{i}0)^2(k_0-\mathrm{i}0)} \to \frac{1}{(k_0\pm\mathrm{i}0)^3}$$

of the definition of the product involved will lead to results with opposite signs.

Thus the potential and soft contributions are individually ill-defined. However, the sum of them is unambiguous. To see this in this particular example, let us decide to evaluate the k_0 integral first in the initial unexpanded integral and, for definiteness, choose to use Cauchy's theorem (a sufficiently fast decrease of the initial integrand at large values of k_0 takes place here) by closing the contour in the upper half-plane. Then, after expanding by regions, we continue to follow this prescription. For the potential contribution, this just means the same recipe for closing the contour in the k_0 plane (with arbitrary behaviour at infinity). For the soft contribution, this also means the same choice of the contour, which apparently is equivalent to nothing but treating all the dangerous products in the integrand as follows:

$$\frac{1}{(q_0k_0+\mathrm{i}0)^{n_1}(q_0k_0-\mathrm{i}0)^{n_2}} \to \frac{1}{(q_0k_0+\mathrm{i}0)^{n_1+n_2}} \,.$$

Evaluating the subleading integral in (7.24) and the leading term in (7.26), with the above prescriptions, we obtain, for general ε,

$$F_{7.3a}^{(p),\,NLO} = i\pi^{d/2} \frac{2^{1+2\varepsilon}\pi}{(-t)^{3/2+\varepsilon}(q^2)^{3/2}} \frac{\Gamma(1/2+\varepsilon)\Gamma(1/2-\varepsilon)}{\Gamma(-\varepsilon)} ,$$

$$F_{7.3a}^{(s),\,LO} = -2i\pi^{d/2} \frac{2^{1+2\varepsilon}\pi}{(-t)^{3/2+\varepsilon}(q^2)^{3/2}} \frac{\Gamma(1/2+\varepsilon)\Gamma(1/2-\varepsilon)}{\Gamma(-\varepsilon)} . \quad (7.27)$$

This result can be confirmed by straightforward evaluation of the given Feynman integral. Starting from the Feynman parameters and introducing Mellin–Barnes (MB) integrals twice, one can arrive at the following twofold MB representation similar to (7.22):

$$F_{7.3a}(q^2, y; d) = \frac{i\pi^{d/2}}{\sqrt{\pi}\Gamma(-1-2\varepsilon)(q^2)^{3+\varepsilon}} \frac{1}{(2\pi i)^2} \int_{-i\infty}^{+i\infty} dz \left(\frac{4y}{q^2}\right)^z$$
$$\times \int_{-i\infty}^{+i\infty} dz' \left(\frac{-t}{4y}\right)^{z'} \frac{\Gamma(3+\varepsilon+z)}{5/2+\varepsilon+z} \Gamma(1+z')^2 \Gamma(-3/2-\varepsilon-z')$$
$$\times \Gamma(-1-\varepsilon-z')\Gamma(-z+z')\Gamma(-z') , \quad (7.28)$$

where the pole of $1/(5/2+\varepsilon+z)$ is to the right of the integration contour.

The leading and subleading asymptotic behaviour is obtained by shifting the integration contour and taking residues, as in the case of (7.22):

$$F_{7.3a}(q^2, y; d) \sim F_{\Gamma}^{(p),\,LO} - \frac{i\pi^{d/2} 2^{1+2\varepsilon}\pi}{(-t)^{3/2+\varepsilon}(q^2)^{3/2}} \frac{\Gamma(1/2+\varepsilon)\Gamma(1/2-\varepsilon)}{\Gamma(-\varepsilon)} . \quad (7.29)$$

The subleading term is indeed given by the sum of $F_{\Gamma}^{(p),\,NLO}$ and $F_{\Gamma}^{(s),\,LO}$. (We are not much interested in the LO term here, because it is obviously unambiguous, so that we keep only the finite part in ε and concentrate on the next-to-leading power for general ε (although it is zero at $\varepsilon = 0$) to illustrate the mixing of the soft and potential contributions.)

Up to now we have encountered non-zero contributions generated by the hard, soft and potential regions. To find a non-trivial ultrasoft contribution, we go to the two-loop level.

Example 7.4. The master two-loop diagram of Fig. 7.4 with four equal non-zero masses and one zero mass in the limit $y = m^2 - q^2/4 \to 0$.

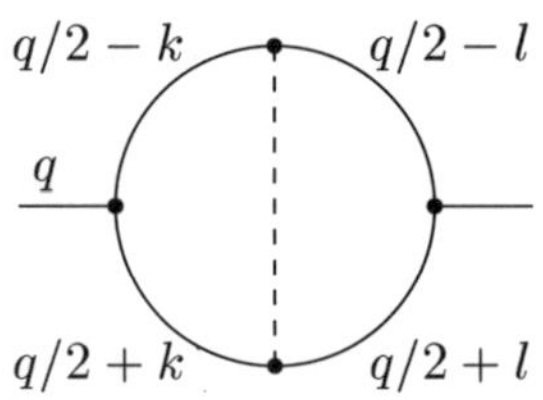

Fig. 7.4. Master two-loop diagram with two non-zero masses in the threshold

The Feynman integral is written as

$$F_{7.4}(q^2, y; d) = \int \frac{\mathrm{d}^d k}{(k^2 + q\cdot k - y)(k^2 - q\cdot k - y)} \times \int \frac{\mathrm{d}^d l}{(l^2 + q\cdot l - y)(l^2 - q\cdot l - y)(k-l)^2} , \tag{7.30}$$

where we again choose the frame $q = \{q_0, \mathbf{0}\}$ and follow our rules for routing the external momentum through the graph. We have already expanded this diagram in the large-momentum limit in Sect. 4.3. Now we have two loop momenta and, when analysing the regions, we consider every loop momentum to be either hard, soft, potential or ultrasoft. When referring to a region, we shall characterize the loop momentum by k in the first place. As we shall see shortly, it is necessary to also consider other ways of introducing the loop momenta, different from that of Fig. 7.4.

The (h–h) region generates a Taylor expansion of the integrand in y. The corresponding integrals are the most difficult in the calculation, among all possible contributions to the expansion, and can be expressed, using partial fractions, through the following family of basic integrals:

$$J_\pm(a_1, \ldots, a_5) = \int\int \frac{\mathrm{d}^d k\, \mathrm{d}^d l}{(k^2)^{a_1}(l^2)^{a_2}[(k-l)^2]^{a_3}(k^2 + q\cdot k)^{a_4}(l^2 \pm q\cdot l)^{a_5}} . \tag{7.31}$$

The integrals of type J_+ can be reduced to gamma functions [129] through recurrence relations derived from IBP [68]. The integrals of type J_- can be expressed in terms of gamma functions and the integrals $J_-(0,0,a_3,a_4,a_5)$. The latter integrals can be reduced to $J_-(0,0,1,1,1)$, using a simplified version of some results in [228]. Then $J_-(0,0,1,1,1)$ (or a more convenient input integral) can be calculated explicitly by means of Feynman parameters as an expansion in ε. We obtain [18]

$$F_{7.4}^{(\mathrm{h-h})} = \frac{\left(\mathrm{i}\pi^{d/2} \mathrm{e}^{-\gamma_{\mathrm{E}}\varepsilon}\right)^2}{q^2} \left[\pi^2 \left(\frac{1}{\varepsilon} - 2\ln q^2 + 6\ln 2 + 2\right) + 21\zeta(3) - 4(8 + 3\pi^2)\frac{y}{q^2} + O(y^2)\right] . \tag{7.32}$$

When one of the loop momenta is hard and the other loop momentum is potential, soft or ultrasoft, only the following two equal, symmetrical contributions (h–p) and (p–h) are non-zero; the others vanish because they contain scaleless integrals. The leading contribution generated by the (h–p) region is

$$F_{7.4}^{(\mathrm{h-p})} = \int \frac{\mathrm{d}^d l}{(-\mathbf{l}^2 + q_0 l_0 - y)(-\mathbf{l}^2 - q_0 l_0 - y)} \int \frac{\mathrm{d}^d k}{k^2 (k^2 + q\cdot k)(k^2 - q\cdot k)} . \tag{7.33}$$

It follows from power counting that this integral contributes at order $\sqrt{y}$. Keeping also the subleading term in the (h–p) contribution, we obtain [18]

$$F_{7.4}^{(\mathrm{h-p})} = F_{7.4}^{(\mathrm{p-h})} = \frac{\left(\mathrm{i}\pi^{d/2}\mathrm{e}^{-\gamma_{\mathrm{E}}\varepsilon}\right)^2}{q^2}\left[8\pi\left(\frac{1}{\varepsilon} - \ln q^2 - \ln y\right)\left(\frac{y}{q^2}\right)^{1/2}\right.$$
$$\left. -\frac{32\pi}{3}\left(\frac{1}{\varepsilon} - \ln q^2 - \ln y + \frac{7}{3}\right)\left(\frac{y}{q^2}\right)^{3/2} + O(y^{5/2})\right] . \quad (7.34)$$

When l is soft or ultrasoft, the integrand is almost the same as in (7.33) In both cases, the resulting integrals vanish.

When both loop momenta are potential, all propagators are expanded in their zero-order components squared. After the residues from the remaining quark poles in k_0 and l_0 have been picked up, the resulting $(d-1)$-dimensional two-loop integral is

$$\begin{aligned} F_{7.4}^{(\mathrm{p-p})} &= \frac{1}{q^2}\int\int \frac{\mathrm{d}^{d-1}\mathbf{k}\,\mathrm{d}^{d-1}\mathbf{l}}{(\mathbf{k}^2+y)(\mathbf{l}^2+y)(\mathbf{k}-\mathbf{l})^2} + \ldots \\ &= \frac{\left(\mathrm{i}\pi^{d/2}\mathrm{e}^{-\gamma_{\mathrm{E}}\varepsilon}\right)^2}{q^2 y^{2\varepsilon}} \frac{\pi\Gamma(\varepsilon+1/2)\Gamma(\varepsilon-1/2)}{2\varepsilon} + \ldots \\ &= \frac{\left(\mathrm{i}\pi^{d/2+1}\mathrm{e}^{-\gamma_{\mathrm{E}}\varepsilon}\right)^2}{q^2}\left(-\frac{1}{\varepsilon} - 2 + 4\ln 2 + 2\ln y\right) , \end{aligned} \quad (7.35)$$

where the $(d-1)$-dimensional version of (A.38) has been applied. The next terms in the (p–p) contribution vanish for the same reason as in Examples 7.1 and 7.2.

All the regions where one of the loop momenta is soft generate only zero contributions. If we consider two regions when one of the loop momenta, k or l, is potential and the other is ultrasoft, we again obtain a zero contribution. To find a missing (p–us) contribution we have to suppose that the momentum flowing through the central line is ultrasoft. So it is better to choose this momentum to be one of the loop momenta. Denoting this momentum by l and the other loop momentum as k, we obtain the following (p–us) contribution:

$$F_{7.4}^{(\mathrm{p-us})} = \int\int \frac{\mathrm{d}^d k\,\mathrm{d}^d l}{[-\mathbf{k}^2 + q_0(k_0 - l_0/2) - y][-\mathbf{k}^2 - q_0(k_0 - l_0/2) - y]}$$
$$\times \frac{1}{[-\mathbf{k}^2 + q_0(k_0 + l_0/2) - y][-\mathbf{k}^2 - q_0(k_0 + l_0/2) - y]\, l^2} + \ldots . \quad (7.36)$$

The integral can be calculated by examining the poles in k_0 and l_0 and closing the integration contours over k_0, l_0 such that the number of terms is minimized. The resulting $(d-1)$-dimensional integrals can be calculated recursively in terms of gamma functions, and we arrive at [18]

$$F_{7.4}^{(\mathrm{p-us})} = \frac{\left(\mathrm{i}\pi^{d/2}\mathrm{e}^{-\gamma_{\mathrm{E}}\varepsilon}\right)^2}{q^2} 8\pi\left(-\frac{1}{\epsilon} - 8 + 10\ln 2 - \ln q^2 + 3\ln y\right)$$

$$\times \left(\left(\frac{y}{q^2}\right)^{1/2} - \frac{4}{3}\left(\frac{y}{q^2}\right)^{3/2} + O(y^{5/2}) \right) . \tag{7.37}$$

The (h–h) and (p–p) regions contribute only to even powers of $\sqrt{y}$ and the (h–p) and (p–us) regions contribute only to odd powers. Each separate contribution contains poles in ε. The poles cancel between the (h–h) and (p–p) contributions and the (h–p) = (p–h) and (p–us) contributions, leaving logarithms of y. The terms computed above combine to give the finite threshold expansion

$$F_{7.4} \sim \frac{\left(\mathrm{i}\pi^{d/2}\right)^2}{q^2} \left[2\pi^2 \ln\frac{32y}{q^2} + 21\zeta(3) + 16\pi\left(\ln\frac{32y}{q^2} - 4\right)\left(\frac{y}{q^2}\right)^{1/2} \right.$$
$$\left. -4(8+3\pi^2) - \frac{64\pi}{3}\left(\ln\frac{32y}{q^2} - \frac{17}{6}\right)\left(\frac{y}{q^2}\right)^{3/2} \right] + \ldots . \tag{7.38}$$

For this diagram, an analytic result is known [97, 36]:

$$F_{7.4} = \frac{\left(\mathrm{i}\pi^{d/2}\right)^2}{q^2} \left[F(1) + F(z^2) - 2F(z) \right] , \tag{7.39}$$

where

$$F(z) = 6\,\mathrm{Li}_3(z) - 4\ln z\,\mathrm{Li}_2(z) - \ln^2(z)\ln(1-z) , \tag{7.40}$$

$$z = -\frac{1 - \mathrm{i}\sqrt{4y/q^2}}{1 + \mathrm{i}\sqrt{4y/q^2}} . \tag{7.41}$$

If we take care to correctly continue the logarithms in $F(z^2)$ to the second sheet when $y/q^2 < 1/4$, the expansion of the exact result reproduces (7.38).

We have concentrated on the situation of two equal non-zero masses in a threshold because of its great importance, from the physical point of view. For completeness, let us point out that, in the case where there are general masses, $m_i \equiv \xi_i m$ with $\sum \xi_i = 1$, in a threshold at $q^2 = (\sum m_i)^2$, it is reasonable to choose the routing of the external momentum in such a way that a portion $\xi_i q$ flows through the ith line. After that, the dependence of the integrand on the expansion variable $y = m^2 - q^2$ becomes explicit. For example, the master sunset diagram of Fig. 2.2, with such a canonical routing, is represented as

$$F_\Gamma(q^2, y; d) = \int\int \frac{\mathrm{d}^d k\, \mathrm{d}^d l}{(k^2 - m_1^2)(l^2 - m_2^2)[(q-k-l)^2 - m_3^2]}$$
$$\equiv \int\int \frac{\mathrm{d}^d k\, \mathrm{d}^d l}{(k^2 + 2\xi_1 q\cdot k - \xi_1^2 - y)(l^2 + 2\xi_2 q\cdot l - \xi_2^2 y)}$$
$$\times \frac{1}{(k+l)^2 - 2\xi_3(q\cdot k + q\cdot l) - \xi_3^2 y} . \tag{7.42}$$

The expansion in the case of three non-zero masses consists of (h–h) and (p–p) contributions – see [93]. If one mass is zero and the other two masses are non-zero, there are (h–h) and (p–us) contributions, and when there is one non-zero mass, one has contributions generated by the (h–h) and (us–us) regions.

For the sunset diagram and its generalization to the case where the diagram consists of $L+1$ lines between two external vertices, the contribution which is most complicated to evaluate is (h–h–...–h). In particular, the (p–p–...–p) contribution to the expansion of this L-loop diagram in the case of all non-zero masses can be expressed in terms of gamma functions for general ε. Its leading term is [93]

$$\mathrm{i}^L \pi^{L(d+1)/2} \left(\prod_{j=1}^{L+1} \xi_j^{1/2-\varepsilon} \right) \frac{\Gamma(1+L(\varepsilon-3/2))}{(q^2)^{L/2} y^{1+L(\varepsilon-3/2)}} \,. \tag{7.43}$$

7.2 General Prescriptions

The prescriptions of the threshold expansion were confirmed by known analytic results for the one- and two-loop examples presented above. Let us now believe in these prescriptions and apply them to any diagram in the threshold limit where, owing to kinematics, some other parameters, such as the Mandelstam variable t (see the above examples of box diagrams), can be small. With the physical orientation towards diagrams consisting of massive fermion and massless boson lines, we suppose that a given diagram contains two paths of massive particles, which can be disconnected, as in the case of box diagrams, or joined together, as in the case of vertex diagrams. In accordance with the strategy of expansion by regions, any given diagram can be expanded at threshold by the following prescriptions [18]:

- Choose the canonical routing for the flow of the external momenta for the given threshold. In particular, when there are two equal non-zero masses in the threshold, let one half of the external momentum flow through one of the massive lines and the other half of it through the other massive line. In the general situation with masses $m_i \equiv \xi_i m$ with $\sum \xi_i = 1$, let a portion $\xi_i q$ flow through line i.
- Choose the frame[2] $q = \{q_0, \mathbf{0}\}$.
- Consider the various regions where any loop momentum can be of one of the following four types:

$$\begin{aligned}
&(\mathrm{h}), \quad k_0 \sim \sqrt{q^2}\,, \quad \mathbf{k} \sim \sqrt{q^2}\,, \\
&(\mathrm{s}), \quad k_0 \sim \sqrt{y}\,, \quad \mathbf{k} \sim \sqrt{y}\,, \\
&(\mathrm{p}), \quad k_0 \sim y/\sqrt{q^2}\,, \quad \mathbf{k} \sim \sqrt{y}\,, \\
&(\mathrm{us}), \quad k_0 \sim y/\sqrt{q^2}\,, \quad \mathbf{k} \sim y/\sqrt{q^2}\,.
\end{aligned}$$

[2] A manifestly Lorentz-invariant treatment of the threshold expansion is possible but very cumbersome.

- Try various choices of the loop momenta (and at the same time avoid double counting).
- In accordance with the general strategy of expansion by regions, extend the integration to the whole space and set scaleless integrals to zero.

In addition, one has to take care of possible soft–potential mixing and introduce consistent prescriptions for treating contributions from regions where the same loop momentum k is either soft or potential (as in Example 7.3a), i.e. one has to close the integration contour over k_0 in the same half-plane in *both* contributions. When considering various possibilities of treating the momenta of the lines, one has to take into account conservation laws in the vertices. In particular, both hard and soft lines always form 1PI subgraphs, two ultrasoft lines cannot generate a potential or a soft line, etc. Note also that massive lines can never be ultrasoft. Although we are considering various possibilities for the lines of the given graph, the prescriptions are not formulated in a pure graph-theoretical language, because, as in the case of the limits treated in Chap. 6, we are stuck to the fixed canonical routing of the external momenta. We shall come back to our 'incompletely' graph-theoretical description of the threshold expansion in Sect. 7.4.

Let us emphasize that the leading order of the contributions of the regions can be obtained in a very simple way, by power counting, without explicit evaluation of any given contribution. The hard contribution always gives y^0. For the other three types of contributions, it is necessary to take account of the following powers from the differentials: y^d from the soft, y^{1+d} from the potential and y^{2d} from the ultrasoft region. We can then estimate the leading behaviour of the propagators. In particular, the quark QCD propagator gives $1/\sqrt{y}$ and $1/y$ from the soft and potential regions, respectively. The gluon QCD propagator gives $1/y$ from the soft and potential regions and $1/y^2$ from the ultrasoft region.

7.3 Two-Loop Examples

Let us turn to examples where explicit analytic results are not (yet?) known and follow the prescriptions formulated in the previous subsection.

Example 7.5. The master two-loop vertex diagram of Fig. 7.5 with $p_1^2 = p_2^2 = m^2$ in the limit $y = m^2 - q^2/4 \to 0$.

The Feynman integral is written as

$$F_{7.5}(q^2, y; d) = \int \frac{\mathrm{d}^d k}{(k^2 + q\cdot k - y)(k^2 - q\cdot k - y)} \times \int \frac{\mathrm{d}^d l}{(l^2 + q\cdot l - y)(l^2 - q\cdot l - y)(k-l)^2(l-p)^2}\,, \tag{7.44}$$

where k and l are chosen to be the loop momenta of the triangle and box subgraphs, respectively.

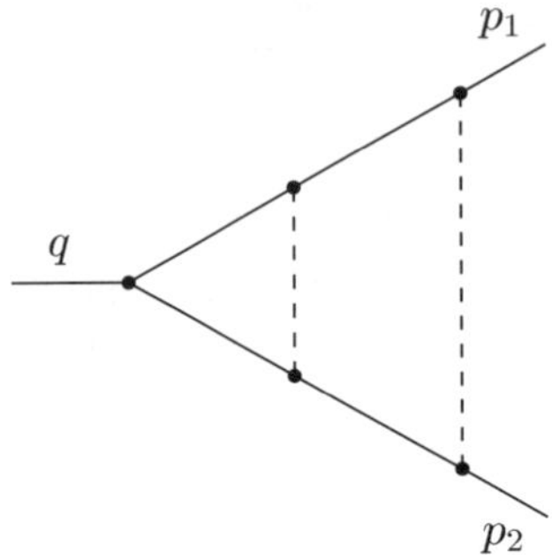

Fig. 7.5. Master two-loop vertex diagram with two non-zero masses in the threshold

This is the list of regions that generate non-zero contributions, together with their leading order [18]:

(h–h) y^0
(h–p) $1/\sqrt{y}$
(p–h) $\sqrt{y}$
(p–p) $1/y$
(p–s) $1/\sqrt{y}$
(p–us) $1/\sqrt{y}$.

Here the loop momentum k is indicated in the first position and l in the second position; the only exception is that (p–us) means that the momentum of the central line is ultrasoft and the momenta of the other lines are potential.

These contributions can be evaluated as in the previous examples. For instance, after partial fractions have been used, the integrals needed to evaluate the (h–h) contribution reduce to $J_\pm$, given by (7.31) (with additional scalar products in the numerator), and the complexity of the integrals is essentially the same as for the self-energy diagram in Example 7.4. The leading singular behaviour at threshold comes from the (p–p) region.

Evaluation of all terms up to order y^0 gives [18]

$$\begin{aligned} F_{7.5}(q^2, y; d) \sim \frac{\left(\mathrm{i}\pi^{d/2}\mathrm{e}^{-\gamma_{\mathrm{E}}\varepsilon}\right)^2 4^{2+2\varepsilon}}{(q^2)^{2+2\varepsilon}} & \left[\frac{1}{\varepsilon^2}\left(\frac{\pi^2 q^2}{128y} - \frac{\pi\sqrt{q^2}}{16\sqrt{y}} + \frac{1}{8}\right)\right. \\ & + \frac{1}{\epsilon}\left(-\frac{\pi^2 q^2}{64y}\ln(16y/q^2) + \frac{\pi\sqrt{q^2}}{8\sqrt{y}}(2-\ln 2) - \frac{1}{4}\right) \\ & + \frac{\pi^2 q^2}{64y}\left(\ln^2(16y/q^2) + \frac{7\pi^2}{12}\right) \\ + \frac{\pi\sqrt{q^2}}{8\sqrt{y}} & \left(\ln^2(16y/q^2) + (3\ln 2 - 4)\ln(16y/q^2) + \ln^2 2 - 2\ln 2 + \frac{\pi^2}{4}\right) \\ & \left. - \frac{29\pi^2}{48} - \frac{3}{2}\right] + O(y^{1/2}) . \end{aligned} \tag{7.45}$$

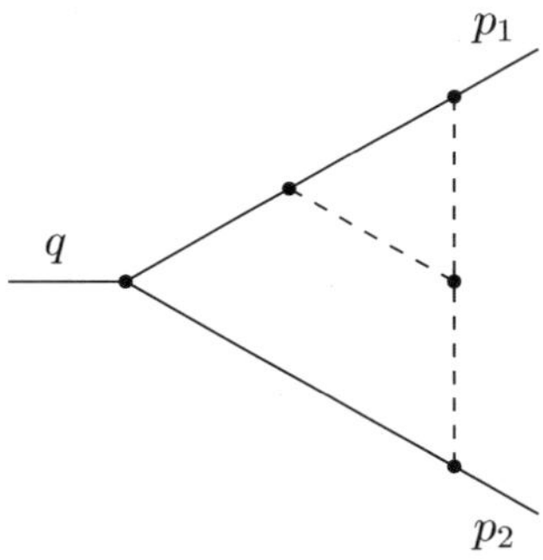

Fig. 7.6. Master two-loop non-Abelian vertex diagram with two non-zero masses in the threshold

The coefficient of the double pole of these first three terms of the expansion agrees [18] with an explicit result for the double-pole part of the given diagram, which is IR divergent:

$$\left(\mathrm{i}\pi^{d/2}\mathrm{e}^{-\gamma_{\mathrm{E}}\varepsilon}\right)^2 \frac{4^{2\varepsilon}}{2\epsilon^2(q^2)^{1+2\varepsilon}y}\left(\frac{\pi}{2} - \arctan\sqrt{4y/q^2}\right)^2 . \tag{7.46}$$

For the next diagram, explicit results are unknown.

Example 7.6. The master two-loop non-Abelian vertex diagram of Fig. 7.6 with $p_1^2 = p_2^2 = m^2$ in the limit $y = m^2 - q^2/4 \to 0$.

The Feynman integral is

$$F_{7.6}(q^2, y; d) = \int \frac{\mathrm{d}^d k}{(k^2 + q\cdot k - y)(k^2 - q\cdot k - y)} \times \int \frac{\mathrm{d}^d l}{(l^2 + q\cdot l - y)(k-l)^2(k-p)^2(l-p)^2} , \tag{7.47}$$

where k and l are the loop momenta of the box and triangle subgraphs, respectively.

The list of regions that generate non-zero contributions is

(h–h) y^0
(p–h) $1/\sqrt{y}$
(p–s) $1/y$
(p–us) $1/\sqrt{y}$,

where the loop momentum k is indicated in the first position and l in the second position; the only exception is that (p–us) means that the momentum of the line that is common to the box and triangle subgraphs is ultrasoft and the momenta of other lines are potential.

The (p–p) region that could give a $1/y$ behaviour does not contribute here, because the poles in l_0 lie on only one side of the real axis, so that the l_0 integral vanishes. The contribution from the potential–soft region is algebraically the most complicated. Since the soft triangle subgraph is less symmetric than the box subgraph in the diagram of Fig. 7.5, the expansion in this region is in powers of $\sqrt{y}$ and it is necessary to compute the first two subleading terms to obtain an accuracy $O(y^0)$ at threshold. Other contributions are evaluated as in the previous example.

Evaluation of terms up to order y^0 gives [18]

$$F_{7.6}(q^2, y; d) \sim \frac{\left(\mathrm{i}\pi^{d/2}\mathrm{e}^{-\gamma_{\mathrm{E}}\varepsilon}\right)^2 4^{2+2\varepsilon}}{(q^2)^{2+2\varepsilon}} \left[\frac{1}{\varepsilon^2}\left(-\frac{\pi\sqrt{q^2}}{32\sqrt{y}} + \frac{1}{8}\right)\right.$$
$$+\frac{1}{\varepsilon}\left(\frac{\pi^2 q^2}{64y} - \frac{\pi^2}{16} - \frac{1}{4}\right) - \frac{\pi^2 q^2}{32y}\left[\ln(16y/q^2) + 2\right]$$
$$+\frac{\pi\sqrt{q^2}}{32\sqrt{y}}\left(\ln^2(16y/q^2) + 8\ln(32y/q^2) + \frac{5\pi^2}{6}\right) + \frac{\pi^2}{8}\ln(16y/q^2)$$
$$\left. -\frac{23\pi^2}{48} - \frac{3}{2}\right] + O(y^{1/2})\,. \tag{7.48}$$

As in the previous example, the double-pole part of this expansion can be checked against an explicit result for the double-pole part of the unexpanded diagram:

$$-\frac{\left(\mathrm{i}\pi^{d/2}\mathrm{e}^{-\gamma_{\mathrm{E}}\varepsilon}\right)^2 4^{2\varepsilon}}{\varepsilon^2 (q^2)^{3/2+2\varepsilon}(1+4y/q^2)\sqrt{y}}\left(\frac{\pi}{2} - \arctan\sqrt{4y/q^2}\right)\,. \tag{7.49}$$

In the expansion of the similar non-planar diagram of Fig. 7.7 near threshold, the non-vanishing contributions are the same as for the non-Abelian diagram, except that the (h–p) contribution scales as $\sqrt{y}$ in the leading order and the (p–s) region as $1/\sqrt{y}$, as for the planar diagram of Fig. 7.6. Furthermore, equal contributions (h–p) = (p–h) and (s–p) = (p–s) are present. The (p–p) contribution vanishes again, for the same reason as in the previous example.

Any integral present in the (h–h) contribution of any of the three basic types of vertex diagrams can be reduced to either $J_\pm$, given by (7.31), or

$$L_\pm(a_1,\ldots,a_5)$$
$$= \int\int \frac{\mathrm{d}^d k\,\mathrm{d}^d l}{(k^2)^{a_1}(l^2)^{a_2}[(k+l)^2 + q\cdot(k+l)]^{a_3}(k^2+q\cdot k)^{a_4}(l^2 \pm q\cdot l)^{a_5}}\,. \tag{7.50}$$

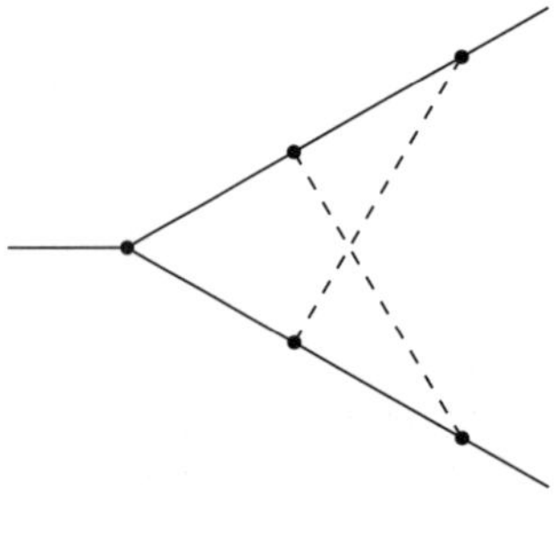

Fig. 7.7. Master two-loop non-planar vertex diagram with two non-zero masses in the threshold

Using recurrence relations that follow from IBP, one can reduce these integrals first to $L_+(0,0,a_3,a_4,a_5) = L_-(0,0,a_3,a_4,a_5)$ and then to a single integral such as $L_+(0,0,2,2,1)$. An algorithm for L_+ has been formulated in [129, 114][3]. Complete algorithms[4] for the evaluation of the $J_\pm$ and $L_\pm$ integrals have been developed for the purpose of performing the calculations in [84, 16]. Furthermore, the hard parts of all possible two-loop three-point diagrams relevant to QCD calculations can be reduced to $J_\pm$ or $L_\pm$ and, consequently, can be evaluated analytically.

7.4 Threshold Expansion and (P)NRQCD

Let us remember that we identified the potential region by studying Example 7.1, where we supposed that the loop momentum was such that $k = (k_0, \mathbf{k})$ with $|k_0|, |\mathbf{k}| \lesssim \sqrt{y}$. Let us not distinguish between the soft, potential and ultrasoft momenta which satisfy these relations. For such loop momenta k, we have to neglect k_0^2 with respect to $q_0 k_0$ and therefore expand the massive propagators in k_0^2. Considering the various possible ways in which the loop momenta may be either hard (large) or small (here in the sense that they are not large), we see that the resulting expansion can be written as a sum of contributions labelled by the lines with hard loop momenta, which form subgraphs consisting of 1PI connectivity components. Thus the expansion near threshold takes the form

$$F_\Gamma \sim \sum_\gamma \mathcal{T}_{k_0^2} F_{\Gamma/\gamma} \circ \mathcal{M}_\gamma F_\gamma \;, \tag{7.51}$$

where the operator $\mathcal{M}_\gamma$, as before, performs Taylor expansion of the integrand of F_γ in its small external momenta and other small parameters (y, the Mandelstam variable t, small masses). The first Taylor operator $\mathcal{T}_{k_0^2}$ expands propagators with a dependence of type $1/(k^2 \pm q_0 k_0 - y)$ in k_0^2, where k symbolizes any loop momentum of the reduced graph Γ/γ.

This expansion is almost identical to the threshold expansion (6.17) when there is one heavy mass in the threshold. Formally, the only difference is that we now have a Taylor expansion in k_0^2 rather than in k^2. A crucial point is that expansion (7.51) is not yet homogeneous with respect to the expansion parameter, while (6.17) already possesses the desired homogeneity property.

It is already clear how to make (7.51) homogeneous, i.e. to identify the small momenta as soft, potential or ultrasoft and expand further. But let us stay for some time at this level and study what the expansion (7.51) gives at the Lagrangian level in the case of QCD. We suppose for simplicity that there is one quark flavour with a heavy mass m and that other quarks can be regarded as massless. If a diagram possesses two paths of quark lines,

[3] No details are given in these references.

[4] Descriptions of the two algorithms are still unpublished.

we follow our conventions for the routing of external momenta. If a quark propagator corresponds to such a line, let its momentum be $q/2 + p - k$, where $q = (q_0, \mathbf{0})$ is the momentum that is typically written as the sum $p_1 + p_2$ of the external momenta of two incoming quarks with mass m and tends to the threshold $q^2 \equiv s \to 4m^2$. Moreover, k is a loop momentum (or, in a more general situation, a linear combination of loop momenta), and p is an external momentum whose square is of order $y = m^2 - s/4$. When k is supposed to be small, we expand the denominator in k_0^2 and neglect p and k with respect to q. So the quark propagator becomes

$$\begin{aligned} S(q/2+p+k) &= \mathrm{i}\frac{\not{q}/2+\not{p}+\not{k}+m}{(q/2+p+k)^2-m^2+\mathrm{i}0} \\ &\sim \mathrm{i}\frac{2m/q_0+\gamma^0}{2}\frac{1}{k_0-(\mathbf{p}+\mathbf{k})^2/q_0-y/q_0+\mathrm{i}0} \,. \end{aligned} \tag{7.52}$$

The threshold limit is a non-relativistic limit when the relative velocities of the quarks are small. In practice, one uses, instead of the 'pure' difference $y = -(s - 4m^2)/4$, the parameter $\beta \equiv \sqrt{-4y/s}$ given by (4.69), which is called the 'velocity' because it is connected with the 'real' non-relativistic velocity $v = \sqrt{E/m}$ by

$$\beta = \sqrt{1 - \frac{4}{(v^2+2)^2}} = v - \frac{3}{8}v^3 + \ldots \,, \tag{7.53}$$

where $E = s - 2m$ is the residual energy of the quarks.

Changing the variables (y, s) in (7.52) to (v, m), we obtain

$$\mathrm{i}\frac{1+\gamma^0}{2}\frac{1}{p_0+k_0-(\mathbf{p}+\mathbf{k})^2/(2m)+\mathrm{i}0} \,, \tag{7.54}$$

where $p_0 = \mathbf{p}^2/(2m) = mv^2/2 \sim -y/q_0$ is the zero component of the four-vector $p = (p_0, \mathbf{p})$, restricted by the non-relativistic on-shell condition. The first factor is a projector that distinguishes quarks from antiquarks and, thus, we arrive at the propagator of NRQCD [231, 164, 25] (which was constructed analogously to NRQED [43]):

$$\tilde{S}(p) = \frac{\mathrm{i}}{p_0 - \mathbf{p}^2/(2m) + \mathrm{i}0} \,. \tag{7.55}$$

The antiquark propagator differs in the overall sign and the sign of p_0. The other Feynman rules of NRQCD can be found in [231, 164, 25]. These rules are determined by the NRQCD Lagrangian [231, 164, 25]

$$\mathcal{L}_{\mathrm{NRQCD}} = \mathcal{L}_{\mathrm{heavy}} + \mathcal{L}_{\mathrm{light}} \,, \tag{7.56}$$

$$\mathcal{L}_{\mathrm{light}} = \sum_i \bar{q}_i \mathrm{i}\not{D} q_i - \frac{1}{4}G^a_{\mu\nu}G^{a\mu\nu} + \mathcal{L}^{\mathrm{GF}} + \mathcal{L}^{\mathrm{FP}} \,, \tag{7.57}$$

$$\begin{aligned}\mathcal{L}_{\text{heavy}} &= \psi^\dagger \left(\mathrm{i}D^0 + \frac{\mathbf{D}^2}{2m} \right) \psi + \frac{1}{8m^3}\, \psi^\dagger \mathbf{D}^4 \psi \\ &\quad - \frac{d_1\, g}{2m} \psi^\dagger \sigma \cdot \mathbf{B} \psi + \frac{d_2\, g}{8m^2} \psi^\dagger \left(\mathbf{D} \cdot \mathbf{E} - \mathbf{E} \cdot \mathbf{D} \right) \psi \\ &\quad + \frac{d_3\, \mathrm{i}g}{8m^2} \psi^\dagger \sigma \cdot \left(\mathbf{D} \times \mathbf{E} - \mathbf{E} \times \mathbf{D} \right) \psi + \dots\,, \end{aligned} \tag{7.58}$$

where the σ_i are the Pauli matrices. Colour indices are omitted in the fermion fields, in the gluon potential $A^a_\mu \equiv A^a_\mu t^a$ and in the electric and magnetic fields $\mathbf{E} \equiv \mathbf{E}^a_\mu t^a$, $\mathbf{B} \equiv \mathbf{B}^a_\mu t^a$, where t^a is the SU(3) generator.

The light part describes gluons and light (massless) quarks exactly as in QCD (2.15). In the heavy part, the bispinor ψ for the heavy-quark field is decoupled, $\psi \to (\psi, \chi)$, and consists of non-relativistic two-spinors ψ and χ describing quarks and antiquarks in the non-relativistic limit. The ellipsis stands for similar antiquark terms bilinear in χ and $\chi^\dagger$, four-fermion operators, and operators with higher dimensions. The quark propagator is determined by the first term and a part of the second term, $\psi^\dagger \left[\mathrm{i}D^0 + \partial^2/(2m) \right] \psi$, in the heavy Lagrangian.

The coefficients $d_1, d_2, d_3, \dots$ are determined by matching conditions which relate the diagrams of QCD and NRQCD evaluated at threshold. Therefore the matching calculations reduce to the evaluation of the hard contributions (more precisely, (h–h–…–h) for a general number of loops) to the diagrammatic threshold expansions of QCD diagrams.

When one transforms to NRQCD from QCD, the composite QCD operators are written in a way similar to the terms of the effective Lagrangian. For example, the vector current is represented as

$$\bar{\psi} \gamma_i \psi = C_0 \psi^\dagger \sigma_i \chi + \frac{C_1}{6m^2} \psi^\dagger \mathbf{D}^2 \sigma_i \chi + \dots\,, \tag{7.59}$$

where the matching coefficients $C_0, C_1, \dots$ are determined similarly to the matching coefficients in the effective Lagrangian, and terms are omitted starting from order v^4.

The matching coefficients include information about the scale of the mass of the heavy quark m (distances of order $1/m$), which corresponds to the relativistic physics of annihilation and production of quarks. NRQCD is designed for the description of phenomena that happen at lower energies (larger distances), i.e. dynamics of bound states (mesons) composed of a quark and antiquark. The velocities of the quark and antiquark, v, are small so that there are three well-separated scales in the problem: m, mv and mv^2, the first two of which have already been separated in the transition to NRQCD. The fact that the last two, a typical momentum and a typical energy, are not yet separated corresponds exactly to the absence of homogeneity of the diagrammatic expansion (7.51).

To continue the process of factorization of the scales, let us come back to the diagrammatic level and distinguish now the soft, potential and ultrasoft regions. The soft region is first in this hierarchy. As in the case of the hard regions, the soft contributions are generated by 1PI subgraphs all of whose lines consist of soft momenta. The corresponding Feynman integral is expanded into a Taylor series with respect to its ultrasoft external momenta but, if an external momentum is potential, the Taylor expansion is performed only in its time component, which has order v^2, in contrast to its vector components, which have the same order, v, as the soft loop momentum. Therefore the soft contributions are described, at the operator level, by insertion into the effective Lagrangian of spatially non-local four-fermion operators, which, on the other hand, are naturally treated as parts of the non-relativistic potential that is used to describe the interaction of quarks by means of quantum mechanics with the Schrödinger equation.

Ultrasoft lines are typically connected by a vertex with two potential lines – see Examples 7.4, 7.5 and 7.6. Let the momentum of an ultrasoft line be l and the momentum of one of the incident quark lines be $k+l$. Then, according to the rules formulated above, the quark–gluon vertex and the quark propagator have to be expanded in the ratio $|\mathbf{l}|/|\mathbf{k}|$, which is of order $v \sim \sqrt{y}$. This procedure corresponds to a multipole expansion.

After all the relevant regions have been distinguished, the soft quarks and gluons, as well as potential gluons, are 'integrated out', in the above sense, and the resulting effective theory, called potential NRQCD (PNRQCD), includes only potential quarks and ultrasoft gluons. The Lagrangian of PNRQCD is of the form[5]

$$
\begin{aligned}
\mathcal{L}_{\mathrm{PNRQCD}} = \psi^\dagger & \left(\mathrm{i}\partial^0 + \frac{\partial^2}{2m} + g_s A^0(t,\mathbf{0}) - g_s x^i E^i(t,\mathbf{0}) \right) \psi \\
& + \chi^\dagger \left(\mathrm{i}\partial^0 - \frac{\partial^2}{2m} + g_s A^0(t,0) - g_s x^i E^i(t,0) \right) \chi \\
& + \int \mathrm{d}^{d-1}\mathbf{x} \left[\psi^\dagger\psi\right](r) \left(-\frac{C_{\mathrm{F}}\alpha_{\mathrm{s}}}{r} \right) \left[\chi^\dagger\chi\right](0) + \frac{1}{8m^3}\psi^\dagger\partial^4\psi \\
& - \frac{1}{8m^3}\chi^\dagger\partial^4\chi + \int \mathrm{d}^{d-1}\mathbf{x} \left[\psi^\dagger\psi\right](r)\, \delta V(r) \left[\chi^\dagger\chi\right](0) + \ldots \,, \qquad (7.60)
\end{aligned}
$$

where $r = |\mathbf{x}|$. The terms containing $A^0(t,\mathbf{0})$ and $E^i(t,\mathbf{0})$ have appeared because of the multipole expansion, e.g. $A^0(t,\mathbf{x}) = A^0(t,\mathbf{0}) + \mathbf{x}\cdot\partial A^0(t,\mathbf{0}) + \ldots$ According to the power-counting rules, the first two terms and the term containing the Coulomb potential have the same order, v^5, so that the Coulomb potential cannot be treated as a perturbation. In contrast, the rest of the terms have to be treated perturbatively. Note that the potential $\delta V(r)$ contains both a local and a non-local part.

[5] Such a construction, obtained by tree-level matching, was first proposed in [193]. The term 'potential', when applied to NRQCD, is taken from [193]. In the context of the threshold expansion, which provides a matching prescription for loop graphs, PNRQCD has been discussed in [17]. A somewhat different, but probably conceptually equivalent, construction has been proposed in [172].

Let us note, for completeness, that it is possible introduce different fields for the soft and potential quarks, as well as for the soft, potential and ultrasoft gluons, and implement the threshold expansion at the operator level by introducing individual effective propagators and vertices for each of these particles [130].

7.5 Two-Loop Correction to the Leptonic Decay of Quarkonium

It turns out that a knowledge of the QCD/NRQCD matching coefficients for the vector current in (7.59) provides the possibility of obtaining corrections to the partial decay rate

$$\Gamma(J/\psi \to l^+ l^-) = \frac{4\pi e_{\rm Q}^2 \alpha f_{J/\psi}^2}{3M_{J/\psi}} \tag{7.61}$$

of the J/ψ meson, whose leptonic decay is described by the interaction with the electromagnetic current. In (7.61), the tiny lepton masses are neglected, $M_{J/\psi}$ is the mass of the J/ψ, α is the fine-structure constant and $e_{\rm Q}$ is the electric charge of the heavy quark in units of the electron charge. The decay constant $f_{J/\psi}$ is defined through the following matrix element of the electromagnetic current:

$$\langle \Psi(p)|\bar{\psi}\gamma_\mu\psi|0\rangle = -{\rm i} f_{J/\psi} M_{J/\psi}\, \epsilon_\mu^*(p) \,, \tag{7.62}$$

where Ψ is a one-particle J/ψ state, $\epsilon_\mu(p)$ is the polarization vector and p is its momentum.

The decay constant parameterizes the strong-interaction effects and contains long- and short-distance contributions. According to the previous section, the transition to NRQCD and, in particular, the use of (7.59) makes it possible to factorize effects connected with short distances, i.e. of order $1/M_{J/\psi}$, from the bound-state effects associated with the next scales, $1/(M_{J/\psi}v)$ and $1/(M_{J/\psi}v^2)$. The matching relation (7.59) is inserted between the vacuum state and the one-particle state $\Psi(p)$ on the left-hand side of (7.62). These matrix elements are defined in the J/ψ rest frame, so that an appropriate Lorentz boost into this frame is introduced.

The matching coefficients C_0 and C_1 are calculated perturbatively as series in the strong coupling α_s and account for the short-distance QCD effects. The matching coefficients and matrix elements depend individually on a factorization scale μ in NRQCD. The coefficient C_1 is defined (as in [33]) such that $C_1 = 1 + O(\alpha_s)$ and [12]

$$C_0\left(\alpha_s, m/\mu\right) = 1 - \frac{2C_{\rm F}\,\alpha_s}{\pi} + c_2(m/\mu)\left(\frac{\alpha_s}{\pi}\right)^2 + \ldots \,. \tag{7.63}$$

The two-loop matching coefficient c_2 has been calculated in [84, 16]. Since the matching coefficient contains only short-distance effects, it can be obtained by projecting (7.59) onto the state consisting of a free quark–antiquark pair of on-shell quarks considered at threshold. In terms of this on-shell matrix element, the matching relation can be rewritten as

$$Z_{2,\mathrm{QCD}}\, G_{\mathrm{QCD}} = C_0\, Z_{2,\mathrm{NRQCD}}\, Z_J^{-1}\, G_{\mathrm{NRQCD}} + O(v^2)\,, \tag{7.64}$$

where $Z_{2,\ldots}$ are the on-shell wave function renormalization constants in QCD and NRQCD, and $G_{\ldots}$ are the amputated, bare electromagnetic annihilation vertices in QCD and NRQCD. The two-loop Feynman diagrams for G_{QCD} are shown in Fig. 7.8. Since the current $J = \psi^\dagger \sigma_i \chi$ need not be conserved in NRQCD, its renormalization, $J^{\mathrm{B}} = Z_J J$, is taken into account on the right-hand side.

Let us stress that only the leading (h–h) contributions, i.e. from the diagrams considered exactly at threshold (in the naive sense, under the sign of the integral over the loop momenta), are needed for the calculation of C_0. Even if an explicit analytic result for the unexpanded diagrams was known, it would be not so easy to extract the desired information about the hard part from it, because the contributions from other regions, e.g. the potential region, which are not relevant here, produce constants which mix with the constants in the hard part.

It should be also pointed out that NRQCD has traditionally been introduced and applied with momentum cut-offs. Matching calculations of the

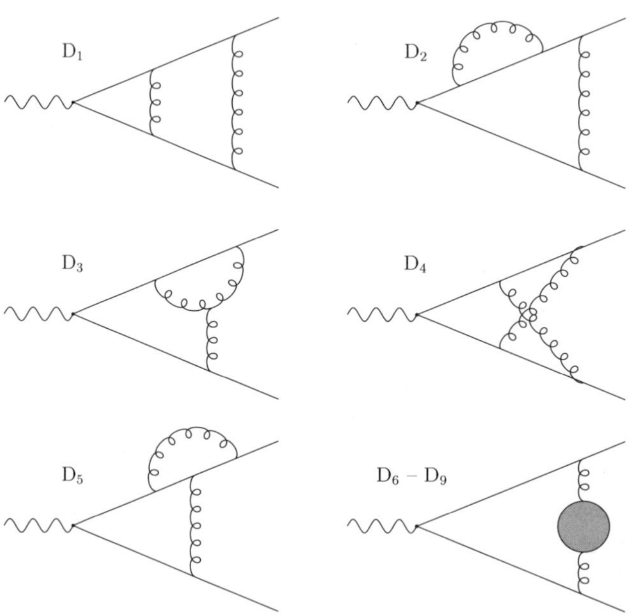

Fig. 7.8. Diagrams that contribute to G_{QCD}. Symmetrically related diagrams exist for $D_{2,3,5}$. The last diagram summarizes vacuum polarization contributions from massless fermions (D_6), gluons (D_7), ghosts (D_8) and a massive fermion with mass m (D_9)

type considered in this section were performed with an additional dimensionful cut-off. In the present approach, within dimensional regularization and without extra cut-offs, the matching calculations are much easier. Furthermore, NRQCD itself is understood in a slightly modified sense: when one writes down the Feynman integrals of NRQCD, according to its Feynman rules, there is an *obligation* to expand every quark propagator by considering its momentum to be either soft or potential and to expand every gluon propagator by considering its momentum to be either soft, potential or ultrasoft. If we do not follow this prescription, we obtain, in particular, non-zero values for NRQCD diagrams at threshold, and the results for the matching coefficients are different.

Let us come back to our matching calculation. The spinor structure of the on-shell matrix element in QCD is conventionally parameterized by two form factors, F_1 and F_2:

$$\mathrm{i}g\bar{u}(p')\left(\gamma_\mu F_1(q^2)+\frac{\mathrm{i}\sigma_{\mu\nu}q^\nu}{2m}F_2(q^2)\right)v(p)\,, \tag{7.65}$$

where $\sigma_{\mu\nu}=(\mathrm{i}/2)\,[\gamma_\mu,\gamma_\nu]$, and $u(p')$ and $v(p)$ denote bispinors for external quarks. Only the combination F_1+F_2 is required for the calculation, which eventually reduces [84, 16] to the integrals (7.31) and (7.50).

After summing up all the diagrams, multiplying by the two-loop QCD on-shell wave function renormalization constant [40] and performing (one-loop) coupling and mass renormalization, the result still contains poles in ϵ. The final result for $c_2(m/\mu)$ in the $\overline{\mathrm{MS}}$ scheme, with separated contributions from the different colour group factors defined by (1.29), takes the following form [84, 16]:

$$c_2(m/\mu)=C_{\mathrm{F}}^2\,c_{2,\mathrm{A}}+C_{\mathrm{F}}C_{\mathrm{A}}\,c_{2,\mathrm{NA}}+C_{\mathrm{F}}T_{\mathrm{F}}\,n_{\mathrm{f}}\,c_{2,\mathrm{L}}+C_{\mathrm{F}}T_{\mathrm{F}}\,c_{2,\mathrm{H}}\,, \tag{7.66}$$

$$\begin{aligned}
c_{2,\mathrm{A}}&=\pi^2\left[\frac{1}{6}\ln\left(\frac{m^2}{\mu^2}\right)-\frac{79}{36}+\ln 2\right]+\frac{23}{8}-\frac{\zeta(3)}{2}\,,\\
c_{2,\mathrm{NA}}&=\pi^2\left[\frac{1}{4}\ln\left(\frac{m^2}{\mu^2}\right)+\frac{89}{144}-\frac{5}{6}\ln 2\right]-\frac{151}{72}-\frac{13\zeta(3)}{4}\,,\\
c_{2,\mathrm{L}}&=\frac{11}{18}\,,\\
c_{2,\mathrm{H}}&=-\frac{2\pi^2}{9}+\frac{22}{9}\,.
\end{aligned} \tag{7.67}$$

Here all fermions with masses less than m are considered massless, which is a good approximation even for $m=m_b$, the bottom quark mass, when the charm quark mass m_c is neglected. The coefficient $c_{2,\mathrm{L}}$, proportional to n_{f}, the number of light fermions, agrees with [33], and the C_{F}^2 term agrees with [138]. The two-loop correction turns out to be very large. See [84, 16] for phenomenological discussions of this result.

7.6 Top Quark Production near Threshold

The total cross-section for $t\bar{t}$ production is expressed through the correlation function of two vector currents by (4.65) and (4.66), where the vector current for the top quark, with mass $m = m_t$ is now implied.

One complication connected with the case of the top quark is that it is an unstable particle, with a decay $t \to bW$. This fact is taken into account by introducing a width Γ_t, which is of order $m_t\alpha_s^2$.

Another complication typical of threshold calculations is that the dominant interaction between the t quark and its antiquark, described by the colour-singlet Coulomb potential, has to be treated non-perturbatively. This means that it is natural to consider the velocity v to be of the same order as the strong coupling α_s, so that summation of the corresponding series in α_s/v is obligatory. In the language of threshold expansion, this is the contribution from the region where *all* the loop momenta are potential to the ladder diagrams generated by exchange gluons[6] that gives the dominant singularity in v, in every perturbative order of QCD. Equivalently, in NRQCD, one considers the same family of diagrams and applies the Coulomb gauge, $\nabla\cdot\mathbf{A} = 0$, which is very useful in non-relativistic problems, where the leading threshold behaviour comes from the exchange of longitudinal gluons with a propagator $i/\mathbf{p}^2$ so that not only the quark propagators but also the gluon propagator takes exactly the same form as in the (p–p–...–p) QCD contribution. In each integral in this family, one first trivially evaluates all the integrals over the zero components of the loop momenta.

The summation of such ladder diagrams can be effectively performed by solving the Schrödinger equation with the Coulomb potential – see, for example, [191] for a proof. The corresponding equation for the Coulomb Green function is

$$\left(-\frac{\nabla^2}{m} - \frac{C_F\alpha_s}{r} - E\right) G_C(\mathbf{r}, \mathbf{r}'; E) = \delta^{(3)}(\mathbf{r} - \mathbf{r}') \,, \tag{7.68}$$

where $E = \sqrt{s} - 2m_t$. The corresponding equation in momentum space is

$$\left(\frac{\mathbf{p}^2}{m_t} - \bar{E}\right) \tilde{G}_C(\mathbf{p}, \mathbf{p}'; E) + \int \frac{d^{d-1}\mathbf{k}}{(2\pi)^{d-1}} \left(\frac{-4\pi C_F\alpha_s}{\mathbf{k}^2}\right) \tilde{G}_C(\mathbf{p} - \mathbf{k}, \mathbf{p}'; E)$$
$$= (2\pi)^{d-1}\, \delta^{(d-1)}(\mathbf{p} - \mathbf{p}') \,. \tag{7.69}$$

Graphically, the momentum space Coulomb Green function

$$\tilde{G}_C(\mathbf{p}, \mathbf{p}'; E) = (2\pi)^3 \delta^{(3)}(\mathbf{p} - \mathbf{p}')/(E - \mathbf{p}^2/m) + \ldots$$

can be represented by the sum of the ladder diagram shown in Fig. 7.9.

---

[6]The term with one exchange gluon determines the Coulomb potential itself. Although there is no integration over loop momenta, the gluon propagator takes the same form as in the potential contributions because of the kinematics of the present problem.

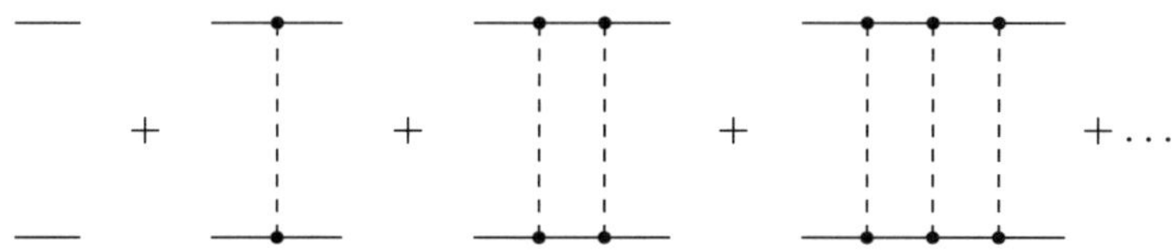

Fig. 7.9. Representation of the Coulomb Green function by a sum of ladder diagrams

So, the summation of the power series in α_s/v reduces to the use of the well-known explicit solution of (7.68). In particular, this is a simple representation of the Coulomb Green function with one argument (in coordinate space) set to zero:

$$G_C(0,r;E) = \frac{-\mathrm{i}m^2 v}{2\pi}\mathrm{e}^{\mathrm{i}mvr}\int_0^\infty \mathrm{d}t\left(1+\frac{1}{t}\right)^\nu \mathrm{e}^{2\mathrm{i}mvrt}\,, \tag{7.70}$$

where $\nu = \mathrm{i}C_F\alpha_s/(2v)$. The Coulomb Green function with two zero coordinate space arguments,

$$G_C(0,0;E) = -\frac{1}{(2\pi)^{d-1}}\int\frac{\mathrm{d}^{d-1}\mathbf{p}}{E-\mathbf{p}^2/m} + \dots\,,$$

is shown graphically in Fig. 7.10. The first two integrals in this series are UV divergent and lead to poles in ε:

$$G_C(0,0;E) = -\frac{m^2\alpha_s}{4\pi}\left\{\mathrm{i}v - C_F\alpha_s\left[\frac{1}{2}\ln\left(\frac{-mE}{\mu^2}\right) - \frac{1}{4\varepsilon} - \frac{1}{2}\right.\right.$$
$$\left.\left. + \ln 2 + \gamma_E + \psi\left(1-\mathrm{i}\frac{C_F\alpha_s}{2v}\right)\right]\right\}\,. \tag{7.71}$$

In fact, the pole does not involve an imaginary part and is irrelevant to the calculation of the leading order (LO) and the next-to-leading order (NLO) cross-sections but it is important in the next-to-next-to-leading order (NNLO) approximation because there are corrections containing $G_C(0,0,E)^2$.

In the double expansion in α_s and v, with the power-counting prescription $v \sim \alpha_s$, the ratio $R(s)$ near threshold is represented by

$$R_{\mathrm{LO}} + R_{\mathrm{NLO}} + R_{\mathrm{NNLO}} + \dots\,,$$

where the LO cross-section in the resummed form is

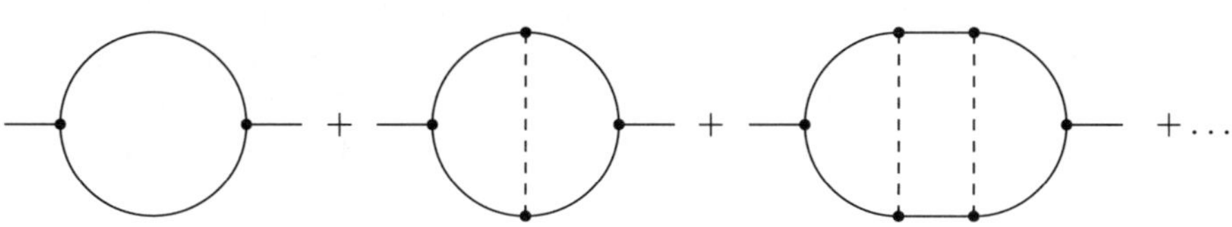

Fig. 7.10. Representation of the Coulomb Green function with two zero coordinate space arguments by a sum of two-point ladder diagrams

$$R_{\rm LO} = \frac{6\pi}{m^2} N e_{\rm Q}^2 \, {\rm Im}\, G_{\rm C}(0,0;E) \; . \tag{7.72}$$

Here N is the number of colours and $e_{\rm Q}$ is the charge of the top quark.

In particular, for $E \equiv s - 2m > 0$, the explicit formula (7.71) gives an expression containing the well-known Sommerfeld–Sakharov factor [218, 199]

$$R_{\rm LO} = \frac{3}{2} N e_{\rm Q}^2 \frac{C_{\rm F} \pi \alpha_{\rm s}}{1 - {\rm e}^{-C_{\rm F}\pi\alpha_{\rm s}/v}} \; . \tag{7.73}$$

When $\alpha_{\rm s} \to 0$, the leading threshold behaviour ($\sim v$) in (4.68) is reproduced.

The LO result (7.72) has been derived in [104] for the case of the top quark under the assumption of a large width (as compared with $\Lambda_{\rm QCD}$), and conclusions about the reliability of perturbative QCD calculations for $R_{\rm LO}$ were drawn. Formally, this result is similar to (7.72), with $E \to E + {\rm i}\Gamma_t$.

The NLO and NNLO contributions to the cross-section have the form

$$R_{\rm NLO} = v \left(R^1_{\rm NLO} \alpha_{\rm s} + R^2_{\rm NLO} v \right) \; , \tag{7.74}$$

$$R_{\rm NNLO} = v \left(R^{11}_{\rm NNLO} \alpha_{\rm s}^2 + R^{12}_{\rm NNLO} \alpha_{\rm s} v + R^{22}_{\rm NNLO} v^2 \right) \; , \tag{7.75}$$

where the functions involved are series of the form $\sum a_n (\alpha_{\rm s}/v)^n$, which are summed by solving the Schrödinger equation. The NLO contribution has been calculated by taking into account the one-loop matching coefficients in (7.59) [151] and the one-loop corrections to the Coulomb potential [108, 22].

The NNLO evaluation is much more non-trivial, not only because it is of the next order but also because complications connected with UV divergences that appear owing to additional terms in the potential in the Schrödinger equation. Nevertheless, one uses again as the input two similar pieces of computationally non-trivial information, which are now of two-loop character: the two-loop matching coefficients in (7.59) given by (7.67) and the two-loop corrections to the Coulomb potential [192, 201] (see below). Similar calculations have been performed by several groups[7] [142, 177, 244, 17, 180, 189], which all used that information. The approach of [17] made maximal use of the strategy of constructing effective field theories by expansion by regions described above. Here is a sketch of main points of this calculation.

The vector currents involved in the polarization function were represented by the matching relations (7.59) and then the correlation function of the non-relativistic currents was computed with the effective Lagrangian (7.58), with a further separation of the scales and an immediate transition to PNRQCD. The presence of the width was taken into account by including the term ${\rm i}\Gamma_t$ in the first expression in parentheses in (7.58) and (7.60). After carrying out a colour and spin projection onto the components relevant to the calculation of the cross-section, the authors of [17] obtained the following result for the heavy-quark potential by matching the on-shell $t\bar{t}$ scattering amplitudes in NRQCD and PNRQCD:

[7] See summarizing publication [141].

$$\tilde{V}(\mathbf{p},\mathbf{q}) = -\frac{4\pi C_{\mathrm{F}}\alpha_s}{\mathbf{q}^2} + \delta\tilde{V}(\mathbf{p},\mathbf{q}) \,, \tag{7.76}$$

$$\begin{aligned}
\delta\tilde{V}(\mathbf{p},\mathbf{q}) = -\frac{4\pi C_{\mathrm{F}}\alpha_s}{\mathbf{q}^2} \Bigg[& \left(a_1 - b_0 \ln\frac{\mathbf{q}^2}{\mu^2}\right)\frac{\alpha_s}{4\pi} \\
& + \left(a_2 - (2a_1 b_0 + b_1)\ln\frac{\mathbf{q}^2}{\mu^2} + b_0^2 \ln^2\frac{\mathbf{q}^2}{\mu^2}\right)\frac{\alpha_s^2}{(4\pi)^2} \\
+\frac{\pi\alpha_s|\mathbf{q}|^{1-2\varepsilon}e^{\gamma_E\varepsilon}\mu^{2\varepsilon}}{4m_t} & \frac{\Gamma(1/2-\varepsilon)^2\Gamma(1/2+\varepsilon)}{\pi^{3/2}\Gamma(1-2\varepsilon)}\left(\frac{C_{\mathrm{F}}}{2}(2\varepsilon-1)+C_{\mathrm{A}}(1-\varepsilon)\right) \\
& +\frac{\mathbf{p}^2}{m_t^2} + \frac{\mathbf{q}^2}{m_t^2}\left(\frac{d^2-7d+10}{4(d-1)}d_1^2 - \frac{1}{4}(1+d_2)\right)\Bigg] \,,
\end{aligned} \tag{7.77}$$

where one can take $d_1 = d_2 = 1$ in the present NNLO approximation, and $b_1 = 102 - 38n_{\mathrm{f}}/3$ is the two-loop coefficient of the QCD β-function. The one- and two-loop corrections to the Coulomb potential are determined by $a_1 = (31C_{\mathrm{A}}/9 - 10n_{\mathrm{f}}/9)$ [108, 22] and

$$\begin{aligned}
a_2 = C_{\mathrm{A}}^2\left(\frac{4343}{162} + \frac{22\zeta(3)}{3} + 4\pi^2 - \frac{\pi^4}{4}\right) - C_{\mathrm{A}}n_{\mathrm{f}}\left(\frac{899}{81} + \frac{28\zeta(3)}{3}\right) \\
- C_{\mathrm{F}}n_{\mathrm{f}}\left(\frac{55}{6} - 8\zeta(3)\right) + \frac{100n_{\mathrm{f}}^2}{81}
\end{aligned} \tag{7.78}$$

[201], respectively.

This result is gauge invariant. To achieve this property it was necessary to combine the contribution from the soft modes with the contribution from potential gluons. (This necessity is connected with the fact, pointed out in Sects. 7.1 and 7.2, that the soft and potential contributions generally mix together.) An 'RG-improved' version of the two-loop matching coefficients (7.67) was used. This allows effective summation of the logarithms of v.

Equation (7.77) defines a dimensionally regularized version of the potential, which is needed here because the terms in the last two lines generate UV divergences. The presence of these divergences leads to a factorization scale dependence, which cancels with the factorization scale dependence in the coefficient function (7.67) of the non-relativistic current. The Coulomb potential itself does not generate divergences, so that $a_{1,2}$ can be evaluated in four dimensions.

The potential (7.77) can be treated perturbatively. In the present NNLO approximation, one needs the following typical integrals:

$$I_1[\tilde{V}_1] = \int \prod_{i=1}^{4} \frac{\mathrm{d}^{d-1}\mathbf{p}_i}{(2\pi)^{d-1}} \tilde{G}_{\mathrm{C}}(\mathbf{p}_1,\mathbf{p}_2)\tilde{V}_1(\mathbf{p}_2-\mathbf{p}_3)\tilde{G}_{\mathrm{C}}(\mathbf{p}_3,\mathbf{p}_4) \,, \tag{7.79}$$

$$\begin{aligned}
I_2[\tilde{V}_2,\tilde{V}_2'] = \int \prod_{i=1}^{6} \frac{\mathrm{d}^{d-1}\mathbf{p}_i}{(2\pi)^{d-1}} & \tilde{G}_{\mathrm{C}}(\mathbf{p}_1,\mathbf{p}_2)\tilde{V}_2(\mathbf{p}_2-\mathbf{p}_3) \\
& \times\tilde{G}_{\mathrm{C}}(\mathbf{p}_3,\mathbf{p}_4)\tilde{V}_2'(\mathbf{p}_4-\mathbf{p}_5)\tilde{G}_{\mathrm{C}}(\mathbf{p}_5,\mathbf{p}_6) \,.
\end{aligned} \tag{7.80}$$

Here the $\tilde{V}_i$ are various contributions to the potential (7.77): $\tilde{V}_1$ can be equal to $(\ln^j q^2/\mu^2)/q^2$, $j = 0, 1, 2$, or to $1/q^{1+2\varepsilon}$ or just to a constant, and $\tilde{V}_2 = \tilde{V}_2' = (\ln q^2/\mu^2)/q^2$. By definition, these integrals are understood as dimensionally regularized quantities. Moreover, in $\tilde{V}_1 = 1/q^{1+2\varepsilon}$, the ε dependence of the exponent is kept because the corresponding integrals have stronger UV divergences. For various pieces of the potential (7.77), one needs either one or two iterations (given by (7.79) and (7.80), respectively).

The NNLO corrections have been found [142, 177, 244, 180, 189, 17] to be much larger than those expected from the known NLO calculations, with minor differences between results of these groups. Owing to the large top width, the total cross-section is a smooth function of the energy, with a clearly distinguished maximum corresponding to a toponium 1S resonance. The position of this peak and its shape and height can be used to determine the top quark mass and Γ_t, as well as some other parameters. A detailed phenomenological discussion of the NNLO results is beyond the scope of the book. (See, e.g., [141].)

The evaluation of the next ($\mathrm{N^3LO}$) corrections is a hard problem but the first results have been already appeared. In [152], the leading retardation contributions (which are absent in the NNLO and NLO approximations), i.e. those from the chromoelectric dipole interaction of the heavy quarkonium with virtual ultrasoft gluons, have been obtained, and the corresponding shifts of the quarkonium energy levels and the wave functions at the origin have been calculated. Another group of applications of the threshold expansion is related to the case of the b quark, within the QCD sum rules. The main technical features of the corresponding NNLO calculations are almost the same as for the top quark – see [188, 139, 178, 140, 15].

The threshold expansion can be applied not only in QED and QCD but also in other cases – see, e.g., [121], where an effective non-relativistic theory that describes bound states of $\pi^+\pi^-$ pairs and their hadronic decays was investigated.

8 Sudakov Limits

In this chapter we investigate the Sudakov limit and related limits. We expand diagrams contributing to the quark or electron form factor and four-point amplitudes – see Fig. 8.1. These diagrams are relevant to a large number of physical processes. In the Sudakov limit for the form factor, the square of the momentum transfer $Q^2 \equiv -(p_1 - p_2)^2$ is large compared with the masses, and, for the four-point amplitudes, the Mandelstam variables $s = (p_1 + p_2)^2$, $t = (p_1 - p_3)^2$ are larger than the masses. In the latter case, one has $t = -s(1 - \cos\theta)/2$ in the centre-of-mass frame, so that this limit is connected with scattering at fixed angles.

Fig. 8.1. (**a**) Form factor. (**b**) Four-point amplitude

In the Regge limit, one has $|t| \ll |s|$, and this corresponds to the scattering at small angles. In the crossed channel, this corresponds to backward scattering, with $|u| \ll |s|$ and $u = (p_1 - p_4)^2 = -s(1 + \cos\theta)/2$.

We start from one-loop examples, as usual. Then, as in Chap. 7, we continue with two-loop examples because not all relevant regions arise at the one-loop level. After we have discovered all the typical regions, we present some general prescriptions. Finally, we describe how the leading and subleading logarithms in the Sudakov limit are summed up by use of evolution equations [73, 75, 203, 204, 155] that govern the dynamics of the amplitudes in the Sudakov limit. We shall consider, as examples, the Abelian form factor in the SU(N) gauge theory and the four-fermion amplitude. This analysis is applied to the fermion annihilation process $f'\bar{f}' \to f\bar{f}$ at high energies. The chapter is concluded by summing up the leading logarithms in the Regge limit.

8.1 List of Limits

There are three standard variants of the Sudakov limit $s = (p_1 - p_2)^2 \to -\infty$ [226, 150] for vertex diagrams:

(*i*) On-shell massless fermions, $p_1^2 = p_2^2 = 0$, and gauge bosons with a small non-zero mass, $m^2 \ll -s \equiv Q^2$. Let us choose, for convenience,

$$p_{1,2} = (Q/2, 0, 0, \mp Q/2) \tag{8.1}$$

so that $2p_1 \cdot p_2 = Q^2$.

(*ii*) Massless gauge bosons and off-shell massless fermions $p_1^2 = p_2^2 = -M^2$, $M^2 \ll -s$. We choose

$$p_{1,2} = \tilde{p}_{1,2} - \frac{M^2}{Q^2} \tilde{p}_{2,1} \,, \tag{8.2}$$

where $\tilde{p}_{1,2}$ are defined as $p_{1,2}$ in Limit (*i*). We have $s = -(1 + M^2/Q^2)^2 Q^2$ and $2p_1 \cdot p_2 = (1 + M^4/Q^4)Q^2$. It is more convenient to expand Feynman diagrams in the ratio M^2/Q^2 rather than $M^2/(-s)$. However, in the leading-power approximation, this makes no difference, so that $Q^2 \sim -s \sim 2p_1 \cdot p_2$.

(*iii*) Massless gauge bosons and on-shell massive fermions $p_1^2 = p_2^2 = m^2 \ll -s$. We choose $p_{1,2}$ to be as in (8.2) with M^2 replaced by $-m^2$.

A limit closely related to the Sudakov limit is the following:

(*iv*) Massless gauge bosons and on-shell fermions of two types, with a small and a large mass, $p_1^2 = M^2$, $p_2^2 = m^2$ and $q^2 = 0$, with $m \ll M$.

In fact, this is a particular case, with $q^2 = 0$, of the limit with general q^2 of order M^2 (but $q^2 \neq M^2$), which has a clear phenomenological relevance: it applies to a decay of a heavy particle into a light particle (for example, muon decay). It turns out that the structure of the asymptotic expansion in Limit (*iv*) is the same as in this general limit, and hence we consider this particular case for simplicity.

The Sudakov limits (*i*)–(*iii*) are defined for four-point diagrams, with both Mandelstam variables t and s of the same order and $m^2 \ll |s|, |t|$ (or $M^2 \ll |s|, |t|$).

The Regge limit for the four-point amplitudes, with zero masses, can be formulated as the following limit:

(*v*) $t \ll |s|$.

Another possible variant is to consider $t = 0$ and expand in the ratio m/s.

8.2 One-Loop Examples: Triangles

Consider first the following example.

Example 8.1. The scalar triangle diagram of Fig. 8.2a in Limit (*i*).

Choosing the loop momentum to be the momentum of the massive line, we have

$$F_{8.1}(Q^2, m^2; d) = \int \frac{\mathrm{d}^d k}{(k^2 - 2p_1 \cdot k)(k^2 - 2p_2 \cdot k)(k^2 - m^2)}$$
$$= \int \frac{\mathrm{d}k_+ \, \mathrm{d}k_- \, \mathrm{d}^{d-2}\underline{k}}{(k_+ k_- - \underline{k}^2 - Qk_+)(k_+ k_- - \underline{k}^2 - Qk_-)(k_+ k_- - \underline{k}^2 - m^2)} , \quad (8.3)$$

where we have introduced the light-cone coordinates $k_\pm = k_0 \pm k_3$, $\underline{k} = (k_1, k_2)$, where $2p_{1,2} \cdot k = Qk_\pm$.

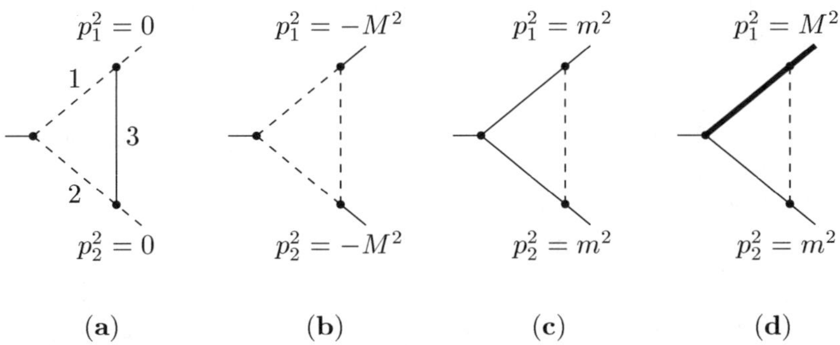

Fig. 8.2a-d. Triangle diagrams in Limits (*i*)–(*iv*)

The diagram is finite at $d = 4$ but we introduce dimensional regularization in advance, as we usually do. An expansion for $m \to 0$ performed naively in the integrand gives the wrong result because it generates IR and collinear divergences starting from the zero-order term. To expand this diagram, let us follow the strategy of expansion by regions. The hard region, $k \sim Q$, gives the expansion of the integrand into a Taylor series in m. The corresponding integrals can be evaluated by means of (A.28):

$$F_{8.1}^{(\mathrm{h})} = \sum_{n=0}^{\infty} (m^2)^n \int \frac{\mathrm{d}^d k}{(k^2 - 2p_1 \cdot k)(k^2 - 2p_2 \cdot k)(k^2)^{n+1}}$$
$$= \mathrm{i}\pi^{d/2} \frac{\Gamma(\varepsilon)\Gamma(1-\varepsilon)}{(Q^2)^{1+\varepsilon}} \sum_{n=0}^{\infty} \frac{\Gamma(-n-\varepsilon)}{\Gamma(1-n-2\varepsilon)} \left(\frac{m^2}{Q^2}\right)^n$$
$$= -\mathrm{i}\pi^{d/2} \frac{1}{(Q^2)^{1+\varepsilon}} \frac{\Gamma(1+\varepsilon)\Gamma(-\varepsilon)^2}{\Gamma(1-2\varepsilon)} + \ldots . \quad (8.4)$$

Let us look for other non-trivial contributions. The soft region, $k \sim m$, gives, in the leading order,

$$F_{8.1}^{(\mathrm{s}),\,\mathrm{LO}} = \int \frac{\mathrm{d}^d k}{(-2p_1\cdot k)(-2p_2\cdot k)(k^2-m^2)}$$
$$= \frac{1}{2Q^2}\int_{-\infty}^{\infty} \mathrm{d}k_+\,\mathrm{d}k_- \int \frac{\mathrm{d}^{d-2}\underline{k}}{(k_+-\mathrm{i}0)(k_--\mathrm{i}0)(k^2-m^2+\mathrm{i}0)}\,. \tag{8.5}$$

Closing the contour of integration over k_+ in the upper half-plane of the complex variable k_+ (only in the case $k_- > 0$, where there is another pole in the lower half-plane; otherwise, both poles would be in the upper half-plane and we would obtain a zero result by closing the contour in the lower half-plane) and taking the residue at $k_+ = 0$, we have

$$F_{8.1}^{(\mathrm{s}),\,\mathrm{LO}} = \frac{\pi\mathrm{i}}{Q^2}\int_0^{\infty}\frac{\mathrm{d}k_-}{k_-}\int\frac{\mathrm{d}^{d-2}\underline{k}}{\underline{k}^2+m^2}\,. \tag{8.6}$$

According to our convention, we set the above integral over k_- to zero because it is an integral without scale. Note that this integral is not regularized by dimensional regularization; this has already happened with integrals occurring in the threshold expansion in Chap. 7. All other terms of the soft contribution vanish for the same reason.

The ultrasoft region, $k \sim m^2/Q$, has even less chance of giving a non-zero contribution because in this case we have to neglect k^2 with respect to m^2 and obtain, in the leading order, the integral

$$F_{8.1}^{(\mathrm{us}),\,\mathrm{LO}} = \frac{1}{m^2}\int\frac{\mathrm{d}^d k}{(-2p_1\cdot k)(-2p_2\cdot k)}\,, \tag{8.7}$$

which is immediately set to zero because it is an integral of a homogeneous function.

It looks reasonable to pay attention to regions where k^2 is of order m^2 and the last propagator is not expanded. Let us consider the following pair of regions:[1]

$$\textit{1-collinear}\ (1\mathrm{c}),\quad k_+\sim m^2/Q,\ k_-\sim Q\,,\ \underline{k}\sim m\,; \tag{8.8}$$
$$\textit{2-collinear}\ (2\mathrm{c}),\quad k_+\sim Q,\ k_-\sim m^2/Q\,,\ \underline{k}\sim m\,. \tag{8.9}$$

When $m \to 0$, d-vectors from the (1c) and (2c) region become collinear with p_1 and p_2, respectively. In both regions, k^2 is indeed of order m^2. In the (1c) region, the first propagator is not expanded, while the second one is expanded in k^2:

$$\frac{1}{k^2-2p_2\cdot k} \to \sum_{n=0}^{\infty}\frac{(k^2)^n}{(-2p_2\cdot k)^{n+1}}\,.$$

A similar expansion should be performed in the (2c) region. It turns out, however, that, when taken alone, the (1c) and (2c) contributions contain divergences that are not dimensionally regularized. To overcome this obstacle,

[1] Collinear regions have been introduced within the 'standard' method based on regions [225, 166, 179] and momentum cut-offs.

we introduce an auxiliary analytic regularization by considering the powers of the first two propagators to be $1+\lambda_1$ and $1+\lambda_2$, with $\lambda_1 \neq \lambda_2$, evaluate both contributions and then take the limit $\lambda_2 \to \lambda_1 \to 0$ in the sum of these contributions. Observing that we can replace factors of k^2 by factors of m^2 in the numerators (because cancelling the denominator $k^2 - m^2$ produces scaleless integrals) and using (A.30), we obtain

$$
\begin{aligned}
F_{8.1}^{(1c)}(\lambda_1,\lambda_2) &= -\frac{1}{\Gamma(1+\lambda_2)} \sum_{n=0}^{\infty} \frac{\Gamma(1+\lambda_2+n)}{n!} (m^2)^n \\
&\quad \times \int \frac{\mathrm{d}^d k}{(-k^2+2p_1\cdot k)^{1+\lambda_1} (2p_2\cdot k)^{n+1+\lambda_2} (m^2-k^2)} \\
&= -\mathrm{i}\pi^{d/2} \frac{\Gamma(\lambda_1+\varepsilon)\Gamma(1-\lambda_1-\varepsilon)}{\Gamma(1+\lambda_1)\Gamma(1+\lambda_2)(m^2)^{\lambda_1+\varepsilon}(Q^2)^{1+\lambda_2}} \\
&\quad \times \sum_{n=0}^{\infty} \frac{\Gamma(\lambda_1-\lambda_2-n)\Gamma(1+\lambda_2+n)}{\Gamma(1-\lambda_2-n-\varepsilon)} \left(\frac{m^2}{Q^2}\right)^n \\
&= -\mathrm{i}\pi^{d/2} \frac{\Gamma(\lambda_1-\lambda_2)\Gamma(\lambda_1+\varepsilon)\Gamma(1-\lambda_1-\varepsilon)}{\Gamma(1+\lambda_1)\Gamma(1-\lambda_2-\varepsilon)(m^2)^{\lambda_1+\varepsilon}(Q^2)^{1+\lambda_2}} + \ldots . \qquad (8.10)
\end{aligned}
$$

The (2c) contribution is obtained from (8.10) by the permutation $\lambda_1 \leftrightarrow \lambda_2$. Now we observe that the pole in $\lambda_1 - \lambda_2$ drops out in the sum of the two collinear contributions and, in the limit $\lambda_2 \to \lambda_1 \to 0$, we obtain

$$
\begin{aligned}
F_{8.1}^{(c)} &= \lim_{\lambda_2 \to \lambda_1 \to 0} \left(F_{8.1}^{(1c)}(\lambda_1,\lambda_2) + F_{8.1}^{(2c)}(\lambda_1,\lambda_2) \right) \\
&= -\mathrm{i}\pi^{d/2} \frac{1}{Q^2 (m^2)^{\varepsilon}} \Gamma(\varepsilon)\Gamma(1-\varepsilon) \sum_{n=0}^{\infty} \frac{(-1)^n}{n!\,\Gamma(1-n-\varepsilon)} \left(\frac{m^2}{Q^2}\right)^n \\
&\times \left[\ln(Q^2/m^2) + \psi(\varepsilon) + \psi(n+1) - \psi(1-\varepsilon) - \psi(1-n-\varepsilon) \right] \\
&= -\mathrm{i}\pi^{d/2} \frac{\Gamma(\varepsilon)}{Q^2 (m^2)^{\varepsilon}} \left[\ln(Q^2/m^2) + \psi(\varepsilon) - \gamma_{\mathrm{E}} - 2\psi(1-\varepsilon) \right] + \ldots . \qquad (8.11)
\end{aligned}
$$

We fail in further attempts to find a region with a non-zero contributions and are left with (8.4) and (8.11). For example, if we choose the loop momentum to be the momentum of the first or the second line, we find no new non-trivial contributions but the description of the previous regions is more complicated. Still, it is useful to think of different choices of the loop momenta – we have seen this in the previous chapters and shall see this again later.

Although the initial quantity is UV and IR finite, we have poles in ε in the individual contributions. Now, the poles are up to the second order, and there are not only UV and IR but also collinear divergences on the right-hand side. Nevertheless, the poles are successfully cancelled and, in the limit $\varepsilon \to 0$, the sum of the (h) and (c) contributions provides, after summing up the power series, the following result at $\varepsilon = 0$:

$$F_{8.1} \sim \frac{\mathrm{i}\pi^2}{Q^2}\left(\mathrm{Li}_2(x) - \frac{1}{2}\ln^2 x + \ln x \ln(1-x) - \frac{\pi^2}{3}\right) , \tag{8.12}$$

where $\mathrm{Li}_2(x)$ is the dilogarithm (see (A.57)) and $x = m^2/Q^2$. This result coincides with an explicit result which can be obtained, for example, from an expression for a triangle with general squares of the external momenta [30, 89]. This means that we have indeed missed no other contributions. We can, however, check the result obtained by an alternative approach based on the Mellin–Barnes representation. Using (2.74) for the massive propagator and evaluating the resulting momentum integral by means of (A.28), we have

$$F_{8.1} = -\frac{\mathrm{i}\pi^{d/2}}{(Q^2)^{1+\varepsilon}}$$
$$\times \frac{1}{2\pi\mathrm{i}}\int_{-\mathrm{i}\infty}^{+\mathrm{i}\infty} \mathrm{d}z \left(\frac{m^2}{Q^2}\right)^z \frac{\Gamma(1+z)\Gamma(1+\varepsilon+z)\Gamma(-\varepsilon-z)^2\Gamma(-z)}{\Gamma(1-2\varepsilon-z)} . \tag{8.13}$$

Setting $\varepsilon = 0$, closing the integration contour on the right and taking a series of residues at the points $z_n = 0, 1, \ldots$, we reproduce (8.12). But before we set ε to zero we can distinguish poles at $z_n^{(\mathrm{h})} = n$ and $z_n^{(\mathrm{c})} = -\varepsilon + n$, which correspond exactly to the hard and collinear contributions.

It appears, in this concrete example, that the Mellin–Barnes technique is simpler than expansion by regions. In particular, we encounter no trouble at all with dimensional regularization in (8.13). (To be more precise, the presence of double poles at $z = z_n^{(\mathrm{c})}$ is connected with those troubles which we had with the individual (1c) and (2c) contributions and cured by analytic regularization.) But we are aiming at a general strategy of expansion by regions which will work for any diagram with any number of loops by producing a result in terms of products of integrals depending on a smaller number of scales, while the Mellin–Barnes technique is applied to each Feynman integral in its own way; at two loops, the optimal choice of Mellin–Barnes integrations is not obvious and, for three or more loops, the problem becomes, generally, almost impossible to solve by this method.

For Limit (*i*), it is possible to formulate prescriptions within the method of expansion by subgraphs [210, 211]. In this one-loop example, the remainder of the asymptotic expansion can be defined in the usual way, as $\mathcal{R}^a F_\Gamma$, where

$$\mathcal{R}^a = (1 - \mathcal{M}_0^a)(1 - \mathcal{M}_1^a - \mathcal{M}_2^a) \tag{8.14}$$

and the pre-subtraction operators $\mathcal{M}_i^a$ correspond to the subgraphs $\{1\}$, $\{2\}$ and the graph Γ itself, with corresponding pre-subtraction operators $\mathcal{M}_1^a, \mathcal{M}_2^a$ and $\mathcal{M}_0^a$. The first two of these operators perform expansion in k^2 of the first and the second propagator, respectively, while the operator $\mathcal{M}_0$ expands the integrand in m. The remainder is UV and IR finite, for any a. Its asymptotic behaviour is $(m^2)^{a+1}/(Q^2)^{a+2}$ modulo logarithms. There is an interplay between various divergences: the operator $\mathcal{M}_0$ generates IR and collinear divergences, which are removed by $\mathcal{M}_1$ and $\mathcal{M}_2$. In turn, the operators $\mathcal{M}_{1,2}$

generate UV and collinear divergences, which are removed by $\mathcal{M}_0$. The terms of the expansion are obtained from the Zimmermann identity:

$$1 = (1 - \mathcal{R}) + \mathcal{R} = \mathcal{M}_0 + \mathcal{M}_1 + \mathcal{M}_2 + \ldots \,, \tag{8.15}$$

where we have dropped products of different operators equal to zero and taken the limit $a \to \infty$ (with $\mathcal{M}_i = \mathcal{M}_i^{\infty}$).

Although the expansion of the triangle diagram in Example 8.1 consists of hard and collinear contributions and does not involve a soft contribution, collinear and soft contributions can sometimes transform into each other. To illustrate this phenomenon, let us consider the Feynman integral $F_{8.1;\mu}$, with an additional factor k_μ in the numerator. Its expansion can be obtained as before and consists of hard and collinear contributions. Let us, however, calculate the leading order of the expansion in another way. The integral is represented as $(p_1 + p_2)_\mu C(q^2, m^2)$, where C is a scalar function which equals $1/q^2$ times an integral with the same denominators as in (8.3) and a numerator $-2p_2 \cdot k \equiv -(k^2 - m^2) + (k^2 - 2p_2 \cdot k) - m^2$. The third term here gives a contribution that starts only from the next-to-leading power, m^2. The first term, $-(k^2 - m^2)$, cancels one of the factors in the denominator and produces exactly the leading order of the hard contribution (8.4). Finally, the second term, $(k^2 - 2p_2 \cdot k)$, cancels another factor in the denominator and produces the integral

$$\int \frac{\mathrm{d}^d k}{(k^2 - 2p_1 \cdot k)(k^2 - m^2)} \,,$$

which depends only on one scale, m^2. Of course, if we dealt with this integral from the beginning, it would be useless to think about its expansion because there would be no expansion parameter. But we obtain such an integral in the problem of expansion when $m^2/q^2 \to 0$ when we have to think about expansion of all the quantities involved. In the method of expansion by regions, only the contribution of the soft region, $k \sim m$, is non-zero, and we observe that the initial collinear contributions have been transformed into soft ones.

Now we turn to the next example.

Example 8.2. The scalar triangle diagram of Fig. 8.2b in Limit (*ii*).

With the same choice of loop momentum, we have

$$F_{8.2}(Q^2, M^2; d) = \int \frac{\mathrm{d}^d k}{(k^2 - 2p_1 \cdot k - M^2)(k^2 - 2p_2 \cdot k - M^2) k^2} \,, \tag{8.16}$$

where $p_{1,2}$ are given by (8.2).

The (h) region generates a contribution obtained from the following expansion of the first two propagators:

$$\frac{1}{k^2 - 2p_{1,2} \cdot k - M^2} \to \frac{1}{k^2 - 2\tilde{p}_{1,2} \cdot k} + \ldots \,.$$

The leading-order term of the expansion equals the term in the last line of (8.4).

As in the previous example, we have non-zero collinear contributions. The (1c) contribution is obtained by expanding the second propagator as

$$\frac{1}{k^2 - 2p_2\cdot k - M^2} \to \sum_{n=0}^{\infty} \frac{\left[M^2 - k^2 - 2(M^2/Q^2)\tilde{p}_1\cdot k\right]^n}{(-2\tilde{p}_2\cdot k)^{n+1}} \, .$$

The resulting integrals can be evaluated by means of (A.33). We obtain

$$F_{8.2}^{(1c)} = \mathrm{i}\pi^{d/2} \frac{1}{Q^2(M^2)^\varepsilon} \frac{\Gamma(1+\varepsilon)\Gamma(-\varepsilon)^2}{\Gamma(1-2\varepsilon)} + \ldots \, . \tag{8.17}$$

There are no troubles with dimensional regularization, so that the (2c) contribution is identical to (1c).

If we sum up the (h) and (c) contributions, we see that the poles are not cancelled. Indeed, we have forgotten to look at the (us) region, $k \sim M^2/Q$, which generates a simultaneous expansion of the first two propagators:

$$\frac{1}{k^2 - 2p_{1,2}\cdot k - M^2} \to \sum_{n=0}^{\infty} (-1)^n \frac{\left[k^2 + 2(M^2/Q^2)\tilde{p}_2\cdot k\right]^n}{(-2\tilde{p}_1\cdot k - M^2)^{n+1}} \, .$$

Evaluating the resulting integrals by means of (A.32), we obtain

$$F_{8.2}^{(\mathrm{us})} = -\mathrm{i}\pi^{d/2} \frac{1}{(Q^2)^{1-\varepsilon}(M^2)^{2\varepsilon}} \Gamma(1-\varepsilon)\Gamma(\varepsilon)^2 + \ldots \, . \tag{8.18}$$

Then the sum of the (h), (1c), (2c) and (us) contributions gives, at $\varepsilon = 0$,

$$F_{8.2} \sim -\frac{\mathrm{i}\pi^2}{Q^2} \left[\ln^2 x + \frac{\pi^2}{3} + 2x \left(\ln^2 x + 2\ln x + \frac{\pi^2}{3} - 2 \right) \right.$$
$$\left. + x^2 \left(6\ln^2 x + 14\ln x + 2\pi^2 - 11 \right) \right] + \ldots \, , \tag{8.19}$$

where $x = M^2/Q^2$. This expansion is in agreement with an explicit result, which can be derived, for example, from the value of the massless triangle diagram at general values of the external momenta [29].

In the next example we consider an expansion of a diagram which is IR divergent.

Example 8.3. The scalar triangle diagram of Fig. 8.2c in Limit (*iii*).

The Feynman integral is of the form

$$F_{8.3}(Q^2, m^2; d) = \int \frac{\mathrm{d}^d k}{(k^2 - 2p_1\cdot k)(k^2 - 2p_2\cdot k)k^2} \, , \tag{8.20}$$

where $p_{1,2}$ are given by (8.2) with M^2 replaced by $-m^2$. We have already expanded this diagram at threshold – see Example 7.2 in Sect. 7.1.

This time, the expansion consists of contributions from the hard region and two collinear regions. The hard contribution is obtained by the expansion of the first two propagators in m^2. In the leading order, the result is the same as in the previous two examples – see the last line of (8.4).

The (1c) and (2c) contributions are equal to each other and can be successfully regularized by dimensional regularization. The (1c) contribution is obtained by the replacement

$$\frac{1}{k^2 - 2p_2 \cdot k} \to \sum_{n=0}^{\infty} \frac{\left[2(m^2/Q^2)\tilde{p}_1 \cdot k - k^2\right]^n}{(-2\tilde{p}_2 \cdot k)^{n+1}} .$$

After evaluation of the resulting integrals by means of (A.37), we have

$$F_{8.3}^{(1c)} = \mathrm{i}\pi^{d/2} \frac{\Gamma(\varepsilon)}{\varepsilon Q^2 (m^2)^\varepsilon} + \ldots . \tag{8.21}$$

The sum of the hard and collinear contributions gives

$$\begin{aligned} F_{8.3} \sim \frac{\mathrm{i}\pi^{d/2} \mathrm{e}^{-\gamma_{\mathrm{E}}\varepsilon}}{(Q^2)^{1+\varepsilon}} & \left\{ -\ln x \frac{1}{\varepsilon} + \frac{1}{2} \ln^2 x + \frac{\pi^2}{6} \right. \\ & + x \left[(\ln x + 1) \frac{2}{\varepsilon} - \left(\ln^2 x - 2 \ln x + \frac{\pi^2}{3} + 2 \right) \right] \\ & \left. + x^2 \left[-(6 \ln x + 7) \frac{1}{\varepsilon} + \left(3 \ln^2 x - 9 \ln x + \pi^2 + \frac{5}{2} \right) \right] \right\} + \ldots , \end{aligned} \tag{8.22}$$

where $x = m^2/Q^2$. The expansion obtained is in agreement with an explicit result for general ε:

$$F_{8.3} = \mathrm{i}\pi^{d/2} \frac{\Gamma(\varepsilon)}{2(m^2)^{1+\varepsilon}} \, {}_2F_1 \left(1, 1+\varepsilon; \frac{3}{2}; -\frac{Q^2}{4m^2} \right) , \tag{8.23}$$

which can be obtained from (7.14) using (A.55) and can be rewritten, using (A.54), in terms of hypergeometric functions depending on the inverse ratio, m^2/Q^2.

The triangle diagram of Fig. 8.2c with general powers of the propagators can be expanded in Limit *(iii)* in the same way, with (h), (1c) and (2c) contributions. It is interesting to analyse the limiting case where the power of the third propagator tends to zero (or to some negative integer) and the resulting diagram is Fig. 2.1, with two masses m. For this diagram, there is no sense in considering collinear regions because the diagram depends only on one external momentum, q. The expansion in the limit $m^2/q^2 \to 0$ is of Euclidean nature when there are contributions from the region of large/hard momentum (i.e. the contribution of the whole graph) and two contributions corresponding to the regions where the momentum of one of the two lines is small/soft (i.e. the contributions of the two subgraphs) – see Example 4.2. It turns out that when the third line is reduced to a point, the (1c) and (2c) contributions are transformed into the corresponding soft contributions. In particular,

$$F_{8.3}^{(1c)}(Q^2, m^2; d, \lambda) = \int \frac{\mathrm{d}^d k}{(k^2 - 2p_1 \cdot k)(-2\tilde{p}_2 \cdot k)(k^2)^\lambda}$$

$$\to \int \frac{\mathrm{d}^d k}{(k^2 - 2p_1 \cdot k)(-2\tilde{p}_2 \cdot k)} = \int \frac{\mathrm{d}^d k}{(k^2 - m^2)(q^2 - 2\tilde{p}_2 \cdot k)}$$
$$= \frac{1}{q^2} \int \frac{\mathrm{d}^d k}{k^2 - m^2} = F^{(1s)}_{8.3}(Q^2, m^2; d, 0) , \tag{8.24}$$

when $\lambda \to 0$. The symbol (1s) means that the momentum of the first line is considered to be soft. Here the change of variables $k \to k + p_1$ has been used. Observe also that $q^2 - 2\tilde{p}_2 \cdot k$ in the denominator has been replaced by q^2 because the value of the integral is in fact the same both for $\tilde{p}_2 \neq 0$ with $\tilde{p}_2^2 = 0$ and for $\tilde{p}_2 = 0$.

The last triangle example is the following.

Example 8.4. The scalar triangle diagram of Fig. 8.2d in Limit (*iv*).

Let us choose the external momenta as follows: $p_1 = (M, \mathbf{0})$, $p_2 = Mn_1 + (m^2/M)n_2$, where $n_{1,2} = (1/2, 0, 0, \mp 1/2)$. We have $2n_{1,2} \cdot k = k_\pm$ for any d-vector k. We use the same characterization of the regions in terms of the components $k_\pm$ and $\underline{k}$ as in Limit (*iii*), with the substitution $Q \to M$. The Feynman integral can be written exactly as the right-hand side of (8.20) but with another assignment of the external momenta.

In the hard region, the first and the third propagator are not expanded, and the second propagator is expanded as

$$\frac{1}{k^2 - 2p_2 \cdot k} \to \frac{1}{k^2 - 2P_1 \cdot k} + \ldots ,$$

where $P_1 = Mn_1$. The resulting integrals can be evaluated by means of (A.35). In the summed form, we obtain

$$F^{(h)}_{8.4} = \mathrm{i}\pi^{d/2} \frac{\Gamma(\varepsilon)}{2\varepsilon (M^2)^{1+\varepsilon}} \frac{1}{1 - m^2/M^2} . \tag{8.25}$$

The (2c) region does not contribute, because it produces integrals without scale:

$$F^{(2c)}_{8.4} = \int \frac{\mathrm{d}^d k}{(-2P_1 \cdot k)^2 k^2} + \ldots . \tag{8.26}$$

The contribution of the (1c) region is obtained by use of the following prescriptions:

(a) Expand the propagator $1/(k^2 - 2p_1 \cdot k)$ into a Taylor series in k^2.
(b) Expand each resulting term, which is a function of three kinematical invariants, $p_1^2 = M^2, p_2^2 = m^2$ and $2p_1 \cdot p_2 = m^2 + M^2$, into a Taylor series at the point $p_1^2 = 0$ and $2p_1 p_2 = M^2$ (do not touch $p_2^2 = m^2$).

It might seem that we expand in the large mass M, but this is just an illusion! The integrals involved can be calculated by means of (A.37) and we obtain

$$F^{(1c)}_{8.4} = -\mathrm{i}\pi^{d/2} \frac{\Gamma(\varepsilon)}{2\varepsilon M^2 (m^2)^{\varepsilon}} \frac{1}{1 - m^2/M^2} . \tag{8.27}$$

Summing (8.25) and (8.27), we reproduce the explicit result (2.73).

8.3 One-Loop Examples: Boxes

We now turn to box diagrams.

Example 8.5. The massless on-shell box diagram of Fig. 8.3 in Limit (*v*), i.e. with $p_i^2 = 0,\ i = 1,2,3,4$ and $t \ll |s|$.

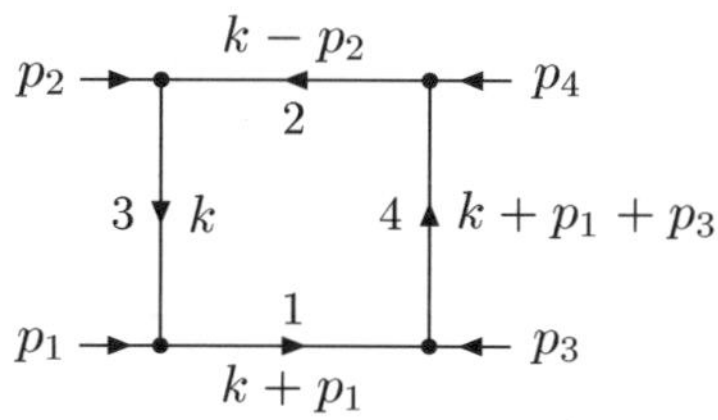

Fig. 8.3. Box diagram

With the loop momentum chosen as in Fig. 8.3, the Feynman integral takes the form

$$F_{8.5}(s,t;d) = \int \frac{\mathrm{d}^d k}{(k^2+2p_1\cdot k)(k^2-2p_2\cdot k)k^2(k+r)^2}\,, \tag{8.28}$$

where $r = p_1 + p_3$. Let us choose the external momenta as follows:

$$p_{1,2} = (\mp Q/2, 0, 0, Q/2),\ r = (T/Q, 0, \sqrt{T + T^2/Q^2}, 0)\,, \tag{8.29}$$

where $s = -Q^2$ and $t = -T$. As in the above examples, the regions that are typical in this limit are hard and collinear. They are defined in the same way, the Mandelstam variable t playing the role of m^2. We have, in particular, the following region:

$$\textit{1-collinear}\ (1c),\quad k_+ \sim T/Q,\ k_- \sim Q\,,\ \underline{k} \sim \sqrt{T}\,. \tag{8.30}$$

The (2c) region is defined by replacing k_+ and k_- (i.e. p_1 and p_2).

The hard region generates a Taylor expansion of the integrand in t but, owing to the kinematics, this is a Taylor expansion in the four-vector $r = p_1 + p_3$. The leading hard term contributes in the next-to-leading-order, $1/s^2$. This term is given by the forward-scattering box, with $p_3 = -p_1$ and $p_4 = -p_2$, and can be evaluated by means of alpha parameters, with the following result:

$$F_{8.5}^{(\mathrm{h}),\,\mathrm{LO}} = \mathrm{i}\pi^{d/2} \frac{\Gamma(-\varepsilon)^2 \Gamma(1+\varepsilon)}{(1+\varepsilon)\Gamma(-2\varepsilon)(-s)^{2+\varepsilon}}\,. \tag{8.31}$$

In the (2c) region, k^2 and $p_2\cdot k$ are of order T, while $p_1\cdot k$ is of order Q^2. Moreover, $(k+r)^2 \equiv k^2 + 2k\cdot r - T \sim (l + \tilde{r})^2$, where

$$\tilde{r} = (T/(2Q), \sqrt{T}, 0, -T/(2Q))\,, \tag{8.32}$$

with $2p_1 \cdot \tilde{r} = 0,\ 2p_2 \cdot \tilde{r} = \tilde{r}^2 = -T$.

Thus the (2c) contribution is obtained by expanding the propagator $1/(k^2 + 2p_1 \cdot k)$ into a Taylor series in k^2, and by expansion also into a

Taylor series in $2p_1 \cdot r$. (Note that we are dealing with a function of three kinematical variables, $2p_1 \cdot r$, $2p_2 \cdot r$ and r^2. So we expand the integrand (e.g. in the alpha representation) in $2p_1 \cdot r$ and then set $2p_1 \cdot r = T$.) Only the leading term in the Taylor expansion in k^2 is non-zero, because, starting from the next order, the factor k^2 cancels the propagator $1/k^2$ and we obtain a zero scaleless integral.

As in Example 8.1, the individual collinear contributions are not regularized dimensionally, so that we introduce an auxiliary analytic regularization by considering the powers of the first two propagators to be $1+\lambda_1$ and $1+\lambda_2$, with $\lambda_1 \neq \lambda_2$, evaluate both contributions and then take the limit $\lambda_2 \to \lambda_1 \to 0$ in the sum of the two contributions. The leading analytically regularized (2c) contribution is then easily evaluated by means of alpha parameters, with the following result:

$$F_{8.5}^{(2c),\,\mathrm{LO}} = \mathrm{i}\pi^d \frac{\Gamma(\lambda_2-\lambda_1)\Gamma(1+\lambda_2+\varepsilon)\Gamma(-\lambda_2-\varepsilon)^2}{\Gamma(1+\lambda_2)\Gamma(-\lambda_1-\lambda_2-2\varepsilon)(Q^2)^{1+\lambda_1}T^{1+\lambda_2+\varepsilon}} \,. \qquad (8.33)$$

The (1c) contribution is obtained by the change $\lambda_1 \leftrightarrow \lambda_2$. Summing up the two collinear contributions and evaluating the next-to-leading contribution in a similar way, we obtain, in the limit $\lambda_{1,2} \to 0$,

$$F_{8.5}^{(c)} = \mathrm{i}\pi^{d/2} \frac{\Gamma(-\varepsilon)^2\Gamma(1+\varepsilon)}{\Gamma(-2\varepsilon)s(-t)^{1+\varepsilon}} \Bigg[\ln(t/s) + \psi(-\varepsilon) - 2\psi(1+\varepsilon) + \gamma_{\mathrm{E}}$$
$$+\varepsilon \frac{t}{s}\left(\ln\frac{t}{s} + \psi(-\varepsilon) - 2\psi(1+\varepsilon) + \gamma_{\mathrm{E}} - \frac{1}{\varepsilon} - 1 \right) + \ldots \Bigg] \,. \qquad (8.34)$$

Combining (8.31) and (8.34) in the limit $\varepsilon \to 0$. we see that, up to the finite part in ε, the hard and collinear terms of the next-to-leading order cancel each other. In fact, this phenomenon takes place in an arbitrary order of the expansion starting from the NLO, so that we are left with only the leading collinear contribution:

$$F_{8.5} = \frac{\mathrm{i}\pi^{d/2}\mathrm{e}^{-\gamma_{\mathrm{E}}\varepsilon}}{st} \left(\frac{4}{\varepsilon^2} - [\ln(-s) + \ln(-t)]\frac{2}{\varepsilon} \right.$$
$$\left. +2\ln(-s)\ln(-t) - \frac{4\pi^2}{3} \right) + O(\varepsilon) \,. \qquad (8.35)$$

It turns out that, in this example, straightforward evaluation is simpler than evaluation by expanding in the limit $t/s \to 0$ with the help of expansion by regions. Indeed, using Feynman parameters, we obtain

$$F_{8.5} = \mathrm{i}\pi^{d/2}\Gamma(2+\varepsilon) \int_0^1 \int_0^1 \int_0^1 \mathrm{d}\xi_1\, \mathrm{d}\xi_2\, \mathrm{d}\eta\, \eta(1-\eta)$$
$$\times \left[-s\xi_1(1-\xi_1)(1-\eta)^2 - t\xi_2(1-\xi_2)\eta^2 - \mathrm{i}0 \right]^{-2-\varepsilon} \,. \qquad (8.36)$$

The Mellin–Barnes representation (2.74) is then used to replace the two terms in the square brackets by a product of some powers of these terms, and, after evaluating parametric integrals in terms of gamma functions, we obtain

$$F_{8.5} = \frac{\mathrm{i}\pi^{d/2}}{(-s)^{2+\varepsilon}\Gamma(-2\varepsilon)} \frac{1}{2\pi\mathrm{i}} \int_{-\mathrm{i}\infty}^{+\mathrm{i}\infty} \mathrm{d}z \left(\frac{t}{s}\right)^z \times \Gamma(2+\varepsilon+z)\Gamma(1+z)^2\Gamma(-1-\varepsilon-z)^2\Gamma(-z) , \quad (8.37)$$

where the standard prescription for dealing with poles has been used. The poles at $z = z_n^{(\mathrm{h})} = n$ and $z = z_n^{(\mathrm{c})} = -1-\varepsilon+n$, with $n = 0,1,\ldots,$ correspond to the hard and collinear contributions, respectively. Our result (8.35), written up to ε^0, is reproduced starting from (8.37) by taking just one residue (with a minus sign, of course) at the first collinear pole $z = -1-\varepsilon$.

We shall see later in this section that the double box provides an even more striking example of this kind.

Example 8.6. The box diagram of Fig. 8.3 with $m_1 = m_2 = m$, $m_3 = m_4 = 0$ and $p_i^2 = 0$, $i = 1,2,3,4$, in Limit (i), i.e. $m^2 \ll |s|, |t|$.

With the loop momentum shown in Fig. 8.3, the Feynman integral takes the form

$$F_{8.6}(s,t,m^2;d) = \int \frac{\mathrm{d}^d k}{(k^2+2p_1\cdot k-m^2)(k^2-2p_2\cdot k-m^2)} \times \frac{1}{k^2[k^2+2(p_1+p_3)\cdot k+t]} . \quad (8.38)$$

Let us choose the external momenta $p_{1,2}$ as in (8.1) and confine ourselves to the leading order of expansion. We obtain the contribution from the hard region, which is nothing but the massless box (8.35). In the (1c) region, we have $k^2 \sim m^2$, $2p_1\cdot k \sim m^2$, $2p_2\cdot k \sim Q^2$ and $2p_3\cdot k \sim (p_3)_+k_- \equiv (-t/Q^2)(2p_2\cdot k)$. Thus

$$F_{8.6}^{(1\mathrm{c}),\,\mathrm{LO}} = -\frac{s}{t}\int \frac{\mathrm{d}^d k}{(k^2+2p_1\cdot k-m^2)(-2p_2\cdot k)k^2(2p_2\cdot k+Q^2)} . \quad (8.39)$$

The (2c) contribution is obtained by interchanging p_1 and p_2. There are, however, two more collinear contributions, which are defined more naturally with another choice of the loop momentum. But it is simpler to think of the interchanges $p_1 \leftrightarrow p_3$ and $p_2 \leftrightarrow p_4$. In this way we obtain the (3c) and (4c) contributions, which correspond to regions where k is almost parallel to p_3 and p_4, respectively. The individual collinear contributions are, again, not regularized dimensionally. Using analytic regularization and evaluating the integrals involved by means of (A.34), we obtain the following result for the sum of all four collinear contributions:

$$F_{8.6}^{(\mathrm{c}),\,\mathrm{LO}} = -\mathrm{i}\pi^{d/2}\frac{2\Gamma(\varepsilon)}{st(m^2)^\varepsilon}\left[\ln\left(\frac{m^2}{-t}\right)+2\psi(-\varepsilon)-\psi(\varepsilon)+\gamma_{\mathrm{E}}\right] . \quad (8.40)$$

The ultrasoft contribution is also non-zero here. Moreover, there are two equal (us) contributions: one of them corresponds to the choice of the loop momentum in Fig. 8.3, and the other corresponds to the loop momentum being the momentum flowing through the fourth line. Using (A.32) we have, in the leading order,

$$F_{8.6}^{(\mathrm{us}),\,\mathrm{LO}} = \frac{2}{t}\int \frac{\mathrm{d}^d k}{(2p_1\cdot k - m^2)(-2p_2\cdot k - m^2)k^2}$$
$$= -2\mathrm{i}\pi^{d/2}\frac{\Gamma(1-\varepsilon)\Gamma(\varepsilon)^2}{(-s)^{1-\varepsilon}t(m^2)^{2\varepsilon}}\,. \tag{8.41}$$

Collecting all the contributions, we arrive at the following result for the leading power of the expansion when $m^2 \ll |s|, |t|$, written at $\varepsilon = 0$:

$$F_{8.6}^{\mathrm{LO}} = \frac{\mathrm{i}\pi^{d/2}}{st}\left[\ln^2\left(\frac{m^2}{-s}\right) + 2\ln\left(\frac{m^2}{-s}\right)\ln\left(\frac{m^2}{-t}\right) - \frac{\pi^2}{3}\right]. \tag{8.42}$$

8.4 Two-Loop Examples: Vertex Diagrams

It turns out that, as in Chap. 7 and in contrast to the limits analysed in Chaps. 4–6, we have not encountered at the one-loop level all the regions that are characteristic of the Sudakov and Regge limits. Thus we continue by analysing examples at the two-loop level. Consider now the following example.

Example 8.7. The diagram of Fig. 8.4 with masses $m_1 = m_2 = m_3 = m_4 = 0$, $m_5 = m_6 = m$ in Limit (i).

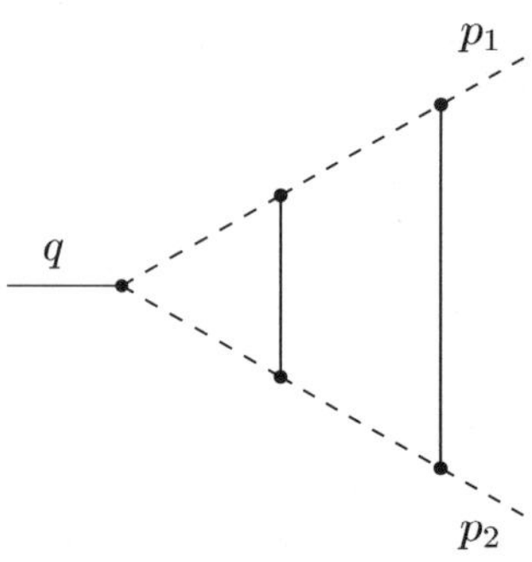

Fig. 8.4. Vertex diagram with two non-zero masses and $p_1^2 = p_2^2 = 0$

This Feynman integral can be written as

$$F_{8.7}(Q^2, m^2; d) = \int \frac{\mathrm{d}^d l}{(l^2 - 2p_1\cdot l)(l^2 - 2p_2\cdot l)}$$
$$\times \int \frac{\mathrm{d}^d k}{(k^2 - 2p_1\cdot k)(k^2 - 2p_2\cdot k)(k^2 - m^2)[(k-l)^2 - m^2]}\,, \tag{8.43}$$

where k and l are the loop momenta of the box and triangle, respectively. In this particular example, we do not encounter new regions. This is the list of all non-trivial contributions: (h–h), (1c–h), (2c–h), (1c–1c) and (2c–2c), where the loop momentum k is characterized in the first position, and the (h–s) contribution, where the momentum flowing through the middle line, $k-l$, is soft.

The (h–h) contribution is obtained by expanding the two massive propagators in a geometric series in m^2. In the leading order, this is (8.43) with pure zero masses, which was first calculated in [125, 182, 156]:

$$\frac{(\mathrm{i}\pi^{d/2})^2}{(Q^2)^{2+2\varepsilon}}\frac{1}{\varepsilon}\left[\frac{1}{2\varepsilon}G_2(2,2)G_3(2+\varepsilon,1,1)\right.$$
$$\left.-G_2(2,1)\left(\frac{1}{\varepsilon}G_3(2,1,1+\varepsilon)+G_3(1,1,1)\right)\right]$$
$$=\frac{(\mathrm{i}\pi^{d/2}\mathrm{e}^{-\gamma_{\mathrm{E}}\varepsilon})^2}{(Q^2)^{2+2\varepsilon}}\left(\frac{1}{4\varepsilon^4}+\frac{5\pi^2}{24\varepsilon^2}+\frac{29\zeta(3)}{6\varepsilon}+\frac{3\pi^4}{32}+O(\varepsilon)\right), \qquad (8.44)$$

where $G_2(\lambda_1,\lambda_2)$ is given by (2.55) and $G_3(\lambda_1,\lambda_2,\lambda_3)$ is given by the ratio of gamma functions on the right-hand side of (A.28).

All the resulting Feynman integrals in the (h–h) contribution, in an arbitrary order of the expansion, are special cases of the integral

$$\int\frac{\mathrm{d}^d l}{(l^2-2p_1\cdot l)^{a_1}(l^2-2p_2\cdot l)^{a_2}}$$
$$\times\int\frac{\mathrm{d}^d k}{(k^2-2p_1\cdot k)^{a_3}(k^2-2p_2\cdot k)^{a_4}(k^2)^{a_5}[(k-l)^2]^{a_6}}, \qquad (8.45)$$

with integer powers of the propagators. These integrals can be evaluated by integration by parts [68] (as in [156]) and can be expressed, for general ε, in terms of gamma functions. In fact, one can use the triangle rule (2.69) for the triangle subgraph of Fig. 8.4 with zero masses and reduce either a_3, a_4 or a_6 to zero. The integrals with $a_6=0$ factorize and can be evaluated in terms of gamma functions by means of (A.7) and (A.28). If $a_4=0$, the fourth line is contracted and then (2.69) is applied again to the triangle with lines $3, 5$ and 6 to reduce either a_1 or a_3 to zero; the resulting integrals can be expressed in terms of gamma functions.

Using another choice of the loop momenta, the (h–s) contribution can be written as

$$F_{8.7}^{(\mathrm{h-s})}=\int\frac{\mathrm{d}^d l}{l^2-m^2}\int\frac{\mathrm{d}^d k}{(k^2-2p_1\cdot k)(k^2-2p_2\cdot k)}\mathcal{T}_m\frac{1}{k^2-m^2}$$
$$\times\mathcal{T}_l\frac{1}{[(k+l)^2-2p_1\cdot(k+l)][(k+l)^2-2p_2\cdot(k+l)]}. \qquad (8.46)$$

The calculation of this contribution is rather simple and reduces to successive application of the one-loop integration formulae (A.7) and (A.28), with the G-functions G_2 and G_3 in the results. The (h–s) contribution is more typical of off-shell limits. This contribution starts from the next-to-leading order, $m^2/(Q^2)^3$.

Consider now the (1c–h) and (2c–h) contributions. According to the general prescription, the (1c–h) contribution is written in terms of the product of the expansion of the propagator $1/(k^2 - 2p_2 \cdot k)$ in a geometric series with respect to k^2 and an expansion of the triangle subdiagram, which is a function $f(q_1^2, q_2^2, q_3^2)$ of the following external momenta squared: $q_1^2 = k^2 - 2p_1 \cdot k$, $q_2^2 = k^2 - 2p_2 \cdot k$ and $q_3^2 = q^2$. So, this second expansion of f is a Taylor series (again in the sense of an expansion under the integral sign) in q_2^2, with subsequent expansion of the result, as a function of $k^2 - 2p_2 \cdot k$, in k^2. All the derivatives of that triangle with respect to q_2^2 at $q_2^2 = 0$ can be calculated by means of recurrence relations based on integration by parts (see, e.g., the results for the first two derivatives in [92]).

This product of the expansion of the fourth line and the lower triangle is then integrated with the product of the other propagators (numbers 4 and 6), using the one-loop integration formulae (A.31). However, the (1c–h) contribution taken alone is not regularized dimensionally, so that it is natural to consider the sum of the (1c–h) and (2c–h) contributions. To handle these terms individually, one can introduce an auxiliary analytic regularization into lines 3 and 4, calculate the terms and switch off the analytic regularization in the sum.

Following this procedure, the (1c–h) and (2c–h) contributions can be evaluated in terms of gamma functions in every order of the expansion, for arbitrary ε. For example, the leading, $1/(Q^2)^2$, contribution takes the form

$$\frac{(\mathrm{i}\pi^{d/2})^2}{(Q^2)^{2+\varepsilon}(m^2)^\varepsilon} \frac{\Gamma(\varepsilon)^2\Gamma(1-\varepsilon)^2}{\varepsilon\Gamma(1-2\varepsilon)} \times \left(-2\frac{\Gamma(-\varepsilon)^2}{\Gamma(-2\varepsilon)} - \ln x + \psi(\varepsilon) - 2\psi(-\varepsilon) - \gamma_{\mathrm{E}} \right) , \quad (8.47)$$

where again $x = m^2/Q^2$.

To calculate the (1c–1c) and (2c–2c) contributions, it is reasonable to use the Mellin–Barnes representation for the fifth propagator and reduce the problem to calculation of the corresponding contribution from the diagram with $m_5 = 0$ in the case where the corresponding propagator is analytically regularized. In particular, the leading, $1/(Q^2)^2$, contribution is obtained as the following Mellin–Barnes integral:

$$-\frac{(\mathrm{i}\pi^{d/2})^2}{(Q^2)^{2+\varepsilon}(m^2)^\varepsilon} \frac{1}{2\pi\mathrm{i}} \int_{-\mathrm{i}\infty}^{\mathrm{i}\infty} \mathrm{d}z\, \Gamma(2\varepsilon+z)\Gamma(\varepsilon+z)\Gamma(-\varepsilon-z)\Gamma(-z) \times [-\ln x + \psi(2\varepsilon+z) + \psi(1+\varepsilon+z) - 2\psi(1+z)] . \quad (8.48)$$

The integration contour is chosen in the standard way, with the qualification that the pole at $z = -\varepsilon$ is to the right of it, i.e. it is considered to be UV. The pole and the finite part of (8.48) are evaluated separately and produce the following result, given up to ε^0:

$$-\frac{(\mathrm{i}\pi^{d/2})^2}{(Q^2)^2(m^2)^{2\varepsilon}} \left(-2\zeta(3)\ln x + \frac{\pi^4}{30} + \Gamma(-\varepsilon)\Gamma(\varepsilon)\Gamma(2\varepsilon) \right.$$

$$\times \left[-\ln x + \psi(1-\varepsilon) - 2\psi(1-2\varepsilon) - \psi(-\varepsilon) + \psi(\varepsilon) + \psi(2\varepsilon)\right] \Big) . \quad (8.49)$$

Using the method of integration by parts [68], one can arrive at recurrence relations that provide all the integrals contributing to an arbitrary power of the expansion [211].

Note that the (h–h) contribution involves IR and collinear divergences, the (c–c) contribution involves UV and collinear divergences, and the (c–h) and (s–h) contributions possess all three kinds of divergence. All the poles, up to $1/\varepsilon^4$, are cancelled, however, and, summing up all the contributions, we have [211, 209]

$$F_{8.7}(Q^2, m^2) \sim \frac{-\pi^4}{(Q^2)^2} \left[\frac{1}{24} \ln^4 x + \frac{\pi^2}{3} \ln^2 x + 6\zeta(3) \ln x + \frac{31\pi^4}{180} \right.$$
$$+ x \left(\frac{1}{6} \ln^3 x + \frac{3}{2} \ln^2 x + 6 \ln x + \frac{2\pi^2}{3} \ln x - 18 + \frac{\pi^2}{3} + 6\zeta(3) \right)$$
$$+ x^2 \left(\frac{1}{12} \ln^3 x + \frac{35}{8} \ln^2 x + \frac{29}{4} \ln x + \frac{\pi^2}{3} \ln x \right.$$
$$\left. \left. - \frac{153}{8} + \frac{19\pi^2}{12} + 3\zeta(3) \right) \right] + \ldots . \quad (8.50)$$

We now turn to the next example.

Example 8.8. The non-planar vertex diagram of Fig. 8.5 with $p_1^2 = p_2^2 = 0$ and masses $m_1 = m_2 = m_3 = m_4 = 0$, $m_5 = m_6 = m$ in Limit (*i*).

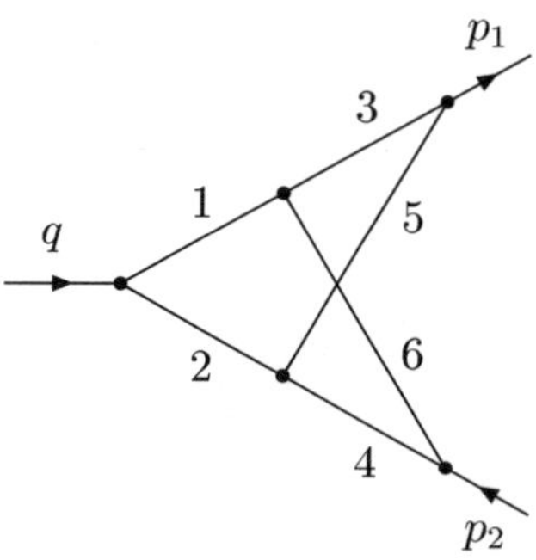

Fig. 8.5. Non-planar vertex diagram

The Feynman integral can be written as

$$F_{8.8}(Q^2, m^2; d) = \int\int \frac{\mathrm{d}^d k \, \mathrm{d}^d l}{[(k+l)^2 - 2p_1 \cdot (k+l)][(k+l)^2 - 2p_2 \cdot (k+l)]}$$
$$\times \frac{1}{(k^2 - 2p_1 \cdot k)(l^2 - 2p_2 \cdot l)(k^2 - m^2)(l^2 - m^2)} , \quad (8.51)$$

where the loop momenta are chosen as the momenta flowing through lines 5 and 6. We shall also use a second choice of the loop momenta, where k and l

are chosen as the momenta of lines 3 and 4, respectively; this is obtained by permutation of the masses and corresponds to (8.51) with $m_1 = m_2 = m_5 = m_6 = 0$, $m_3 = m_4 = m$.

Let us restrict ourselves to the leading order of expansion. In this case, non-zero contributions are generated by the following regions: (h–h), (h–1c), (1c–h), (1c–1c), (2c–2c), (1c–2c), (1c–1c)′, (2c–2c)′ and (us–us)′. As above, we indicate the region of the loop momentum k in the first position and that of l in the second position. We indicate by primes the regions for the second natural choice of the loop momenta. The (h–h) contribution is given by the massless non-planar diagram. The result, as an expansion in ε, can be found in [125]:

$$
\begin{aligned}
F_{8.8}^{(\mathrm{h-h}),\,\mathrm{LO}} &= \int\int \frac{\mathrm{d}^d k\, \mathrm{d}^d l}{[(k+l)^2 - 2p_1\cdot(k+l)][(k+l)^2 - 2p_2\cdot(k+l)]} \\
&\quad\times \frac{1}{(k^2 - 2p_1\cdot k)(l^2 - 2p_2\cdot l)k^2 l^2} \\
&= \frac{\left(\mathrm{i}\pi^{d/2}\mathrm{e}^{-\gamma_{\mathrm{E}}\varepsilon}\right)^2}{(Q^2)^{2+2\varepsilon}} \left(\frac{1}{\varepsilon^4} - \frac{\pi^2}{\varepsilon^2} - \frac{83\zeta(3)}{3\varepsilon} - \frac{59\pi^4}{120} \right) + O(\varepsilon) \, . \qquad (8.52)
\end{aligned}
$$

An algorithm for the evaluation of general integrals of type (8.52) with arbitrary numerators and integer powers of propagators has been presented in [9, 3].

The (us–us)′ contribution is easily evaluated in terms of gamma functions by means of alpha parameters:

$$
\begin{aligned}
F_{8.8}^{(\mathrm{us-us})',\,\mathrm{LO}} &= \int\int \frac{\mathrm{d}^d k\, \mathrm{d}^d l}{[-2p_1\cdot(k+l)][-2p_2\cdot(k+l)]} \\
&\quad\times \frac{1}{(-2p_1\cdot k + m^2)(-2p_2\cdot l + m^2)k^2 l^2} \\
&= \frac{\left(\mathrm{i}\pi^{d/2}\right)^2}{(Q^2)^{2-2\varepsilon}(m^2)^{4\varepsilon}} \left[\Gamma(\varepsilon)\Gamma(2\varepsilon)\Gamma(1-2\varepsilon)\right]^2 \, . \qquad (8.53)
\end{aligned}
$$

The (1c–h) contribution is given by

$$
\begin{aligned}
F_{8.8}^{(\mathrm{1c-h}),\,\mathrm{LO}} &= \int\int \frac{\mathrm{d}^d k\, \mathrm{d}^d l}{l^2 - 2p_1\cdot l + (2p_2\cdot k)(2p_1\cdot l)/Q^2} \\
&\quad\times \frac{1}{l^2 - 2p_2\cdot(k+l) + (2p_2\cdot k)(2p_1\cdot l)/Q^2} \\
&\quad\times \frac{1}{(k^2 - 2p_1\cdot k)(l^2 - 2p_2\cdot l)(k^2 - m^2) l^2} \, , \qquad (8.54)
\end{aligned}
$$

and the same leading-order (h–1c) contribution is obtained by permutation of k and l. Using alpha parameters and Mellin–Barnes representation (twice), we obtain

$$
\begin{aligned}
F_{8.8}^{(\mathrm{h-1c}),\,\mathrm{LO}} &= F_{8.8}^{(\mathrm{1c-h}),\,\mathrm{LO}} \\
&= \frac{\left(\mathrm{i}\pi^{d/2}\mathrm{e}^{-\gamma_{\mathrm{E}}\varepsilon}\right)^2}{(Q^2)^{2+\varepsilon}(m^2)^{\varepsilon}} \left(-\frac{3}{\varepsilon^4} + \frac{\pi^2}{\varepsilon^2} + \frac{22\zeta(3)}{\varepsilon} + \frac{16\pi^4}{45} \right) + O(\varepsilon) \,.
\end{aligned} \tag{8.55}
$$

The (1c–1c) contribution is given by

$$
\begin{aligned}
F_{8.8}^{(\mathrm{1c-1c}),\,\mathrm{LO}} &= \int\int \frac{\mathrm{d}^d k\,\mathrm{d}^d l}{[(k+l)^2 - 2p_1\cdot(k+l)][-2p_2\cdot(k+l)]} \\
&\quad \times \frac{1}{(k^2-2p_1\cdot k)(-2p_2\cdot l)(k^2-m^2)(l^2-m^2)} \,,
\end{aligned} \tag{8.56}
$$

and the (2c–2c) contribution is obtained by permutation of k and l. We should also consider the similar (1c–1c)$'$ and (2c–2c)$'$ contributions with the second choice of the loop momenta. The corresponding expressions are obtained by a permutation of the masses (see above). The fifth non-zero contribution of collinear–collinear type originates from the (1c–2c) region. It turns out that these five contributions are dimensionally regularized only in the sum. It is convenient to introduce an auxiliary analytic regularization into lines 3 and 4 by inserting the factor $(k^2-2p_1k)^{-x_1}(l^2-2p_2l)^{-x_2}$. In contrast to Example 8.7, we encounter, in this example, poles in x_i up to the second order. In particular, the (1c–2c) contribution can be evaluated in terms of gamma functions, for general ε:

$$
\begin{aligned}
F_{8.8}^{(\mathrm{1c-2c}),\,\mathrm{LO}}(x_1,x_2) &= \int\int \frac{\mathrm{d}^d k\,\mathrm{d}^d l}{(k^2-2p_1\cdot k)^{1+x_1}(l^2-2p_2\cdot l)^{1+x_2}} \\
&\quad \times \frac{1}{[-2p_1\cdot l + (2p_2\cdot k)(2p_1\cdot l)/Q^2][-2p_2\cdot k + (2p_2\cdot k)(2p_1\cdot l)/Q^2]} \\
&\quad \times \frac{1}{(k^2-m^2)(l^2-m^2)} \\
&= \left(\mathrm{i}\pi^{d/2}\right)^2 \frac{\Gamma(-x_1-\varepsilon)\Gamma(-x_2-\varepsilon)\Gamma(x_1+\varepsilon)\Gamma(x_2+\varepsilon)}{x_1x_2\Gamma(-\varepsilon)^2(-m^2)^{x_1+x_2+2\varepsilon}(Q^2)^2} \,.
\end{aligned} \tag{8.57}
$$

Using the technique of alpha parameters and the Mellin–Barnes representation for the other four (c–c) contributions, we obtain, for each of them, a result in the form of an expansion in x_i. We then switch off the analytic regularization (first $x_2 \to x_1$ and then $x_1 \to 0$), observe that in the sum of all five contributions, the singular dependence on x_i drops out and obtain the following result as an expansion in ε:

$$
\begin{aligned}
& F_{8.8}^{(\mathrm{1c-1c}),\,\mathrm{LO}} + F_{8.8}^{(\mathrm{2c-2c}),\,\mathrm{LO}} \\
&\quad + F_{8.8}^{(\mathrm{1c-1c})',\,\mathrm{LO}} + F_{8.8}^{(\mathrm{2c-2c})',\,\mathrm{LO}} + F_{8.8}^{(\mathrm{1c-2c}),\,\mathrm{LO}} \\
&= \frac{\left(\mathrm{i}\pi^{d/2}\mathrm{e}^{-\gamma_{\mathrm{E}}\varepsilon}\right)^2}{(Q^2)^2} \left[\frac{19}{4\varepsilon^4} - \frac{9}{2\varepsilon^3}L + \left(L^2 - \frac{11\pi^2}{4}\right)\frac{1}{2\varepsilon^2} \right. \\
&\quad \left. - \left(\frac{3\pi^2}{4}L + \frac{97\zeta(3)}{6}\right)\frac{1}{\varepsilon} + \frac{\pi^2}{12}L^2 + 9\zeta(3)L - \frac{23\pi^4}{32} \right] + O(\varepsilon) \,,
\end{aligned} \tag{8.58}
$$

where $L = \ln(Q^2/m^2)$, and we have set $m = 1$ for brevity (to drop the $\ln m^2$ that occurs with L).

Collecting all the leading-order contributions, we see that the poles in ε are cancelled and we arrive at the following result:

$$F_{8.8}(Q^2, m^2) \sim \frac{\left(\mathrm{i}\pi^{d/2}\right)^2}{(Q^2)^2} \left(\frac{7}{12} L^4 - \frac{\pi^2}{2} L^2 + 20\zeta(3) L - \frac{31\pi^4}{180} \right) . \tag{8.59}$$

At the cost of computer algebra, it is possible to extend this result to any order in $1/Q^2$.

We have found no new types of regions in this example. However, we have observed a rich structure of regions, and poles up to the second order in the auxiliary analytic regularization which are cancelled in the sum of the five contributions. We have also checked that the poles in ε are cancelled. There is, however, no check of the finite part because an explicit result for this diagram is not available. In the next example we shall be able to check the expansion against a known result.

Example 8.9. The massless planar vertex diagram of Fig. 5.3 with $p_1^2 = p_2^2 = -M^2$ in Limit (*ii*), i.e. $M^2 \ll -s$.

The Feynman integral can be written as

$$F_{8.9}(Q^2, M^2; d) = \int\int \frac{\mathrm{d}^d k \, \mathrm{d}^d l}{(l^2 - 2p_1 \cdot l - M^2)(l^2 - 2p_2 \cdot l - M^2)} \times \frac{1}{(k^2 - 2p_1 \cdot k - M^2)(k^2 - 2p_2 \cdot k - M^2) k^2 (k-l)^2} . \tag{8.60}$$

We again restrict ourselves to the leading order, $1/Q^4$, where we obtain contributions from the following nine regions: (h–h), (1c–h), (2c–h), (1c–1c), (2c–2c), (us–h), (us–1c), (us–2c) and (us–us). The (h–h) region generates terms obtained by Taylor expanding the integrand in the expansion parameter, M. In the leading order, the contribution from this region is again, as in Example 8.7, the massless master vertex diagram given by (8.44).

All the contributions connected with ultrasoft regions can be easily evaluated in terms of gamma functions by use of alpha parameters. In the leading order, we have

$$\begin{aligned} F_{8.9}^{(\mathrm{us-us}),\,\mathrm{LO}} &= \int\int \frac{\mathrm{d}^d k \, \mathrm{d}^d l}{(-2\tilde{p}_1 \cdot l - M^2)(-2\tilde{p}_2 \cdot l - M^2)} \\ &\quad \times \frac{1}{(-2\tilde{p}_1 \cdot k - M^2)(-2\tilde{p}_2 \cdot k - M^2) k^2 (k-l)^2} \\ &= \frac{\Gamma(\varepsilon)^2 \Gamma(2\varepsilon)^2}{(M^2)^{4\varepsilon} (Q^2)^{2-2\varepsilon}} , \end{aligned} \tag{8.61}$$

$$F_{8.9}^{(\mathrm{us-h}),\,\mathrm{LO}} = \int\int \frac{\mathrm{d}^d k \, \mathrm{d}^d l}{(l^2 - 2\tilde{p}_1 \cdot l)(l^2 - 2\tilde{p}_2 \cdot l)(-2\tilde{p}_1 \cdot k - M^2)(-2\tilde{p}_2 \cdot k - M^2) k^2 l^2}$$

$$= \frac{\Gamma(1+\varepsilon)\Gamma(1-\varepsilon)\Gamma(\varepsilon)^2\Gamma(-\varepsilon)^2}{\Gamma(1-2\varepsilon)(M^2)^{2\varepsilon}(Q^2)^2}\,, \tag{8.62}$$

$$\begin{aligned} F_{8.9}^{(\mathrm{us-1c}),\,\mathrm{LO}} &= \int\int \frac{\mathrm{d}^d k\, \mathrm{d}^d l}{(l^2 - 2p_1\cdot l - M^2)(-2\tilde{p}_2\cdot l)(-2\tilde{p}_1\cdot k - M^2)} \\ &\quad\times \frac{1}{(-2\tilde{p}_2\cdot k - M^2)k^2[l^2 - (2\tilde{p}_1\cdot k)(2\tilde{p}_2\cdot l)/Q^2]} \\ &= \frac{\Gamma(1-\varepsilon)^2\Gamma(\varepsilon)\Gamma(2\varepsilon)\Gamma(-\varepsilon)}{\varepsilon\Gamma(1-2\varepsilon)(M^2)^{3\varepsilon}(Q^2)^{2-\varepsilon}} \equiv F_{8.9}^{(\mathrm{us-2c}),\,\mathrm{LO}}\,. \end{aligned} \tag{8.63}$$

Using alpha parameters, the rest of the contributions can be evaluated in terms of onefold Mellin–Barnes integrals:

$$\begin{aligned} F_{8.9}^{(\mathrm{1c-1c}),\,\mathrm{LO}} &= \int\int \frac{\mathrm{d}^d k\, \mathrm{d}^d l}{(-2\tilde{p}_1\cdot l)(l^2 - 2p_2\cdot l - M^2)(-2\tilde{p}_1\cdot k)} \\ &\quad\times \frac{1}{(k^2 - 2p_2\cdot k - M^2)k^2(k-l)^2} \\ &= \frac{\Gamma(\varepsilon)\Gamma(-\varepsilon)\Gamma(2\varepsilon)}{\Gamma(1+\varepsilon)(M^2)^{2\varepsilon}(Q^2)^2} \\ \times \frac{1}{2\pi\mathrm{i}}\int_{-\mathrm{i}\infty}^{+\mathrm{i}\infty} \mathrm{d}z\, & \frac{\Gamma(z-3\varepsilon)\Gamma(z+1-2\varepsilon)\Gamma(1+z-\varepsilon)\Gamma(\varepsilon-z)\Gamma(-z)}{\Gamma(1+z-3\varepsilon)} \\ &\equiv F_{8.9}^{(\mathrm{2c-2c}),\,\mathrm{LO}}\,, \end{aligned} \tag{8.64}$$

$$\begin{aligned} F_{8.9}^{(\mathrm{1c-h}),\,\mathrm{LO}} &= \int\int \frac{\mathrm{d}^d k\, \mathrm{d}^d l}{(l^2 - 2\tilde{p}_1\cdot l)(l^2 - 2\tilde{p}_2\cdot l)(-2\tilde{p}_1\cdot k)(k^2 - 2p_2\cdot k - M^2)} \\ &\quad\times \frac{1}{k^2[l^2 - (2\tilde{p}_1\cdot k)(2\tilde{p}_2\cdot l)/Q^2]} \\ &= \frac{\Gamma(\varepsilon)\Gamma(-\varepsilon)\Gamma(1-\varepsilon)}{\Gamma(1-2\varepsilon)(M^2)^{\varepsilon}(Q^2)^{2+\varepsilon}} \\ \times \frac{1}{2\pi\mathrm{i}}\int_{-\mathrm{i}\infty}^{+\mathrm{i}\infty} \mathrm{d}z\, & \frac{\Gamma(1+z)\Gamma(z-\varepsilon)\Gamma(1+z+\varepsilon)\Gamma(-\varepsilon-z)\Gamma(-z)}{\Gamma(1+z-2\varepsilon)} \\ &\equiv F_{8.9}^{(\mathrm{2c-h}),\,\mathrm{LO}}\,, \end{aligned} \tag{8.65}$$

where the contours are chosen in the standard way.

The following is the result for the sum of all four (c–c) and (c–h) contributions [216]:

$$\begin{aligned} & F_{8.9}^{(\mathrm{1c-1c}),\,\mathrm{LO}} + F_{8.9}^{(\mathrm{2c-2c}),\,\mathrm{LO}} + F_{8.9}^{(\mathrm{1c-h}),\,\mathrm{LO}} + F_{8.9}^{(\mathrm{2c-h}),\,\mathrm{LO}} \\ &= \frac{\left(\mathrm{i}\pi^{d/2}\mathrm{e}^{-\gamma_{\mathrm{E}}\varepsilon}\right)^2}{(Q^2)^2}\left[-\frac{1}{2\varepsilon^4} + \left(L^2 - \frac{\pi^2}{2}\right)\frac{1}{2\varepsilon^2}\right. \\ &\left. + \left(\frac{1}{2}L^3 - \frac{\pi^2}{6}L - \frac{17\zeta(3)}{3}\right)\frac{1}{\varepsilon} + \frac{7}{24}L^4 - 4\zeta(3)L - \frac{\pi^4}{144}\right] + O(\varepsilon)\,, \end{aligned} \tag{8.66}$$

where $L = \ln(Q^2/M^2)$ and we have dropped terms containing $\ln M^2$.

Collecting all nine contributions together, we observe that the poles in ε, which turn out to be of very different (UV, IR and collinear) nature, cancel, with the following result:

$$F_{8.9}(Q^2, M^2) \sim \frac{\left(\mathrm{i}\pi^{d/2}\right)^2}{(Q^2)^2} \left(\frac{1}{4} L^4 + \frac{\pi^2}{2} L^2 + \frac{7\pi^4}{60} \right) . \tag{8.67}$$

This is in agreement with the leading-order expansion of the well-known explicit result [233].

It is in the next example where we shall finally encounter a new region.

Example 8.10. The planar vertex diagram of Fig. 7.5 with $m_1 = m_2 = m_3 = m_4 = m$, $m_5 = m_6 = 0$ and $p_1^2 = p_2^2 = m^2$ in Limit (*iii*), i.e. when $m^2 \ll -s$.

We have already expanded this diagram at threshold, $q^2 - 4m^2 \sim 0$, in Example 7.5. Now we write down the Feynman integral as

$$F_{8.10}(Q^2, m^2; d) = \int\int \frac{\mathrm{d}^d k\, \mathrm{d}^d l}{(l^2 - 2p_1 \cdot l)(l^2 - 2p_2 \cdot l)(k^2 - 2p_1 \cdot k)(k^2 - 2p_2 \cdot k) k^2 (k - l)^2} , \tag{8.68}$$

where $p_{1,2}$ are given by (8.2) with M^2 replaced by $-m^2$, so that $2\tilde{p}_1 \cdot \tilde{p}_2 = 2\tilde{p}_1 \cdot p_2 = Q^2$.

In the leading order, $1/Q^4$, we obtain five contributions generated by the following regions: (h–h), (1c–h), (2c–h), (1c–1c) and (2c–2c). In contrast to Example 8.8, every term can now be considered separately, and we have, symbolically, (2c–h) = (1c–h), (2c–2c) = (1c–1c).

The (h–h) region generates terms obtained by Taylor expanding the integrand in the expansion parameter, m. In the leading order, this is again the value of the massless planar diagram at $p_1^2 = p_2^2 = 0$ given by (8.44).

We then have

$$F_{8.10}^{(1c-1c),\, \mathrm{LO}} = \int\int \frac{\mathrm{d}^d k\, \mathrm{d}^d l}{(l^2 - 2p_1 \cdot l)(-2\tilde{p}_2 \cdot l)(k^2 - 2p_1 \cdot k)(-2\tilde{p}_2 \cdot k) k^2 (k - l)^2} , \tag{8.69}$$

$$F_{8.10}^{(1c-h),\, \mathrm{LO}} = \int\int \frac{\mathrm{d}^d k\, \mathrm{d}^d l}{(l^2 - 2\tilde{p}_1 \cdot l)(l^2 - 2\tilde{p}_2 \cdot l)(k^2 - 2p_1 \cdot k)(-2\tilde{p}_2 \cdot k)} \times \frac{1}{k^2 [l^2 - (2\tilde{p}_1 \cdot l)(2\tilde{p}_2 \cdot k)/Q^2]} . \tag{8.70}$$

Using the technique of alpha parameters the and Mellin–Barnes representation, we obtain the following results up to the finite part in ε:

$$F_{8.10}^{(1c-1c),\, \mathrm{LO}} = \frac{\left(\mathrm{i}\pi^{d/2} \mathrm{e}^{-\gamma_{\mathrm{E}}\varepsilon}\right)^2}{(Q^2)^2 (m^2)^{2\varepsilon}} \left(-\frac{\pi^2}{24\varepsilon^2} + \frac{5\zeta(3)}{4\varepsilon} + \frac{\pi^4}{48} \right) , \tag{8.71}$$

$$F_{8.10}^{(1c-h),\,LO} = \frac{-\left(i\pi^{d/2}e^{-\gamma_E\varepsilon}\right)^2}{(Q^2)^{2+\varepsilon}(m^2)^{\varepsilon}}\left(\frac{1}{6\varepsilon^4}+\frac{\pi^2}{6\varepsilon^2}+\frac{41\zeta(3)}{9\varepsilon}+\frac{37\pi^4}{180}\right). \quad (8.72)$$

If we combine all these five contributions we shall obviously obtain the wrong result because the poles of the third and fourth order fail to cancel. It turns out that the following non-standard region has to be included in our list [214]:

$$\textit{1-ultracollinear (1uc)}, \quad k_+ \sim \frac{m^4}{Q^3},\ k_- \sim \frac{m^2}{Q},\ \underline{k} \sim \frac{m^3}{Q^2}. \quad (8.73)$$

The missing contributions are (1uc–2c) and (2uc–1c), which are equal to each other and can easily be evaluated by means of alpha parameters for general ε. In the leading order, we have

$$\begin{aligned} F_{8.10}^{(1uc-2c),\,LO} &= \int\int \frac{d^dk\, d^dl}{(-2\tilde{p}_1\cdot l)(l^2-2p_2\cdot l)(-2p_1\cdot k)(-2\tilde{p}_2\cdot k)k^2} \\ &\quad\times \frac{1}{l^2-(2\tilde{p}_1\cdot l)(2\tilde{p}_2\cdot k)/Q^2} \\ &= -\frac{\left(i\pi^{d/2}\right)^2}{(Q^2)^{2-2\varepsilon}(m^2)^{4\varepsilon}}\Gamma(\varepsilon)\Gamma(2\varepsilon)\Gamma(3\varepsilon)\Gamma(-4\varepsilon) \\ = \frac{\left(i\pi^{d/2}e^{-\gamma_E\varepsilon}\right)^2}{(Q^2)^{2-2\varepsilon}(m^2)^{4\varepsilon}}&\left(\frac{1}{24\varepsilon^4}+\frac{5\pi^2}{48\varepsilon^2}+\frac{7\zeta(3)}{18\varepsilon}+\frac{493\pi^4}{2880}\right)+O(\varepsilon). \quad (8.74) \end{aligned}$$

Collecting all seven contributions together, we observe that the poles of the third and fourth order in ε cancel, and we arrive at the following result:

$$\begin{aligned} &F_{8.10}(Q^2,m^2;d) \sim \\ &= \frac{\left(i\pi^{d/2}e^{-\gamma_E\varepsilon}\right)^2}{(Q^2)^{2+2\varepsilon}}\left[\ln^2\frac{m^2}{Q^2}\frac{1}{2\varepsilon^2} - \left(\frac{5}{6}\ln^3\frac{m^2}{Q^2}+\frac{\pi^2}{3}\ln\frac{m^2}{Q^2}+\zeta(3)\right)\frac{1}{\varepsilon}\right. \\ &\left. +\frac{7}{8}\ln^4\frac{m^2}{Q^2}+\frac{4\pi^2}{3}\ln^2\frac{m^2}{Q^2}+\zeta(3)\ln\frac{m^2}{Q^2}+\frac{\pi^4}{15}\right]+O(\varepsilon), \quad (8.75) \end{aligned}$$

which has the proper coefficient of the double pole, given by (7.46).

Note that in the case of the non-planar diagram in Limit (*iii*), with $p_1^2 = p_2^2 = m^2$, there are exactly the same problems with dimensional regularization as in case of the Limit (*i*) with $p_1^2 = p_2^2 = 0$ – see Example 8.7.

In the above example we encountered the ultracollinear regions which can be symbolically described as 1uc $= (m^2/Q^2) \times$ 1c and 2uc $= (m^2/Q^2) \times$ 2c. A *1-ultra-...-ultracollinear* region 1uc $= (m^2/Q^2)^h \times$ 1c can occur for an arbitrary h. Consider the h-loop ladder diagram in Limit (*iii*). In this case the contributions from the (2u^{h-1}c–1u^{h-2}c–...–1c) and (1u^{h-1}c–2u^{h-2}c–...–2c) regions are non-zero (here h is supposed to be even, for definiteness).

For completeness, we consider also the master planar diagram in Limit (*iv*):

Example 8.11. The planar vertex diagram of Fig. 5.3 with $m_1 = m_3 = M$, $m_2 = m_4 = m$, $m_5 = m_6 = 0$ in Limit (*iv*).

The corresponding Feynman integral $F_{8.11}(M^2, m^2; d)$ has the same form as (8.68), but the assignments for the external momenta are chosen as in one-loop Example 8.4.

No new regions are relevant here. In the leading order, there are four non-zero contributions, corresponding to the (h–h), (1c–h),(1c–1c) and (us–1c) regions. The (h–h) contribution is obtained by expanding the integrand into a Taylor series in m^2. Using again the technique of alpha parameters and Mellin–Barnes representation, we arrive at the following result as an expansion in ε up to the finite part:

$$F_{8.11}^{(\text{h}-\text{h}),\,\text{LO}} = \frac{\left(\mathrm{i}\pi^{d/2}\mathrm{e}^{-\gamma_{\mathrm{E}}\varepsilon}\right)^2}{(M^2)^{2+2\varepsilon}} \left(\frac{1}{12\varepsilon^4} + \frac{\pi^2}{12\varepsilon^2} + \frac{91\zeta(3)}{36\varepsilon} + \frac{179\pi^4}{1440} \right) . \tag{8.76}$$

The contribution of the (1c–1c) region is obtained by almost the same prescriptions (a) and (b) as in the corresponding one-loop Example 8.4; in addition, the propagator $1/(l^2 - 2p_1 \cdot l)$ is expanded into a Taylor series in l^2. The (1c–h) contribution is of an intermediate character. In the leading order, we have

$$F_{8.11}^{(\text{1c}-\text{1c}),\,\text{LO}} \\ = \int\int \frac{\mathrm{d}^d k\, \mathrm{d}^d l}{(-2P_2 \cdot l)(l^2 - 2p_2 \cdot l)(-2P_2 \cdot k)(k^2 - 2p_2 \cdot k)k^2(k-l)^2} , \tag{8.77}$$

$$F_{8.11}^{(\text{1c}-\text{h}),\,\text{LO}} = \int\int \frac{\mathrm{d}^d k\, \mathrm{d}^d l}{(l^2 - 2p_1 \cdot l)(l^2 - 2P_1 \cdot l)} \\ \times \frac{1}{(-2P_2 \cdot k)(k^2 - 2p_2 \cdot k)k^2[l^2 - (2n_1 \cdot l)(2P_2 \cdot k)/M]} , \tag{8.78}$$

where $P_1 = Mn_1$, $P_2 = Mn_2$. These contributions can be evaluated by the same techniques as before, with the following results as expansions in ε:

$$F_{8.11}^{(\text{1c}-\text{1c}),\,\text{LO}} = \frac{\left(\mathrm{i}\pi^{d/2}\mathrm{e}^{-\gamma_{\mathrm{E}}\varepsilon}\right)^2}{(M^2)^2(m^2)^{2\varepsilon}} \left(-\frac{\pi^2}{24\varepsilon^2} + \frac{5\zeta(3)}{4\varepsilon} + \frac{\pi^4}{48} \right) + O(\varepsilon) , \tag{8.79}$$

$$F_{8.11}^{(\text{2c}-\text{h}),\,\text{LO}} \\ = \frac{-\left(\mathrm{i}\pi^{d/2}\mathrm{e}^{-\gamma_{\mathrm{E}}\varepsilon}\right)^2}{(M^2)^{2+\varepsilon}(m^2)^{\varepsilon}} \left(\frac{1}{8\varepsilon^4} + \frac{7\pi^2}{48\varepsilon^2} + \frac{31\zeta(3)}{6\varepsilon} + \frac{871\pi^4}{2880} \right) + O(\varepsilon) . \tag{8.80}$$

The (us–1c) contribution

$$F_{8.11}^{(\text{us}-\text{1c}),\,\text{LO}} = \int\int \frac{\mathrm{d}^d k\, \mathrm{d}^d l}{(-2P_2 \cdot l)(l^2 - 2p_2 \cdot l)} \\ \times \frac{1}{(-2p_1 \cdot k)(-2P_1 \cdot k)k^2[l^2 - (2n_1 \cdot k)(2P_2 \cdot l)/M]} \tag{8.81}$$

can be evaluated in terms of gamma functions, with a result which is closely related to the (2uc–1c) and (1uc–2c) contributions in Example 8.10:

$$F_{8.11}^{(\mathrm{us-2c}),\,\mathrm{LO}}$$
$$= \frac{\left(\mathrm{i}\pi^{d/2}\mathrm{e}^{-\gamma_{\mathrm{E}}\varepsilon}\right)^2}{(Q^2)^{2-\varepsilon}(m^2)^{3\varepsilon}}\left(\frac{1}{24\varepsilon^4}+\frac{5\pi^2}{48\varepsilon^2}+\frac{7\zeta(3)}{18\varepsilon}+\frac{493\pi^4}{2880}\right)+O(\varepsilon)\,. \tag{8.82}$$

Collecting all four contributions together, we observe that the poles of the third and fourth order in ε cancel, and we arrive the following result:

$$F_{8.11}(M^2,m^2;d)$$
$$\sim \frac{\left(\mathrm{i}\pi^{d/2}\mathrm{e}^{-\gamma_{\mathrm{E}}\varepsilon}\right)^2}{(M^2)^{2+2\varepsilon}}\left[\ln^2 x\frac{1}{8\varepsilon^2}-\left(\frac{1}{6}\ln^3 x+\frac{\pi^2}{12}\ln x+\zeta(3)\right)\frac{1}{\varepsilon}\right.$$
$$\left.+\frac{13}{96}\ln^4 x+\frac{5\pi^2}{16}\ln^2 x+\frac{3\zeta(3)}{2}\ln x+\frac{\pi^4}{72}\right]+O(\varepsilon)\,, \tag{8.83}$$

where $x = m^2/M^2$, with the proper coefficient of the double pole, which can be evaluated starting from the full diagram.

8.5 Two-Loop Example: the Double Box in the Regge Limit

Our last example in this chapter is the following.

Example 8.12. The massless on-shell (i.e. with $p_i^2 = 0$) double box diagram of Fig. 8.6 in Limit (*v*).

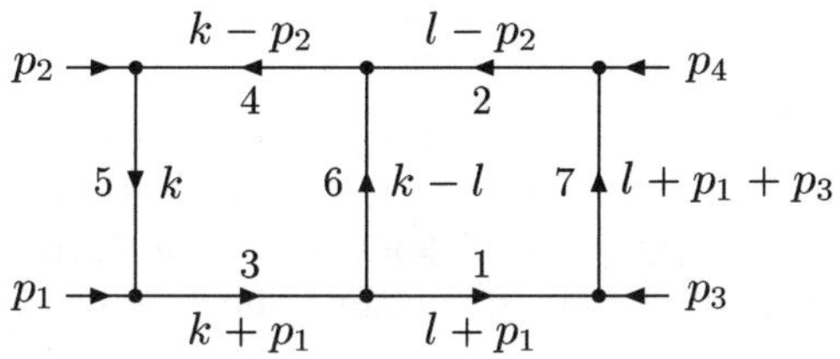

Fig. 8.6. Double box diagram

We consider the master double box, i.e. with all powers of propagators equal to one. With the assignments of the loop momenta shown in Fig. 8.6, the Feynman integral takes the form

$$F_{8.12} = \int\int \frac{\mathrm{d}^d k\,\mathrm{d}^d l}{(l^2+2p_1\cdot l)(l^2-2p_2\cdot l)(k^2+2p_1\cdot k)(k^2-2p_2\cdot k)}$$
$$\times\frac{1}{k^2(k-l)^2(l+r)^2}$$
$$\equiv \frac{\left(\mathrm{i}\pi^{d/2}\mathrm{e}^{-\gamma_{\mathrm{E}}\varepsilon}\right)^2}{(-s)^{2+2\varepsilon}(-t)}\,K(t/s;\varepsilon)\,. \tag{8.84}$$

We again choose the external momenta (8.29). It turns out that the (h–h), (1c–1c) and (2c–2c) contributions are the only non-zero contributions to the asymptotic expansion in the limit $t/s \to 0$. In particular, both of the (c–h) contributions and all the contributions with soft or ultrasoft momenta are zero because they generate scaleless integrals.

The (h–h) region generates a contribution given by Taylor expansion of the integrand in the vector r. Every diagram from this contribution corresponds to the forward-scattering configuration, i.e. $p_3 = -p_1$ and $p_4 = -p_2$. Such diagrams, with arbitrary numerators and integer powers of propagators, can be evaluated for general ε in terms of gamma functions by resolving recurrence relations following from integration by parts [68]. The first step of this procedure is to reduce the index a_5 or a_7 to zero and thereby obtain vertex massless diagrams. The latter reduction, in the scalar case [211], has been described in the previous section – see Example 8.7. (In the case without numerators, the reduction of the forward-scattering double box diagram was presented in [91].) In [217], two different procedures for the evaluation of the (h–h) contribution were used: one procedure along the lines of this standard recursion, and another that expands the integrand of the alpha representation in the variable t and uses tricks based on a shift of the dimension [228].

The (h–h) contribution starts from the next-to-leading order of the expansion, $1/s^3$, with the first term equal to the forward-scattering double box diagram. Although this contribution can be expressed in terms of gamma functions for general ε, we present it here as an expansion up to ε^0:

$$K^{(\mathrm{h-h}),\,\mathrm{NLO}}(x;\varepsilon) = x\left(\frac{1}{\varepsilon^3} - \frac{5}{\varepsilon^2} + \frac{5\pi^2+96}{6\varepsilon} + \frac{58\zeta(3)}{3} - \frac{\pi^2}{6} - 44\right), \qquad (8.85)$$

where $K(x;\varepsilon)$ is given by (8.84).

The (1c–1c) and (2c–2c) contributions are defined by generalizing the arguments used in Example 8.5 for the box. The (2c–2c) contribution is obtained by expanding the propagators $1/(k^2+2p_1{\cdot}k)$ and $1/(l^2+2p_1{\cdot}l)$ into Taylor series in k^2 and l^2, respectively, and also by expanding the integrands into Taylor series in $2p_1{\cdot}r$.

Thus, we need integrals of the type

$$J(a_1,\ldots,a_9;d,s,t) = \frac{1}{a_8!}\left(\frac{\partial}{\partial X}\right)^{a_8} \int\int \frac{\mathrm{d}^d k\,\mathrm{d}^d l}{(2p_1{\cdot}l)^{a_1}(l^2-2p_2{\cdot}l)^{a_2}(2p_1{\cdot}k)^{a_3}} \times \left.\frac{(l^2)^{a_9}}{(k^2-2p_2{\cdot}k)^{a_4}(k^2)^{a_5}[(k-l)^2]^{a_6}[(l+r)^2]^{a_7}}\right|_{X=0}, \qquad (8.86)$$

where $X = 2p_1 \cdot r$, for integer a_i. However, this integral, taken alone, is not generally regularized dimensionally. Only if we add the corresponding symmetrically related contribution (i.e. if we make the replacements $a_1 \leftrightarrow a_2$, $a_3 \leftrightarrow a_4$) do we obtain a result that exists within dimensional regularization. As usual, we introduce an auxiliary analytic regularization which enables us to consider the above terms separately. Let us introduce this regularization into lines 1 and 2 (although we could just as well choose 3 and 4, or all these four lines), i.e. $a_1 \to a_1 + x_1, a_2 \to a_2 + x_2$ plus a symmetrically related contribution which is given by interchanging $a_1 + x_1$ and $a_2 + x_2$. Only in the sum may we switch off this regularization, i.e. let $x_1 \to 0, x_2 \to 0$.

The integrals with $a_9 > 0$ can be reduced [217] to those with $a_9 = 0$ by recurrence relations following from IBP [68]. Then the alpha representation can be used for a general integral (8.86) with $a_9 = 0$. By introducing Mellin–Barnes integrations in an appropriate way, any resulting integral can be written in terms of a threefold Mellin–Barnes integral representation of a ratio of gamma functions. These integrals can be evaluated, as an expansion in ε, by the standard technique of shifting contours and expansion in MB integrals. First, the singularities in $x_1 - x_2$ are localized. Second, the same technique is applied for picking up the singularities in ε – see [217] for details.

The collinear contributions start from the order $1/(s^2 t)$. The following are the results for the LO ($1/(s^2 t)$) and NLO ($1/(s^3)$) contributions [217]:

$$
\begin{aligned}
K^{(\mathrm{c-c}),\,\mathrm{LO+NLO}}(x;\varepsilon) = x^{-2\varepsilon} \Bigg[& -\frac{4}{\varepsilon^4} - \frac{3\ln x}{\varepsilon^3} + \frac{5\pi^2}{2\varepsilon^2} \\
& - \left(\frac{\pi^2}{2}\ln x - \frac{65}{3}\zeta(3)\right)\frac{1}{\varepsilon} + 14\zeta(3)\ln x + \frac{29}{30}\pi^4 \Bigg] \\
& + x^{1-2\varepsilon} \left[\frac{1}{\varepsilon^3} + (2\ln x - 5)\frac{1}{\varepsilon^2} - \left(6\ln x + \frac{7\pi^2}{6} - 12\right)\frac{1}{\varepsilon} \right. \\
& \left. + \pi^2 \ln x + \frac{46}{3}\zeta(3) + \frac{13\pi^2}{2} \right] .
\end{aligned} \tag{8.87}
$$

Combining the result above with the order $1/s^3$ (h–h) contribution given by (8.85), we obtain

$$
\begin{aligned}
K(x,\varepsilon) = & -\frac{4}{\varepsilon^4} + \frac{5\ln x}{\varepsilon^3} - \left(2\ln^2 x - \frac{5}{2}\pi^2\right)\frac{1}{\varepsilon^2} \\
& - \left(\frac{2}{3}\ln^3 x + \frac{11}{2}\pi^2 \ln x - \frac{65}{3}\zeta(3)\right)\frac{1}{\varepsilon} \\
& + \frac{4}{3}\ln^4 x + 6\pi^2 \ln^2 x - \frac{88}{3}\zeta(3)\ln x + \frac{29}{30}\pi^4 \\
& + 2x \left(\frac{1}{\varepsilon}\left(\ln^2 x - 2\ln x + \pi^2 + 2\right) \right. \\
& \left. - \frac{1}{3}\left\{4\ln^3 x + 3\ln^2 x + (5\pi^2 - 36)\ln x + 2[33 + 5\pi^2 - 3\zeta(3)]\right\} \right) \\
& + O(x^2 \ln^3 x) .
\end{aligned} \tag{8.88}
$$

The expansion in the limit $s/t \to 0$ has the same structure as in the previous case: only the (h–h) and (c–c) contributions are non-zero. However, in this case, there are three (c–c) contributions, and the poles with respect to the auxiliary parameter of analytic regularization are of up to second order, as in Example 8.8.

The above initial terms of the expansion are in agreement with an explicit analytic result [213] obtained by use of alpha parameters and an intermediate fivefold Mellin–Barnes representation:

$$
\begin{aligned}
K(x,\varepsilon) = & -\frac{4}{\varepsilon^4} + \frac{5\ln x}{\varepsilon^3} - \left(2\ln^2 x - \frac{5}{2}\pi^2\right)\frac{1}{\varepsilon^2} \\
& - \left(\frac{2}{3}\ln^3 x + \frac{11}{2}\pi^2 \ln x - \frac{65}{3}\zeta(3)\right)\frac{1}{\varepsilon} \\
& + \frac{4}{3}\ln^4 x + 6\pi^2 \ln^2 x - \frac{88}{3}\zeta(3)\ln x + \frac{29}{30}\pi^4 \\
& - \left[2\,\mathrm{Li}_3\left(-x\right) - 2\ln x\,\mathrm{Li}_2\left(-x\right) - \left(\ln^2 x + \pi^2\right)\ln(1+x)\right]\frac{2}{\varepsilon} \\
& - 4\left[S_{2,2}(-x) - \ln x S_{1,2}(-x)\right] + 44\,\mathrm{Li}_4\left(-x\right) \\
& - 4\left[\ln(1+x) + 6\ln x\right]\mathrm{Li}_3\left(-x\right) \\
& + 2\left(\ln^2 x + 2\ln x\ln(1+x) + \frac{10}{3}\pi^2\right)\mathrm{Li}_2\left(-x\right) \\
& + \left(\ln^2 x + \pi^2\right)\ln^2(1+x) \\
& - \frac{2}{3}\left[4\ln^3 x + 5\pi^2 \ln x - 6\zeta(3)\right]\ln(1+x)\;, \qquad (8.89)
\end{aligned}
$$

where, in addition to the polylogarithms (A.57), the generalized polylogarithms defined by (A.58) also occur.

The very existence of this analytic result provides a new, curious two-loop example (in addition to Example 8.8) of a diagram whose analytic evaluation happens to be possible while the terms of its expansion (here in the limit $t/s \to 0$) cannot be evaluated so easily. It turns out that even a general massless on-shell double box with arbitrary numerators and integer powers of the propagators can be evaluated analytically – see an algorithm developed in [217]. Let us note, for completeness, that in the case of the second two-loop non-trivial scattering graph, the non-planar double box shown in Fig. 8.7, the master diagram has been evaluated analytically in [230]. This result is expressed in terms of the same class of functions as for the master planar double box. A reduction of general massless on-shell non-planar double box diagrams with arbitrary numerators and integer powers of the propagators

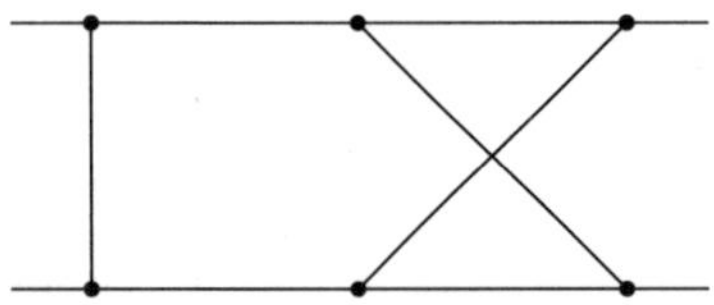

Fig. 8.7. Non-planar double box

has been presented in [3]. For a simpler scattering graph, called the pentabox, the problem of its evaluation has been solved in [2].

It is possible to analytically evaluate [215, 123] double boxes even with one leg off shell, $p_1^2 = q^2 \neq 0, p_2^2 = p_3^2 = p_4^2 = 0$. In [215], an asymptotic expansion, obtained within the strategy of expansion by regions, in the leading power when $q^2 \to 0$ has been used for crucial checks.

8.6 General Prescriptions

Now we are ready to present the list of regions which are involved in the Sudakov and Regge limits:

(h), (s), (us), (1c), (2c), (1uc), (2uc), (1uuc), (2uuc) , ...

When considering various regions, there is a risk of double counting. To illustrate this fact consider again Examples 8.10 and 8.11. There are contributions from the (h–1c) region (in both examples) and the (h–2c) region (in Example 8.10) which can be evaluated in terms of gamma functions for general ε and start from the next-to-leading order, m^2. Consider now another choice of the loop momenta, where the momentum l is chosen as the momentum of the second line. In this case we can recognize a non-zero contribution from the (h–s) region (where k is hard and this new momentum l is soft). However, we obtain nothing but double counting because these contributions coincide with the previous ones: this can be seen by analysing this shift of the variables $l \to l - p_2$. This example shows that one has to be careful when testing different choices of the loop momenta. Of course, there is no sense in choosing different assignments of the loop momenta for the hard regions. But for other types of regions, certain choices of loop momenta are crucial – see, for example, the two-loop examples in the case of the threshold expansion in Chapter 7, particularly in the case of the ultrasoft regions. There is a possibility of completely avoiding such double counting by using the alpha representation and a similar strategy of expansion by regions in the language of alpha parameters – see Sect. 9.1.

In the Sudakov and Regge limits, individual (1c), (2c), (3c), . . . , (1c–h), (2c–h), . . . contributions sometimes are not regularized by dimensional regularization, in which case a natural procedure is to introduce an auxiliary analytic regularization, which is switched off after picking up spurious poles. The structure of regions can be very rich. To illustrate this point let us consider one more example: the scalar master box diagram of Fig. 8.8 with three massless lines, four lines with mass m, and on-shell external momenta $p_i^2 = m^2$, $i = 1, 2, 3, 4$, in the leading order, m^0, of expansion in the (Sudakov) limit of fixed-angle scattering, $m^2 \ll |s|, |t|$.

The following family of seventeen regions participates here:

(h–h), (1c–h), . . . , (4c–h), (1c–1c), . . . , (4c–4c),

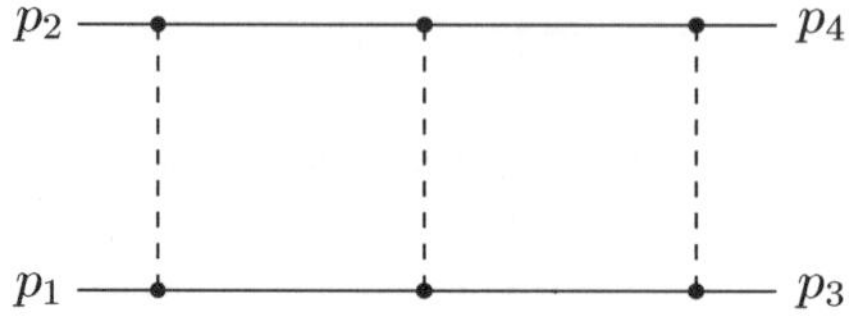

Fig. 8.8. Double box diagram with external momenta $p_i^2 = m^2$

(1c–3c), (2c–4c), (1c–4c), (2c–3c),
(1uc–2c), (2uc–1c), (3uc–4c), (4uc–3c).

The contributions (1c–1c), (3c–3c), (1c–3c), as well as the symmetrical contributions (2c–2c), (4c–4c) and (2c–4c), are not individually regularized by dimensional regularization. The poles in the auxiliary analytic regularization turn out to be of the second order and are cancelled in the sum. After adding the rest of the contributions, the poles of the third and fourth order in ε are cancelled and, using the techniques described above, one can observe that the coefficient of the double pole of the initial diagram in ε (which can be evaluated simply) is reproduced in the limit $m \to 0$ in the sum of the contributions on the right-hand side.

Let us point out that the expansion in k^2 in the collinear contributions (as well as in the soft contributions in the case of Limit (A) considered in Chap. 6) reminds us of the prescriptions of the eikonal[2] approximation (see, e.g., [100]). There are, however, essential differences. In the eikonal approximation, both propagators containing $k^2 - 2p_1 \cdot k$ or $k^2 - 2p_2 \cdot k$ would be expanded in k^2, while we expand only one of them. Moreover, a (momentum) cut-off is implied when the eikonal approximation is used, while we systematically work only with dimensional (and, sometimes, analytic) regularization without any other cut-off and have manifest homogeneity of the terms of the expansion from the beginning. For us, such a contribution with both propagators expanded in k^2 is zero because it involves a scaleless integral.

8.7 Summing up Sudakov Logarithms

8.7.1 Form Factor

As we saw in the examples, there is a double power of a logarithm per loop in the Sudakov limit. Indeed, the *leading* logarithms in QED/QCD enter at the n-loop level in the combination $\alpha^n \ln^{2n} Q^2$ (with α_s in place of α in QCD). Starting from the pioneering work of Sudakov [226], it has become well known how to sum up these leading logarithms. The corresponding nice

[2]The bad news is that there are doubts whether the eikonal approximation is always valid. But the good news is that the method of expansion by regions can be successfully applied. At least one can take the diagrams considered in [100] and expand them in accordance with the prescriptions formulated in this section.

combinatorial arguments, which are the same as those used in the summation of IR and collinear divergences, can be found in many books (see, e.g., [190]) and are not presented here. In Limit (*ii*), the leading logarithms contributing to the electron form factor can be summed up [226] to give

$$\exp\left(-\frac{\alpha}{4\pi}\ln^2\frac{Q^2}{M^2}\right) , \tag{8.90}$$

and, in Limit (*i*), one has [150]

$$\exp\left(-\frac{\alpha}{2\pi}\ln^2\frac{Q^2}{m^2}\right) . \tag{8.91}$$

The summation of the leading and *subleading* (i.e. $\alpha^n \ln^{2n-1} Q^2$) logarithms was studied in [226, 150, 77, 120] and [206, 73, 203, 204, 154, 155, 173], respectively.

Let us rely on evolution equations obtained by means of non-trivial diagrammatic analyses within standard methods [73, 75, 203, 204, 155] and insert, as an input, the leading-power behaviour of diagrams obtained within the strategy of expansion by regions following [159].

Further, let us consider the vector form factor in the SU(N) non-Abelian gauge theory which determines the amplitude of fermion scattering in an external Abelian field – see Fig. 8.1a. The form factor is parameterized according to (7.65). The contribution of F_2 is, however, suppressed by a power of Q^2 so that we are left with F_1. At the tree level (in the Born approximation), the form factor is

$$F_{\mathrm{B},\mu} = \bar{\psi}(p_2)\gamma_\mu\psi(p_1) . \tag{8.92}$$

In Limit (*i*), the evolution equation is of the form [72, 203]

$$\frac{\partial}{\partial \ln Q^2} F = \left[\int_{m^2}^{Q^2} \frac{\mathrm{d}x}{x}\gamma(\alpha(x)) + \zeta(\alpha(Q^2)) + \xi(\alpha(m^2))\right] F , \tag{8.93}$$

with the solution

$$\begin{aligned} F^{(i)} &= F_0(\alpha(m^2)) \\ &\times \exp\left\{\int_{m^2}^{Q^2} \frac{\mathrm{d}x}{x}\left[\int_{m^2}^{x} \frac{\mathrm{d}x'}{x'}\gamma(\alpha(x')) + \zeta(\alpha(x)) + \xi(\alpha(m^2))\right]\right\} . \end{aligned} \tag{8.94}$$

In Limit (*ii*), one has [155]

$$\begin{aligned} &\frac{\partial}{\partial \ln Q^2} F \\ &= \left[\int_{M^4/Q^2}^{Q^2} \frac{\mathrm{d}x}{x}\gamma(\alpha(x)) + \zeta(\alpha(Q^2)) + \zeta_1(\alpha(M^4/Q^2)) + \xi(\alpha(M^2))\right] F , \end{aligned} \tag{8.95}$$

with the solution

$$F^{(ii)} = F_0(\alpha(M^2)) \times \exp\left\{ \int_{M^2}^{Q^2} \frac{\mathrm{d}x}{x} \left[\int_{M^2}^{x} \frac{\mathrm{d}x'}{x'} \gamma(\alpha(x')) + \zeta(\alpha(x)) + \xi(\alpha(M^2)) \right] + \int_{M^4/Q^2}^{M^2} \frac{\mathrm{d}x}{x} \left[\int_{x}^{M^2} \frac{\mathrm{d}x'}{x'} \gamma(\alpha(x')) + \zeta_1(\alpha(x)) \right] \right\} . \tag{8.96}$$

The functions F_0 and ξ are, generally, different in these two limits. We omit the corresponding label ((i) or (ii)) and the Lorentz index μ for brevity.

As we saw in the examples in this chapter, there are logarithms of types $\ln(Q^2/m^2)$ and $\log(Q^2/\mu^2)$, where μ is the massive parameter which is always implicitly present within dimensional regularization. We shall call the former logarithms 'infrared' because they involve logarithms of the masses and are singular in the massless limit. The latter logarithms will be called 'renormalization group' (RG) logarithms because they are connected with the UV properties of the theory and are effectively summed up by solving RG equations [27, 74]. Similarly, the evolution equations (8.93) and (8.95), with information about the first orders of perturbation theory used as an input, gives the possibility of summing up the (sub)leading IR logarithms. Since the RG logarithms also contribute to the asymptotic behaviour at large Q^2, starting from the level of subleading logarithms, we have to keep RG corrections to the leading logarithmic approximation as well as single one-loop IR and RG logarithms. In this approximation, the form factor $F^{(i)}$ takes the form

$$F^{(i)} = F_0(\alpha) \exp\left[\int_{m^2}^{Q^2} \frac{\mathrm{d}x}{x} \int_{m^2}^{x} \frac{\mathrm{d}x'}{x'} \gamma(\alpha(x')) + (\zeta(\alpha) + \xi(\alpha)) \ln\left(Q^2/m^2\right) \right] \tag{8.97}$$

and, in the case of Limit *(ii)*, we have

$$F^{(ii)} = F_0(\alpha) \times \exp\left\{ \int_{M^2}^{Q^2} \frac{\mathrm{d}x}{x} \int_{M^2}^{x} \frac{\mathrm{d}x'}{x'} \gamma(\alpha(x')) + \int_{M^4/Q^2}^{M^2} \frac{\mathrm{d}x}{x} \int_{x}^{M^2} \frac{\mathrm{d}x'}{x'} \gamma(\alpha(x')) + (\zeta(\alpha) + \zeta_1(\alpha) + \xi(\alpha)) \ln\left(Q^2/M^2\right) \right\} . \tag{8.98}$$

All the functions in the exponent have to be computed in one loop, and the one-loop running of α (according to the RG equations) in the argument of the γ function should be taken into account.

We proceed in the covariant gauge, where the self-energy insertions in the external fermion lines are independent of Q. The calculation of the one-loop triangle diagram gives

$$F^a = \frac{\alpha}{2\pi} C_{\mathrm{F}} \left[-V_0^a + 2V_1^a + 2(1-2\varepsilon)V_2^a - \bar{V}_2^a \right] F_{\mathrm{B}} , \tag{8.99}$$

where $a = (i)$ or (ii) and C_{F} is defined by (1.29) The functions V_i^a and $\bar{V}_2^a$ are given by

$$F_{8.1} = -\mathrm{i}\pi^{d/2} \mathrm{e}^{-\gamma_{\mathrm{E}}\varepsilon} \frac{V_0^{(i)}}{Q^2} ,$$

$$F_{8.1;\mu} = -\mathrm{i}\pi^{d/2} \mathrm{e}^{-\gamma_{\mathrm{E}}\varepsilon} (p_1 + p_2)_\mu \frac{V_1^{(i)}}{Q^2} ,$$

$$F_{8.1;\mu\nu} = \mathrm{i}\pi^{d/2} \mathrm{e}^{-\gamma_{\mathrm{E}}\varepsilon} \left(g_{\mu\nu} V_2^{(i)} - \frac{(p_1)_\mu (p_2)_\nu - (p_2)_\mu (p_1)_\nu}{Q^2} \bar{V}_2^{(i)} \right) . \quad (8.100)$$

Here $F_{8.1}$ is given by (8.3), and $F_{8.1;\mu}$ and $F_{8.1;\mu\nu}$ are defined in a similar way with factors k_μ and $k_\mu k_\nu$, respectively, in the numerator. The corresponding functions in Limit (ii) are defined similarly, with $F_{8.1}$ replaced by $F_{8.2}$ – see (8.16). In the representation of the third integral, tensor structures that are power suppressed in the expansion when $Q \to \infty$ are omitted.

Keeping the leading power in (8.4), (8.11), (8.17) and (8.18), as well as in similar expansions for the integrals with numerators, we obtain,[3] in Limit (i),

$$V_0^{(i)\mathrm{h}} = \frac{1}{\varepsilon^2} - \frac{1}{\varepsilon} \ln Q^2 + \frac{1}{2} \ln^2 Q^2 - \frac{\pi^2}{12} ,$$

$$V_0^{(i)\mathrm{c}} = -\frac{1}{\varepsilon^2} + \frac{1}{\varepsilon} \ln Q^2 - (\ln m^2) \ln Q^2 + \frac{1}{2} \ln^2 m^2 + \frac{5\pi^2}{12} ,$$

$$V_0^{(i)} \sim \frac{1}{2} \ln^2 \frac{Q^2}{m^2} + \frac{\pi^2}{3} ; \quad (8.101)$$

$$V_1^{(i)\mathrm{h}} = -\frac{1}{\varepsilon} + \ln Q^2 - 2 , \quad V_1^{(i)\mathrm{c}} = \frac{1}{\varepsilon} - \ln m^2 + 1 ,$$

$$V_1^{(i)} \sim \ln \frac{Q^2}{m^2} - 1 ; \quad (8.102)$$

$$V_2^{(i)} \sim V_2^{(i)\mathrm{h}} = \frac{1}{4} \left(\frac{1}{\varepsilon} - \ln \frac{Q^2}{\mu^2} + 3 \right) , \quad (8.103)$$

$$\bar{V}_2^{(i)} \sim \bar{V}_2^{(i)\mathrm{h}} \sim \frac{1}{2} . \quad (8.104)$$

The pole in (8.103) and the corresponding logarithm are of UV nature. Observe that this RG logarithm contributes only to the function ζ. We explicitly insert there the dependence on the renormalization parameter μ which is usually implied (and omitted) within dimensional regularization. All other poles in the hard contributions are cancelled, together with the poles in the collinear and/or ultrasoft contributions. Correspondingly, all other logarithms $\ln Q^2$ are naturally combined with logarithms of the mass squared and provide the IR logarithms $\ln(Q^2/m^2)$.

[3] In fact, in summing up the leading and next-to-leading logarithms, we do not need the finite parts present in these and similar results. These parts are, however, needed for the summation of NNLL logarithms (see [158]).

The hard contribution (in the leading power) in Limit (*ii*) is the same. The new ingredients are

$$
\begin{aligned}
V_0^{(ii)\mathrm{c}} &= -\frac{2}{\varepsilon^2} + \frac{2}{\varepsilon}\ln M^2 - \ln^2 M^2 + \frac{\pi^2}{6} \,, \\
V_0^{(ii)\mathrm{us}} &= \frac{1}{\varepsilon^2} + \frac{1}{\varepsilon}\left(\ln Q^2 - 2\ln M^2\right) + \frac{1}{2}\ln^2 Q^2 \\
&\qquad -2(\ln M^2)\ln Q^2 + 2\ln M^2 + \frac{\pi^2}{4} \,, \\
V_0^{(ii)} &\sim \ln^2\frac{Q^2}{M^2} + \frac{\pi^2}{3} \,;
\end{aligned}
\tag{8.105}
$$

$$
V_1^{(ii)\mathrm{c}} = \frac{1}{\varepsilon} - \ln M^2 + 2 \,, \quad V_1^{(ii)} \sim \ln\frac{Q^2}{M^2} \,. \tag{8.106}
$$

The strategy of expansion by regions enables us to straightforwardly identify the regions relevant to determining any given function of the evolution equation and to compute them separately to the required accuracy. For example, the anomalous dimensions $\gamma(\alpha)$ and $\zeta(\alpha)$ in an arbitrary order are completely determined by the coefficients of the double and single poles, respectively, of the hard contribution. In contrast, the functions $\xi(\alpha)$ and $F_0(\alpha)$ determine the initial conditions for the evolution equation and depend on the infrared sector of the model. To determine the function $\xi(\alpha)$ one has to know also the singularities of the collinear contribution, while $F_0(\alpha)$ requires complete information about the contributions of all the regions.

From the one-loop result, we find

$$
\gamma(\alpha) = -C_{\mathrm{F}}\frac{\alpha}{2\pi} \,. \tag{8.107}
$$

Moreover, it is clear from the above expressions that in Limit (*i*), the total double logarithm of Q^2 comes from the hard contribution, while in Limit (*ii*) one-half of the double logarithm comes from the ultrasoft contribution. This explicitly determines the scale of the coupling constant in the second order logarithmic derivative of the form factor with respect to Q. This scale is Q^2 in Limit (*i*) and M^2 in Limit (*ii*). Furthermore, all the Q-dependent terms in the ultrasoft contribution of (8.105) are related to the γ term, and therefore $\zeta_1(\alpha) = 0$ and $\xi(\alpha) = 0$. Moreover, we find

$$
\zeta(\alpha) = 3C_{\mathrm{F}}\frac{\alpha}{4\pi} \,. \tag{8.108}
$$

To complete the Q-independent part of the one-loop corrections to the form factor, one has to include the fermion wave function renormalization determined by the self-energy insertions into the external lines, which are nothing but the soft contributions,[4] which, for a 1PI vertex diagram, play a

[4] Such integrals depend on one scale, m^2 or M^2, and it is natural to think of the loop momentum as soft. See a similar situation at the end of the discussion of Example 8.1, where the connection between collinear and soft contributions was exemplified.

minor role in the leading-power approximation, within our strategy of expansion by regions. (They are definitely more important in the standard strategy of expansion by regions, with momentum space cut-offs [75].) In Limit (*i*), these insertions lead the factor

$$1 + C_{\rm F}\frac{\alpha}{4\pi}\left(-\frac{1}{\varepsilon} + \ln\frac{m^2}{\mu^2} + \frac{1}{2}\right) , \tag{8.109}$$

and, in Limit (*ii*),

$$1 + C_{\rm F}\frac{\alpha}{4\pi}\left(-\frac{1}{\varepsilon} + \ln\frac{M^2}{\mu^2} - 1\right) . \tag{8.110}$$

The UV poles of (8.109) and (8.110) cancel the UV pole that arises from (8.103), in accordance with the Ward identity and the non-renormalization property of the conserved vector current. Finally, in the next-to-leading logarithmic approximation, we find the form factor (*i*) to be

$$\begin{aligned} F^{(i)} &= F_{\rm B}\left[1 - C_{\rm F}\frac{\alpha}{4\pi}\left(\frac{7}{2} + \frac{2\pi^2}{3}\right)\right] \\ &\quad \times \exp\left[\frac{C_{\rm F}}{2\pi}\left(-\int_{m^2}^{Q^2}\frac{{\rm d}x}{x}\int_{m^2}^{x}\frac{{\rm d}x'}{x'}\alpha(x') + 3\alpha\ln\left(Q^2/m^2\right)\right)\right] \end{aligned} \tag{8.111}$$

and the form factor (*ii*) to be

$$\begin{aligned} F^{(ii)} &= F_B\left[1 - C_{\rm F}\frac{\alpha}{4\pi}\left(1 + \frac{2\pi^2}{3}\right)\right] \\ &\quad \times \exp\left[\frac{C_{\rm F}}{2\pi}\left(-\int_{M^2}^{Q^2}\frac{{\rm d}x}{x}\int_{M^2}^{x}\frac{{\rm d}x'}{x'}\alpha(x')\right.\right. \\ &\quad \left.\left. -\int_{M^4/Q^2}^{M^2}\frac{{\rm d}x}{x}\int_{x}^{M^2}\frac{{\rm d}x'}{x'}\alpha(x') + 3\alpha\ln\left(Q^2/M^2\right)\right)\right] . \end{aligned} \tag{8.112}$$

Equations (8.111) and (8.112) are in agreement with a similar result of [154, 155].

Let us show, using expansion by regions, that the single logarithmic term in the exponents of (8.111) and (8.112) can be symbolically represented as

$$3\ln\left(Q^2/m^2\right) = 4\ln\left(Q^2/m^2\right)_{\rm (IR)} - \ln\left(Q^2/m^2\right)_{\rm (RG)} , \tag{8.113}$$

where $m \to M$ in the second case. We explicitly separate the IR and renormalization group[5] logarithms, which are of essentially different nature and are related here to the V_1 and V_2 integrals, respectively.

[5]Although this logarithm originates from the integration over the virtual momentum region between the M^2 and Q^2 scales and does not depend on μ, we call it 'RG' because it is directly related to and can be read off the renormalization group properties of the Abelian vertex and the fermion wave function.

It is useful to distinguish between the collinear[6] and IR poles in ε (resulting in logarithms of Q^2) in the hard contribution. The collinear logarithms, in a physical (Coulomb or axial) gauge originate only from the self-energy insertions into the external fermion lines [120, 73, 203, 204, 119, 118] and therefore are universal, i.e. independent of specific processes.

Within expansion by regions, the total pole part in ε of the collinear contribution in Limit (*i*) cancels both the IR and the collinear poles of the hard contribution. Hence it is not straightforward to separate the collinear logarithms. Let us note that the collinear poles (but not the IR ones) can be removed by considering Limit (*iii*), with a finite fermion mass and zero gauge boson mass. This enables us to distinguish the IR and collinear poles. In Limit (*iii*), the collinear poles in the hard contribution are cancelled by the poles of the collinear contribution when the logarithms are obtained, while the IR poles in the hard contribution are not cancelled. Thus one can determine the origin of the poles in the hard contribution and therefore the origin of the logarithms. For example, in the hard contribution to the integral V_1, the pole is collinear since it is cancelled in Limit (*iii*) by

$$V_1^{(iii)\mathrm{c}} = \frac{1}{\varepsilon} - \ln m^2 + 2\,. \tag{8.114}$$

Thus the single infrared logarithm in (8.113) is of collinear origin and therefore is universal. We should emphasize that this is not true for the RG logarithm in (8.113), which depends on a specific amplitude and model. For example, this logarithm is different for a scalar form factor or for a vector form factor in a model containing an additional Yukawa interaction of the fermions with the scalar bosons.

A less trivial example is the Feynman integral (8.20):

$$F_{8.3} = -\mathrm{i}\pi^{d/2} \mathrm{e}^{-\gamma_{\mathrm{E}}\varepsilon} \frac{V_0^{(iii)}}{Q^2}\,. \tag{8.115}$$

In the hard contribution to this integral in Limit (*iii*), the collinear part of the double pole is cancelled by poles in the collinear contribution (8.21) and transforms into the logarithm of Q^2/m^2, but an IR single pole in the leading-power asymptotics is left:

$$V_0^{(iii)} \sim -\frac{1}{\varepsilon} \ln \frac{Q^2}{m^2} + \left(\ln \frac{Q^2}{m^2} \right) \ln Q^2 - \frac{1}{2} \ln^2 \frac{Q^2}{m^2} - \frac{\pi^2}{6}\,. \tag{8.116}$$

8.7.2 Four-Fermion Amplitude

Now we turn to the Sudakov limit for the four-fermion amplitude (following [159]), where all the invariant energy and momentum transfers of the

[6] Although IR divergences are sometimes called 'soft' we prefer not to use this word here, in order not to interfere with the word 'soft' used to denote soft regions/contributions. The same word 'collinear' is, however, used to denote both collinear regions/contributions and collinear divergences.

process are much larger than the typical mass scale of the internal particles, i.e. $|s| \sim |t| \sim |u| \gg M^2$. Besides the extra kinematical variable, the analysis of the four-fermion amplitude is made more complicated by the presence of different colour and Lorentz structures. The Born amplitude, for example, can be expanded in a basis of colour/chiral amplitudes:

$$A_{\mathrm{B}} = \frac{\mathrm{i}g^2}{s} A^\lambda = \frac{\mathrm{i}g^2}{s} T_{\mathrm{F}} \left(-\frac{A^{\mathrm{d}}_{\mathrm{LL}} + A^{\mathrm{d}}_{\mathrm{LR}}}{N} + A^{\mathrm{c}}_{\mathrm{LL}} + A^{\mathrm{c}}_{\mathrm{LR}} + \ldots \right) , \tag{8.117}$$

where

$$\begin{aligned} A^\lambda &= \bar{\psi}_2(p_2) t^a \gamma_\mu \psi_1(p_1) \bar{\psi}_4(p_4) t^a \gamma_\mu \psi_3(p_3) , \\ A^d_{\mathrm{LL}} &= \bar{\psi}^{\,i}_{2\mathrm{L}} \gamma_\mu \psi^{\,i}_{1\mathrm{L}} \bar{\psi}^{\,j}_{4\mathrm{L}} \gamma_\mu \psi^{\,j}_{3\mathrm{L}} , \quad A^c_{\mathrm{LR}} = \bar{\psi}^{\,j}_{2\mathrm{L}} \gamma_\mu \psi^{\,i}_{1\mathrm{L}} \bar{\psi}^{\,i}_{4\mathrm{R}} \gamma_\mu \psi^{\,j}_{3R} , \end{aligned} \tag{8.118}$$

$\psi_{\mathrm{L/R}} = (1 \mp \gamma_5)\psi/2$, T_{F} is defined by (1.29), the ellipsis stands for terms with L$\leftrightarrow$R, and the superscripts d and c denote direct and crossed tree diagrams, respectively. Further, t^a is the SU(N) generator, p_1, p_3 are the incoming momenta, and p_2, p_4 are the outgoing momenta, and the Mandelstam variables are $s = (p_1 - p_2)^2$, $t = (p_1 - p_4)^2$ and $u = (p_1 + p_3)^2 = -(s + t)$. For the moment we consider a parity-conserving theory. Hence only two chiral amplitudes are independent, for example LL and LR. Similarly, only two colour amplitudes are independent, for example λ and d.

Let us first compute the one-loop corrections in Limit (i), keeping the leading-power behaviour. Vertex diagrams are expanded as in the previous subsection, with the following total contribution:

$$\frac{\alpha}{\pi} \left(C_{\mathrm{F}} \left(-V_0 + 2V_1 \right) + \frac{C_{\mathrm{A}}}{2} V_0 + \ldots \right) A_{\mathrm{B}} , \tag{8.119}$$

where the ellipsis stands for the contribution without IR logarithms, and C_{A} is defined by (1.29). In the vertex correction involving the gauge boson self-coupling, a contribution of the form (8.114) appears with m replaced by M, and we have used the fact that the pole (logarithmic) term of V_1^{c} is the same in both (8.102) and (8.106).

When dealing with (crossed) box diagrams, it is convenient to take into account polynomials of the loop momenta in the numerator by shifting the dimension d and the powers of propagators. The box diagrams give

$$\begin{aligned} &\frac{\mathrm{i}g^2}{s} \frac{\alpha}{4\pi} \Bigg\{ (5B_0 + 3B_1 + 4B_2 - 3B_3 - 6B_4 + 2B_5) \\ &\times \left[\left(C_{\mathrm{F}} - \frac{T_{\mathrm{F}}}{N} \right) \bar{\psi}_2 t^a \gamma_\mu \psi_1 \bar{\psi}_4 t^a \gamma_\mu \psi_3 + C_{\mathrm{F}} \frac{T_{\mathrm{F}}}{N} \bar{\psi}_2 \gamma_\mu \psi_1 \bar{\psi}_4 \gamma_\mu \psi_3 \right] \\ &- (3B_0 + B_1 - B_3 - 2B_4 + 2B_5) \\ &\times \left[\left(C_{\mathrm{F}} - \frac{T_{\mathrm{F}}}{N} \right) \bar{\psi}_2 t^a \gamma_\mu \gamma_5 \psi_1 \bar{\psi}_4 t^a \gamma_\mu \gamma_5 \psi_3 + C_{\mathrm{F}} \frac{T_{\mathrm{F}}}{N} \bar{\psi}_2 \gamma_\mu \gamma_5 \psi_1 \bar{\psi}_4 \gamma_\mu \gamma_5 \psi_3 \right] \Bigg\} , \end{aligned} \tag{8.120}$$

where

$$\begin{aligned}
&B_0 = -\mathrm{i}s\,J(1,1,1,1,1)\,, \quad B_1 = st\,J(1,2,1,1,1)\,,\\
&B_2 = st\,J(1,1,2,1,1)\,, \quad B_3 = st\,J(1,3,1,1,2)\,,\\
&B_4 = st\,J(2,1,2,1,2)\,, \quad B_5 = s^2\,J(1,1,2,2,2)\,,
\end{aligned} \tag{8.121}$$

and the functions $J(a_1,a_2,a_3,a_4,n)$ are proportional to scalar box integrals in shifted dimensions:

$$\begin{aligned}
J(a_1,a_2,a_3,a_4,n) &= \mathrm{i}^{-a_1-a_2-a_3-a_4-1+d/2+n}\prod_i (a_i-1)!\\
&\times \int \frac{\mathrm{d}^{d+n}k}{(k^2)^{a_1}(k^2-2p_1\cdot k-m^2)^{a_2}(k^2-2p_2\cdot k-m^2)^{a_3}}\\
&\times \frac{1}{(k^2-2(p_1-p_4)\cdot k+t)^{a_4}}\,.
\end{aligned} \tag{8.122}$$

The expansion of the master integral $J(1,1,1,1,0)$ has been described in Example 8.6. All other integrals (8.122) are expanded similarly. Keeping the leading power in the expansion and the terms containing the leading and subleading logarithms, we have

$$B_0 \sim B_5 \sim B_1 - B_3 - 2B_4 \sim 0\,, \tag{8.123}$$

$$\begin{aligned}
B_2^{\mathrm{h}}(s,t) &= -\frac{1}{\varepsilon^2} + \frac{1}{\varepsilon}\ln(-t) + \frac{1}{2}\ln^2(-s) - \ln(-s)\ln(-t)\,,\\
B_2^{\mathrm{c}}(s,t) &= \frac{1}{\varepsilon^2} - \frac{1}{\varepsilon}\ln(-t) - \ln M^2\ln(-t) - \frac{1}{2}\ln^2 M^2\,,\\
B_2(s,t) &\sim \frac{1}{2}\ln^2\left(\frac{-s}{M^2}\right) + \ln\left(\frac{-s}{M^2}\right)\ln\frac{t}{s}\,,
\end{aligned} \tag{8.124}$$

and the box contribution takes the form

$$-\frac{\mathrm{i}g^2}{s}\frac{\alpha}{\pi}B_2(s,t)\left[\left(C_{\mathrm{F}} - \frac{T_{\mathrm{F}}}{N}\right)A^\lambda + C_{\mathrm{F}}\frac{T_{\mathrm{F}}}{N}A^d\right]\,. \tag{8.125}$$

The crossed box diagrams give

$$\frac{\mathrm{i}g^2}{s}\frac{\alpha}{\pi}B_2(s,u)\left[\left(C_{\mathrm{F}} - \frac{T_{\mathrm{F}}}{N} - \frac{C_{\mathrm{A}}}{2}\right)A^\lambda + C_{\mathrm{F}}\frac{T_{\mathrm{F}}}{N}A^d\right]\,. \tag{8.126}$$

The rest of the one-loop logarithmic contributions from the vertex diagrams and the self-energy insertions are of renormalization group nature. In addition to the vertex and external-fermion self-energy diagrams considered in the previous subsection, the renormalization group logarithms set the scale of g in the Born amplitude as Q^2.

The total one-loop correction in the logarithmic approximation becomes

$$\frac{ig^2(Q^2)}{s}\frac{\alpha}{2\pi}\left(\left\{-C_{\mathrm{F}}\ln^2\left(\frac{-s}{M^2}\right)\right.\right.$$
$$\left.+\left[3C_{\mathrm{F}}-C_{\mathrm{A}}\ln\frac{u}{s}+2\left(C_{\mathrm{F}}-\frac{T_{\mathrm{F}}}{N}\right)\ln\frac{u}{t}\right]\ln\left(\frac{-s}{M^2}\right)\right\}A^{\lambda}$$
$$\left.+2\frac{C_{\mathrm{F}}T_{\mathrm{F}}}{N}\ln\frac{u}{t}\ln\left(\frac{-s}{M^2}\right)A^{d}\right)\,. \tag{8.127}$$

Note that the next-to-leading logarithms do not depend on chirality and are the same for both the LL and LR amplitudes.

The collinear logarithms can be now separated from the total one-loop correction. For each fermion–antifermion pair, they form the exponential factor (8.111). This factor incorporates in addition the renormalization group logarithms which are not absorbed by changing the normalization scale of the gauge coupling. The remaining single logarithms in (8.127) are of IR nature. Let us denote by $\tilde{A}$ the amplitude with the collinear logarithms factored out. This amplitude can be represented as a vector in the basis A^{λ}, A^{d} and satisfies the following matrix evolution equation [204, 31]:

$$\frac{\partial}{\partial\ln Q^2}\tilde{A}=\chi(\alpha(Q^2))\tilde{A}\,, \tag{8.128}$$

where χ is the matrix of the 'soft' (IR) anomalous dimensions. From (8.128), we find the elements of this matrix to be, in units of $\alpha/(4\pi)$,

$$\chi_{\lambda\lambda}=-2C_{\mathrm{A}}\ln\frac{u}{s}+4\left(C_{\mathrm{F}}-\frac{T_{\mathrm{F}}}{N}\right)\ln\frac{u}{t}\,,$$
$$\chi_{\lambda d}=4\frac{C_{\mathrm{F}}T_{\mathrm{F}}}{N}\ln\frac{u}{t}\,,\quad \chi_{d\lambda}=4\ln\frac{u}{t}\,,\quad \chi_{dd}=0\,. \tag{8.129}$$

The solution of (8.128) reads

$$\tilde{A}=A_1^0(\alpha(M^2))\exp\left[\int_{M^2}^{Q^2}\frac{\mathrm{d}x}{x}\chi_1(\alpha(x))\right]$$
$$+A_2^0(\alpha(M^2))\exp\left[\int_{M^2}^{Q^2}\frac{\mathrm{d}x}{x}\chi_2(\alpha(x))\right]\,, \tag{8.130}$$

where the χ_i are the eigenvalues of the χ matrix and the A_i^0 are Q-independent vectors.

Equations (8.124) imply that only the hard contributions are relevant to (8.128). This fixes the scale of α in this equation to be Q. For this reason, the matrix (8.129) is the same in Limit (*ii*). Hence, in the next-to-leading logarithmic approximation, the difference between the corrections to the four-quark amplitude in Limits (*i*) and (*ii*) is explained by the different factors (8.111) and (8.112).

In particular, in the Abelian case, there are no different colour amplitudes and there is only one anomalous dimension,

$$\chi = 4 \ln \frac{u}{t} \; . \tag{8.131}$$

8.8 An Application: Electroweak Processes at High Energies

Let us apply (following [159] where details can be found) the results of the previous section to the fermion annihilation process $f'\bar{f}' \to f\bar{f}$ at high energies. At the tree level (Born approximation), its amplitude is

$$A_{\mathrm{B}} = \frac{ig^2}{s} \sum_{I,J=\mathrm{L,R}} \left(T^3_{f'} T^3_f + t^2_{\mathrm{W}} \frac{Y_{f'} Y_f}{4} \right) A^{f'f}_{IJ} \; . \tag{8.132}$$

Here

$$A^{f'f}_{IJ} = \bar{f}'_I \gamma_\mu f'_I \bar{f}_J \gamma_\mu f_J \; , \tag{8.133}$$

$t_{\mathrm{W}} = \tan\theta_{\mathrm{W}}$, where θ_{W} is the Weinberg angle, and T_f and Y_f are the isospin and hypercharge, respectively, of the fermion, which depend on the fermion chirality.

To analyse the electroweak corrections to the above process we shall proceed in two steps. First, we use an approximation in which the W and Z bosons have the same mass M and the quarks and leptons are massless. Moreover, a fictitious photon mass λ is introduced to regularize the IR divergences, and the equal-mass case $\lambda = M$ is considered. Now we can work in terms of fields of unbroken phase and directly apply the results of the previous section for Limit (i) by projecting onto a relevant initial/final state. For each fermion–antifermion pair, the factor (8.111) takes the form

$$\exp\left[-\left(T_f(T_f+1) + t^2_{\mathrm{W}} \frac{Y^2_f}{4} \right) [L(s) - 3l(s)] \right] , \tag{8.134}$$

where

$$L(s) = \frac{g^2}{16\pi^2} \ln^2\left(\frac{-s}{M^2}\right) , \quad l(s) = \frac{g^2}{16\pi^2} \ln\left(\frac{-s}{M^2}\right) , \tag{8.135}$$

and we neglect the running of the coupling constant in the integral in (8.111) but fix the scale of the couplings g and $t_{\mathrm{W}} g$ in the double logarithmic contribution to be Q. The soft anomalous dimension for I and/or J =R is Abelian and, in units of $\alpha/(4\pi)$, reads

$$\chi = t^2_{\mathrm{W}} Y_{f'} Y_f \ln \frac{u}{t} \; . \tag{8.136}$$

The matrix of the soft anomalous dimensions for $I = J$ =L is the sum of the Abelian and non-Abelian parts:

$$\chi_{\lambda\lambda} = -4 \ln \frac{u}{s} + \left(t_{\mathrm{W}}^2 Y_{f'} Y_f + 2\right) \ln \frac{u}{t} , \quad \chi_{\lambda d} = \frac{3}{4} \ln \frac{u}{t} ,$$
$$\chi_{d\lambda} = 4 \ln \frac{u}{t} , \quad \chi_{dd} = t_{\mathrm{W}}^2 Y_{f'} Y_f \ln \frac{u}{t} . \tag{8.137}$$

Second, it is necessary to switch to the situation of a massless photon, where the corresponding IR-divergent contributions should be accompanied by integration of the real soft-photon radiation over some energy resolution ω_{res} to obtain an infrared-safe cross-section independent of the auxiliary photon mass. At the same time, the massive gauge bosons are supposed to be detected as separate particles. In practice, the energy resolution is much less than the W and Z boson mass so that the soft-photon emission is of QED nature. This cancels the IR singularities of the QED virtual correction. We should therefore separate the QED virtual correction from the complete result computed for a photon of some mass λ and then evaluate the QED virtual correction together with the real soft-photon radiation effects for vanishing λ. It is convenient to subtract the QED contribution computed for a photon of mass M from the result obtained for the virtual corrections and then take the limit $\lambda \to 0$ for the sum of the QED virtual and real photon contributions to the total amplitude. In the language of the approach of [106], this prescription means that we use the auxiliary photon mass λ as a variable of the evolution equation below the scale M, and the subtraction fixes a relevant initial condition for this differential equation. This leads to a modification of the factor (8.134) and the soft anomalous dimensions (8.136), (8.137).

The common factor for each fermion–antifermion pair becomes

$$\exp\left[-\left(T_f(T_f+1) + t_{\mathrm{W}}^2 \frac{Y_f^2}{4} - s_{\mathrm{W}}^2 Q_f^2\right)[L(s) - 3l(s)]\right] , \tag{8.138}$$

where $s_{\mathrm{W}} = \sin\theta_{\mathrm{W}}$. We then have

$$\chi = \left(t_{\mathrm{W}}^2 Y_{f'} Y_f - 4 s_{\mathrm{W}}^2 Q_{f'} Q_f\right) \ln \frac{u}{t} , \tag{8.139}$$

and the matrix of the soft anomalous dimension for $I = J = \mathrm{L}$ is

$$\chi_{\lambda\lambda} = -4 \ln \frac{u}{s} + \left(t_{\mathrm{W}}^2 Y_{f'} Y_f - 4 s_{\mathrm{W}}^2 Q_{f'} Q_f + 2\right) \ln \frac{u}{t} , \quad \chi_{\lambda d} = \frac{3}{4} \ln \frac{u}{t} ,$$
$$\chi_{d\lambda} = 4 \ln \frac{u}{t} , \quad \chi_{dd} = \left(t_{\mathrm{W}}^2 Y_{f'} Y_f - 4 s_{\mathrm{W}}^2 Q_{f'} Q_f\right) \ln \frac{u}{t} . \tag{8.140}$$

We can now estimate the dominant one- and two-loop logarithmic corrections. The renormalization group logarithms which are not included in (8.138) can be taken into account trivially by choosing the relevant scale of the coupling constants in the Born amplitude. At the same time, the remaining logarithmic corrections are the main feature of interest because they are expected to dominate the (still unknown) total two-loop electroweak corrections. The one-loop leading and subleading logarithms can be obtained directly from (8.127). The corresponding corrections to the chiral amplitudes read

$$\left\{\left(T_f(T_f+1)+t_{\mathrm{W}}^2\frac{Y_f^2}{4}-s_{\mathrm{W}}^2Q_f^2+(f\leftrightarrow f')\right)\left(T_{f'}^3T_f^3+t_{\mathrm{W}}^2\frac{Y_{f'}Y_f}{4}\right)\right.$$
$$\times\left[3l(s)-L(s)\right]+\left[\left[-4\ln\frac{u}{s}+2\ln\frac{u}{t}+\left(\ln\frac{u}{t}\right)t_{\mathrm{W}}^2Y_{f'}Y_f\right]T_{f'}^3T_f^3\right.$$
$$+\frac{3}{4}\left(\ln\frac{u}{t}\right)\delta_{IL}\delta_{JL}+\ln\frac{u}{t}$$
$$\left.\left.\times\left(t_{\mathrm{W}}^2Y_{f'}Y_f-4s_{\mathrm{W}}^2Q_{f'}Q_f\right)\left(T_{f'}^3T_f^3+t_{\mathrm{W}}^2\frac{Y_{f'}Y_f}{4}\right)\right]l(s)\right\}A_{IJ}^{f'f}\,,$$

where $\delta_{I\mathrm{L}}=1$ for $I=$L and zero otherwise. The two-loop leading (IR) logarithms are determined by the second-order term of the expansion of the double (soft $\times$ collinear) logarithmic part of the collinear factors (8.138). The corresponding corrections to the chiral amplitudes are

$$\frac{1}{2}\left(T_f(T_f+1)+t_{\mathrm{W}}^2\frac{Y_f^2}{4}-s_{\mathrm{W}}^2Q_f^2+(f\leftrightarrow f')\right)^2$$
$$\times\left(T_{f'}^3T_f^3+t_{\mathrm{W}}^2\frac{Y_{f'}Y_f}{4}\right)L^2(s)A_{IJ}^{f'f}\,. \tag{8.141}$$

The two-loop next-to-leading logarithms are generated by the interference between the first order terms of the expansions of the double (soft $\times$ collinear) and single (soft + collinear + RG) logarithmic exponents. The corresponding corrections to the chiral amplitudes are of the following form:

$$-\left(T_f(T_f+1)+t_{\mathrm{W}}^2\frac{Y_f^2}{4}-s_{\mathrm{W}}^2Q_f^2+(f\leftrightarrow f')\right)$$
$$\times\left[3\left(T_f(T_f+1)+t_{\mathrm{W}}^2\frac{Y_f^2}{4}-s_{\mathrm{W}}^2Q_f^2+(f\leftrightarrow f')\right)\left(T_{f'}^3T_f^3+t_{\mathrm{W}}^2\frac{Y_{f'}Y_f}{4}\right)\right.$$
$$+\left[-4\ln\frac{u}{s}+2\ln\frac{u}{t}+\left(\ln\frac{u}{t}\right)t_{\mathrm{W}}^2Y_{f'}Y_f\right]T_{f'}^3T_f^3+\frac{3}{4}\left(\ln\frac{u}{t}\right)\delta_{I\mathrm{L}}\delta_{J\mathrm{L}}$$
$$\left.+\left(\ln\frac{u}{t}\right)\left(t_{\mathrm{W}}^2Y_{f'}Y_f-4s_{\mathrm{W}}^2Q_{f'}Q_f\right)\left(T_{f'}^3T_f^3+t_{\mathrm{W}}^2\frac{Y_{f'}Y_f}{4}\right)\right]L(s)l(s)A_{IJ}^{f'f}\,.$$

With the expression for the chiral amplitudes to hand, one can compute the leading and subleading logarithmic corrections to the basic observables for $e^+e^-\rightarrow f\bar{f}$ using standard formulae. In [159] one can find numerical estimates which result, in particular, in the conclusion that at energies of 1 and 2 TeV the two-loop corrections are huge and amount to 5% and 7%, respectively.

8.9 An Application: Small-Angle Scattering of Massless Quarks at High Energies

Now we study the Regge limit (v), i.e. the regime of the small-angle scattering of massless quarks when $|s| \approx |u| \gg |t|$. The scattering amplitude can be expanded in the following basis of four independent colour/chiral four-quark operators:

$$\begin{aligned} A_{\mathrm{v}}^{\lambda} &= \bar{\psi}_4(p_4) t^a \gamma_\mu \psi_1(p_1) \bar{\psi}_2(p_2) t^a \gamma_\mu \psi_3(p_3)\,, \\ A_{\mathrm{a}}^{\mathrm{d}} &= \bar{\psi}_4(p_4) \gamma_\mu \gamma_5 \psi_1(p_1) \bar{\psi}_2(p_2) \gamma_\mu \gamma_5 \psi_3(p_3)\,, \end{aligned} \tag{8.142}$$

with similar expressions for A_{a}^{λ} and $A_{\mathrm{v}}^{\mathrm{d}}$. The expansion in the vector/axial basis (symbolized by the indices v and a) is now more convenient than the expansion in the left/right one.

According to Regge theory [76], the behaviour of the amplitude in this limit is determined by the Regge pole with gluon quantum numbers and negative signature. Thus the t channel exchange of the reggeized gluon (reggeon) saturates the amplitude, which takes the form

$$A(q\bar{q} \to q\bar{q}) = \frac{\mathrm{i}}{2t} \left[\left(\frac{s}{-t} \right)^{\omega(t)} + \left(\frac{s}{t} \right)^{\omega(t)} \right] \Gamma_{q\bar{q}}^2(t) A_{\mathrm{v}}^{\lambda}\,, \tag{8.143}$$

where $1 + \omega(t)$ is the reggeon trajectory and $\Gamma_{q\bar{q}}(t)$ is the quark–quark–reggeon vertex. In the Born approximation $\omega(t) = 0$ and $\Gamma_{q\bar{q}}(t) = g$. The reggeon trajectory is of special interest since it enters the kernel of the BFKL equation [167, 105, 11] describing small-x QCD processes (see [168] for a review). Let us consider the one-loop amplitude. The box diagrams give

$$\begin{aligned} \frac{\mathrm{i}g^2}{t} \frac{\alpha}{2\pi} \Bigg\{ (3B_1 + 2B_2) & \left[\left(C_{\mathrm{F}} - \frac{T_{\mathrm{F}}}{N} \right) A_{\mathrm{v}}^{\lambda} + C_{\mathrm{F}} \frac{T_{\mathrm{F}}}{N} A_{\mathrm{v}}^{\mathrm{d}} \right] \\ & -2B_1 \left[\left(C_{\mathrm{F}} - \frac{T_{\mathrm{F}}}{N} \right) A_{\mathrm{a}}^{\lambda} + C_{\mathrm{F}} \frac{T_{\mathrm{F}}}{N} A_{\mathrm{a}}^{\mathrm{d}} \right] \Bigg\}\,, \end{aligned} \tag{8.144}$$

where

$$\begin{aligned} B_1 &= st \left[J(1,3,1,1,2) + J(1,2,2,1,2) + J(1,2,1,2,2) \right. \\ &\quad \left. - J(1,2,1,1,1) - J(1,1,1,2,1) \right]\,, \\ B_2 &= st\, J(1,2,1,1,1)\,, \end{aligned} \tag{8.145}$$

and, as in the previous section, we reduce box integrals with polynomials in the numerator to scalar box integrals with shifted dimensions:

$$\begin{aligned} J(a_1, a_2, a_3, a_4, n) &= \mathrm{i}^{-a_1 - a_2 - a_3 - a_4 - 1 + d/2 + n} \prod_i (a_i - 1)! \\ \times \int & \frac{\mathrm{d}^{d+n} k}{(k^2)^{a_1} (k^2 + 2p_4 \cdot k)^{a_2} (k^2 + 2p_1 \cdot k)^{a_3} [k^2 + 2(p_1 - p_2) \cdot k + s]^{a_4}}\,. \end{aligned} \tag{8.146}$$

This time, the integrals J are massless boxes. The expansion of the master integral $J(1,1,1,1,0)$ has been described in Example 8.5. All other integrals

(8.146) can be expanded similarly and, keeping the leading power in the expansion, we find

$$B_1 \sim 0\,, \quad B_2 \sim \left(\frac{1}{\varepsilon} - \ln(-t)\right) \ln\frac{s}{t}\,, \tag{8.147}$$

so that the box contribution takes the form

$$-\frac{\mathrm{i}g^2}{t}\frac{\alpha}{\pi}B_2(s,t)\left(C_\mathrm{F} - \frac{T_\mathrm{F}}{N}\right)A_\mathrm{v}^\lambda\,. \tag{8.148}$$

The crossed-box contribution is expanded in the same way and gives

$$\frac{\mathrm{i}g^2}{t}\frac{\alpha}{\pi}B_2(-s,t)\left(C_\mathrm{F} - \frac{T_\mathrm{F}}{N} - \frac{C_\mathrm{A}}{2}\right)A_v^o\,. \tag{8.149}$$

The rest of the one-loop contributions are trivial one-scale vertices and self-energy insertions which do not depend on s and do not affect $\omega(t)$. From (8.143), (8.148) and (8.149) we reproduce the well-known result [167]

$$\omega(t) = \frac{\alpha}{2\pi}C_\mathrm{A}\left(\frac{1}{\varepsilon} - \ln(-t)\right)\,. \tag{8.150}$$

In the BFKL kernel, the pole of (8.150) is cancelled by the singular part of the real radiation contribution.

We have seen how the one-loop result for the reggeon trajectory can be reproduced in a simple way using expansion by regions. To reproduce the existing two-loop results in a similar way, the (c–c) contributions to the expansions of double boxes (see Sect. 8.5) have to be used as an input. One may hope that new results will be obtained by this technique.

9 Conclusion

'But one must have some proof...' began Berlioz. 'There's no need for any proof,' answered the professor.[1]

(M.A. Bulgakov, *The Master and Margarita*)

Renormalization theory has a history of egregious errors by distinguished savants. It has a justified reputation for perversity; a method that works up to 13th order in the perturbation series fails in the 14th order. Arguments that sound plausible often dissolve into mush when examined closely. The worst that can happen often happens.

(A.S. Wightman [240])

In this final chapter, various auxiliary techniques related to the expansion of Feynman diagrams in various limits of momenta and masses are first described: summation by Padé approximants, summing up expansions by guess, Taylor expansion in a mass difference, and a method of direct evaluation of general integrals appearing in contributions to the asymptotic expansions by use of IBP. An alternative approach within the strategy of expansion by regions, which is based on the alpha representation and thereby is manifestly independent of the choice of the loop momenta, is then described. I conclude by characterizing the present status of expansion by regions and giving some advice about its application.

9.1 Related Techniques

9.1.1 Summation by Padé Approximants

If a sufficiently large number of terms in the asymptotic expansion of a given Feynman diagram in a limit of momenta and masses is known, one often applies the method of summing up series by the use of Padé approximants. Suppose we have a Feynman diagram in the large-mass (small-momentum) limit and there are no massless thresholds. For example, let us keep in mind the self-energy diagram of Fig. 2.5 or the vertex diagram of Fig. 5.3 with equal masses, with the first threshold at $q^2 = 4m^2$. In this case the combinatorial structure of the expansion is trivial, and the expansion is given by a Taylor series in $z = q^2$,

$$f(z) = \sum_{n=0}^{\infty} a_n z^n \,, \tag{9.1}$$

[1]Translated from the Russian by M. Glenny. Collins and Harvill Press, London, 1967.

with a non-zero radius of convergence z_0, which equals $4m^2$ in our case. The function $f(z)$ is analytically continued from the circle $|z| < z_0$ to the wholer z plane with a cut starting at z_0. To do this, one applies some summation technique (see, e.g., [135]).

In the case of Feynman diagrams, the method of Padé approximants has turned out to be very effective. If the first N coefficients of the series (9.1) are known, one can construct N different Padé approximants, which are denoted by $[N-M/M]$ for $M = 0, 1, \ldots, N-1$ and are defined as rational functions of the form (see, e.g., [10])

$$\frac{b_0 + b_1 z + b_2 z^2 + \ldots + b_J z^J}{1 + c_1 z + c_2 z^2 + \ldots + c_K z^K} . \tag{9.2}$$

Such a function is called a $[J/K]$ Padé approximant if its expansion into a Taylor series in z reproduces the first $J+K$ coefficients of (9.1). Usually one applies 'diagonal' Padé approximants $[J/J]$ or $[J/J \pm 1]$. The coefficients b_i and c_i in (9.2) are uniquely determined by the given Taylor series. In practice, one uses various specific methods to evaluate these coefficients [10].

Summation by Padé approximants can be applied to the initial power series. It turns out, however, that it is worthwhile first to perform a conformal mapping [69]. When the initial function is analytic with a cut starting at $z = z_0$, a very useful trick is to transform to a new variable by means of the conformal mapping

$$\omega = \frac{1 - \sqrt{1 - z/z_0}}{1 + \sqrt{1 - z/z_0}} , \tag{9.3}$$

which maps the whole domain of analyticity of the function $f(z)$ into the interior of the unit circle in the ω plane. By means of this conformal transformation of the z plane, the cut from z_0 to $+\infty$ is mapped onto the boundary of the unit circle. The upper semicircle corresponds to the upper side of the cut. The initial power series is transformed into the following series in ω, which is eventually summed up by Padé approximants:

$$f(z(\omega)) = \sum_{n=0}^{\infty} \phi_n \omega^n , \tag{9.4}$$

where

$$\begin{aligned} \phi_0 &= a_0 , \\ \phi_n &= \sum_{k=1}^{n} \frac{(n+k-1)!(-1)^{n-k}}{(2k-1)!(n-k)!} (4z_0)^k a_k , \quad \text{if } n \geq 1 . \end{aligned} \tag{9.5}$$

As was shown in [115], this procedure works very well for Feynman diagrams without massless thresholds. Suppose now that a given Feynman integral has massless thresholds so that the corresponding large-mass expansion consists of contributions from various subgraphs – see, for instance, Examples 5.4 and 5.5. The expansion is no longer a Taylor series in q^2. It can, however, be represented as a finite sum of powers of the logarithm of q^2 times Taylor series:

$$F_\Gamma(q^2, M^2) = \frac{1}{M^4}\left[f_0(z) + f_1(z)\ln(-z) + f_2(z)\ln^2(-z)\right] . \tag{9.6}$$

(For a similar general h-loop diagram, the maximal power of the logarithm would be h in this typical Euclidean limit.) Remember that in the case without massless thresholds, the value of the radius of convergence of the Taylor series is determined, according to well-known theorems [98], by the position of the lowest threshold. No similar mathematical results are known for the functions f_i in (9.6) that arise after expanding the given Feynman integral in the large-mass limit. It is, however, quite plausible to suppose that the radius of convergence of the power series associated with the functions f_i is determined by the position of the *next* threshold of the initial diagram. (Remember that the first threshold is at $z = 0$.) A natural idea is to apply the technique of Padé approximants to each of the power series f_i on the right-hand side of the asymptotic expansion. This procedure was suggested in [113] and was 'experimentally' confirmed [113, 109] using Examples 5.4 and 5.5 and other typical vertex diagrams. It has been shown that the resulting convergence is very good and gives the possibility to perform precise calculations far beyond the lowest non-zero threshold. The method of summation by Padé approximants combined with the expansion of Feynman integrals in momenta and masses has been applied successfully in phenomenological calculations – see, for example, [58, 52, 66, 101, 134, 110].

9.1.2 Summing up an Expansion by Guess, Taylor Expansion in a Mass Difference and Expansion without Small Parameters

When many terms in an asymptotic expansion of a given Feynman diagram are known, summation by Padé approximants is not the only possibility. An alternative is to carefully analyse the available terms and make a guess about the exact analytic dependence of the general term on the order of expansion, n. An example of such a strategy can be found in [80], where two-loop QCD and electroweak corrections to the hadronic Z boson decay rate have been calculated with the help of asymptotic expansions in the limits of both small and large values of q^2/M_Z^2. A guess about the general structure of the series in q^2/M_Z^2 led to explicit summation in terms of rather complicated functions expressed in terms of tetralogarithms.

Other examples of summing up series by guess are some calculations of QCD corrections to the decay $Z \to Wu\bar{d}$ in [83], with explicit results for summed up series in terms of polylogarithms up to the fourth order, and some calculations of two-loop QED corrections to the muon lifetime in [196], within the large-mass expansion, where the resulting series in $x = -q^2/m_\mu^2$ were summed up with subsequent continuation of the explicit results to the point $x = 1$.

Reference [83] provides another type of expansion used in practice, in powers of a difference between masses, $(M-m)/M$, with the choice $M = M_Z$ and $m = M_W$. As in the case of the small momentum expansion without massless thresholds, such an expansion does not require special techniques and reduces just to expanding the integrand into a Taylor series. Technically, the simplifications arise because one obtains integrals depending on one mass instead of two. An earlier example of an expansion in $(M - m)/M$ can be taken from [78], with $M = m_b$ and $m = m_c$, where complete two-loop QCD corrections to the $b \to c$ transitions have been calculated in the limit of zero recoil.

It is not always clear what the domain of convergence of a given asymptotic expansion is. Sometimes it is possible to determine the domain 'experimentally' and then successfully apply the expansion even if there are no natural small parameters in the problem. The experience accumulated in the course of numerous practical applications shows that asymptotic expansions in momenta and masses can work even for large values of the expansion parameters. To illustrate this point, let us recall Example 4.1, where the expansion in the limit $p^2 \ll q^2, (p+q)^2$ was given by (4.8). Imagine now that we are interested in the value of this triangle diagram at the symmetrical kinematical point $p^2 = q^2 = (p+q)^2$. (This value is of special importance because it is used to define the so-called symmetrical MOM scheme, which is significantly more suited (than $\overline{\text{MS}}$ [13]) to analysing lattice results.) A comparison with the known analytical result (4.9) shows that even a limited number (e.g. four or five) of terms of the expansion (4.8) provides a reasonably accurate approximation at the point where the parameter of expansion p^2/q^2 equals unity, rather than being small. As has been shown in [60] by comparison with known analytical results [233] for two-loop, massless, finite scalar vertex diagrams, this phenomenon takes place also at the two-loop level. Since no analytical results for two-loop vertex diagrams (which are typically UV divergent) with general numerators and integer powers of propagators are available at the moment, the authors of [60] have applied this strategy of expanding in the limit $p^2 \ll q^2, (p+q)^2$ to compute the three-gluon QCD vertex at the symmetrical kinematical point. The result obtained has been applied to construct the first non-universal (three-loop) term in the QCD MOM-scheme beta function, which has already found important applications in lattice calculations [32].

A similar situation is exemplified by the diagram of Fig. 9.1 To calculate this diagram, one can consider one of the three equal masses, M, to be different from the other two, $m \neq M$, i.e. we consider a diagram with masses $(M, M, 0, m)$, and apply the asymptotic expansion in the limit $m/M \to 0$. According to the prescriptions of the large-mass expansion described in Chap. 5, two subgraphs give non-zero contributions: the graph itself and the subgraph consisting of the lines with masses $(M, M, 0)$. Any term in these two series can be evaluated by means of (A.7), (A.38) and (A.3). It turns out that

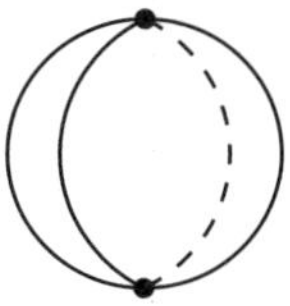

Fig. 9.1. Three-loop vacuum diagram with one massless and three massive lines

the resulting expansion in the ratio m^2/M^2 converges very well at the point $m = M$. The evaluation of the terms of this asymptotic expansion is easily performed by computer. This technique has been applied in [67], where it has been shown, in particular, that the first twenty expansion terms of the diagram of Fig. 9.1 reproduce the first ten digits of the exact result [66, 111], given up to terms of order ε:

$$\frac{1}{\varepsilon^3} + \frac{15}{4\varepsilon^2} + \frac{1}{\varepsilon}\left(\frac{65}{8} + \frac{\pi^2}{4}\right) + \frac{135}{16} + \frac{15\pi^2}{16} - \zeta(3) + 3\sqrt{3}\,\mathrm{Cl}_2\left(\frac{\pi}{3}\right)$$
$$+\varepsilon\left\{-\frac{763}{32} - \frac{9\sqrt{3}\pi}{16}\ln^2 3 - \frac{35\sqrt{3}\pi^3}{48} + \frac{65\pi^2}{32} - \frac{15}{4}\zeta(3) + \frac{19\pi^4}{480}\right.$$
$$\left. + \frac{45\sqrt{3}}{2}\mathrm{Cl}_2\left(\frac{\pi}{3}\right) - 27\sqrt{3}\,\mathrm{Im}\left[\mathrm{Li}_3\left(\frac{\mathrm{e}^{-\mathrm{i}\pi/6}}{\sqrt{3}}\right)\right]\right\} .$$

Here Cl_2 is the Clausen function [165].

This trick can be used for the (semi-)analytic evaluation of diagrams where explicit results are unknown. It is very effective in any situation where Taylor expansion in some of the non-zero masses leads to Feynman integrals that can be evaluated in terms of gamma functions for general powers of the propagators.

9.1.3 A Direct Algorithm to Evaluate Contributions to the Asymptotic Expansion

A standard problem that arises when expanding Feynman diagrams in various limits of momenta and masses is the evaluation of the class of integrals that occur in the contribution from a given specific region. A standard way to solve this problem is based on solving recurrence relations obtained by the method of IBP [68] for these integrals. Any dimensionally regularized Feynman integral that appears in the asymptotic expansions when the strategy of expansion by regions is used, can be represented as

$$F(\underline{n}) \equiv \int \cdots \int \frac{\mathrm{d}^d k_1 \ldots \mathrm{d}^d k_h}{D_1^{n_1} \ldots D_N^{n_N}}, \tag{9.7}$$

where the k_i, $i = 1, \ldots, h$, are loop momenta, the n_i are integer indices and the $D_a = \sum_{i,j=1} A_a^{ij} p_i \cdot p_j - m_a^2$, with $a = 1, \ldots, N$, are denominators. Irreducible numerators which cannot be expressed through a given set of propagators are naturally treated as extra denominators with a negative power. The momenta p_i are either the loop momenta $p_i = k_i$, $i = 1, \ldots, h$, or external momenta of the graph $p_{h+1}, \ldots$, so that the denominators can be either quadratic or linear with respect to the loop momenta k_i. Underlined letters denote collections of variables here, i.e. $\underline{n} = (n_1, \ldots, n_N)$. The IBP identities that prescribe zero values for the integrals containing $(\partial/\partial p_i)\cdot p_k$, for $i = 1, \ldots, h$, can be symbolically written as

$$P_{ik}(I^+, I^-)F(\underline{n}) = 0 \,, \tag{9.8}$$

where P_i are polynomials in the increasing and lowering operators $\mathbf{I}_a^+$ and $\mathbf{I}_a^-$, where

$$\mathbf{I}_a^+ F(\ldots, n_a, \ldots) = F(\ldots, n_a + 1, \ldots) \,,$$
$$\mathbf{I}_a^- F(\ldots, n_a, \ldots) = F(\ldots, n_a - 1, \ldots) \,.$$

One says that these relations can be recursively solved if an integral with a given value of the multi-index $\underline{n}$ can be expressed in terms of simpler integrals corresponding to some finite family of multi-indices $\underline{n} = \underline{n}_i$, where usually $\underline{n}_i = (n_{i1}, \ldots, n_{iN})$ consists of $n_{ia} = 0$ or 1:

$$F(\underline{n}) = \sum_{i=1}^{k} c_i(\underline{n}) F(\underline{n}_i) \,, \tag{9.9}$$

where

$$c_i(\underline{n}_j) = \delta_{ji} \,. \tag{9.10}$$

The integrals $F(\underline{n}_i)$ are called *master*, *basic* or *irreducible* integrals and are usually calculated without using IBP.

In concrete situations, the realization of this procedure turns out to be far from straightforward. Examples of solving IBP recurrence relations have been presented in the previous chapters. Some attempts [228, 229, 122, 123] to make this procedure systematic have been mentioned in Sect. 2.5.6. Another attempt to solve recurrence relations systematically was made in [7]. This procedure is manifestly universal with respect to the form of the integrand and can be applied to a contribution from an arbitrary region on the right-hand side of an asymptotic expansion in momenta and masses. The goal of this method is to find an algorithm for direct evaluation of the coefficients $c_i(\underline{n})$, i.e. to understand

(1) which integrals should be taken as master integrals and
(2) how to calculate the coefficients $c_i(\underline{n})$.

The first problem is connected with the formulation of an adequate criterion for irreducibility of a given Feynman integral to a fixed set of master integrals. See [8], where such a criterion has been suggested.

Consider the second problem and assume that the integrals $F(\underline{n}_i)$ are irreducible. Then the $c_i(\underline{n})$ have to be solutions of the relations (9.8). Indeed, acting with $P_{ik}(I^+, I^-)$ on (9.9), we obtain zero on the left-hand side and linear combinations of $F(\underline{n}_i)$ on the right-hand side. If $F(\underline{n}_i)$ is irreducible, all coefficients in this linear combination have to vanish. This means that the $c_i(\underline{n})$ are solutions of (9.8). So, if one finds $\tilde{c}_i(\underline{n})$ as some solutions of (9.8) one can obtain $c_i(\underline{n})$ as their linear combinations which fit (9.10).

To construct $\tilde{c}_i(\underline{n})$, the following integral representation has been suggested [7]:

$$\tilde{c}(\underline{n}) = \int \frac{\mathrm{d}x_1 \ldots \mathrm{d}x_N}{x_1^{n_1} \ldots x_N^{n_N}} g(\underline{x}) \, . \tag{9.11}$$

This parametric integral can be chosen to be a contour integral or an integral over some appropriately chosen real domain. The action of (9.8) on (9.11), after formal IBP of the integral over x_i, leads to a differential equation for $g(\underline{x})$, which can then be solved [7] for Feynman integrals with an arbitrary number of legs and with arbitrary masses. In some practically important cases (see [7, 8]), (9.11) can be solved in terms of rational combinations of Pochhammer symbols, thus providing the desired direct solutions of (9.8).

9.2 Expansion by Regions in the Alpha Representation

Let us turn our attention to the strategy expansion by regions in the alpha-parameter representation. In this case, the strategy is formulated in the same way as for the integrals over the loop momenta. One has to consider the alpha parameters to be of different orders, measured in terms of the given masses and kinematical invariants. Since the functions involved in the alpha representation (2.42) are homogeneous in α_l, only the relative order of the alpha parameters is relevant when specifying a region. Let us measure this order in terms of the small parameters associated with a given limit. For example, when there are two parameters, $q^2 \gg m^2$, let us characterize the regions in terms of powers of m. Thus the hard region, which participates in every limit, is characterized by $\alpha_l \sim m^0$ for every l, so that the corresponding contribution is defined as a Taylor expansion of the exponent in (2.42) in the small external parameters.

In particular, in the case of the limits typical of Euclidean space, it is enough to consider only the regions of large (hard) and small (soft) alpha parameters, i.e. where $\alpha_l \sim 1/q^2$ and $\alpha_l \sim m^2/(q^2)^2$, respectively, and by q^2 and m^2 we denote the large and small typical parameters, respectively, of the problem.

Although the Feynman parametric representation (2.63) is obtained from the alpha-parameter representation just by one integration over the overall scaling variable, it is not so convenient for expansion by regions as the representation (2.42), because, when one uses regions defined by the use of Feynman parameters, one arrives at terms of the expansion without manifest homogeneity.

Let us illustrate the strategy of expansion by regions in the alpha representation in the case of some limits typical of pseudo-Euclidean space, using Examples 8.10 and 8.11. The function (2.43) is the same in both cases:

$$D = (\alpha_1 + \alpha_2)(\alpha_3 + \alpha_4 + \alpha_5) + \alpha_6(\alpha_1 + \alpha_2 + \alpha_3 + \alpha_4 + \alpha_5) \; . \tag{9.12}$$

The functions (2.44) are

$$\begin{aligned}
&A_{8.10} \\
&= [(\alpha_1 + \alpha_3)(\alpha_2 + \alpha_4)\alpha_6 + (\alpha_1 + \alpha_2)\alpha_3\alpha_4 + \alpha_1\alpha_2(\alpha_3 + \alpha_4 + \alpha_5)]Q^2 \\
&+\{(\alpha_1 + \alpha_2 + \alpha_3 + \alpha_4)^2\alpha_6 \\
&+(\alpha_1 + \alpha_2)[(\alpha_3 + \alpha_4)(\alpha_1 + \alpha_2 + \alpha_3 + \alpha_4) + (\alpha_1 + \alpha_2)\alpha_5]\}m^2 \, , \\
&A_{8.11} = [(\alpha_1 + \alpha_2 + \alpha_3 + \alpha_4)\alpha_6 + (\alpha_1 + \alpha_2)(\alpha_3 + \alpha_4)] \\
&\times [M^2(\alpha_1 + \alpha_3) + m^2(\alpha_2 + \alpha_4)] + (M^2\alpha_1 + m^2\alpha_2)(\alpha_1 + \alpha_2)\alpha_5 \; .
\end{aligned} \tag{9.13}$$

In the (h–h) alpha region all the α_l are of order m^0, so that the function D is not expanded and the part of the exponent proportional to m^2 is expanded into a Taylor series in m.

In Example 8.10, we reproduce the contributions of the previously found momentum space regions as follows:

$$\begin{aligned}
(\text{1c–h}) &\to \{\alpha_{3,5} \sim m^0 \, , \; \alpha_{1,2,4,6} \sim m^2\} \, , \\
(\text{1c–1c}) &\to \{\alpha_{1,3,5,6} \sim m^0 \, , \; \alpha_{2,4} \sim m^2\} \, , \\
(\text{h–1c}) &\to \{\alpha_1 \sim m^0 \, , \; \alpha_{2,3,4,5,6} \sim m^2\} \, , \\
(\text{1uc–2c}) &\to \{\alpha_5 \sim m^0 \, , \; \alpha_3 \sim m^2 \, , \; \alpha_{2,4,6} \sim m^4 \, , \; \alpha_1 \sim m^6\} \, .
\end{aligned} \tag{9.14}$$

The alpha parameters have dimension m^{-2} but, to simplify the relations, we set Q and M to one. The remaining regions are obtained by permutations.

In Example 8.11, we have

$$\begin{aligned}
(\text{1c–h}) &\to \{\alpha_{4,5} \sim m^0 \, , \; \alpha_{1,2,3,6} \sim m^2\} \, , \\
(\text{1c–1c}) &\to \{\alpha_{2,4,5,6} \sim m^0 \, , \; \alpha_{1,3} \sim m^2\} \, , \\
(\text{h–1c}) &\to \{\alpha_2 \sim m^0 \, , \; \alpha_{1,3,4,5,6} \sim m^2\} \, , \\
(\text{us–1c}) &\to \{\alpha_5 \sim m^0 \, , \; \alpha_{2,3,4,6} \sim m^2 \, , \; \alpha_1 \sim m^4\} \, .
\end{aligned} \tag{9.15}$$

Note that here we do not risk performing double counting of the (h–1c) and (h–s) regions (with another choice of the second loop momentum) as was discussed in Chap. 8. Moreover, the description of the regions in the alpha-parameter language is manifestly Lorentz invariant. Some regions are, however, described in a rather non-trivial way: for example, in the case of the (us–1c) and (1uc–2c) regions, the alpha parameters are subdivided into three and four groups, respectively, with different orders measured in powers of the expansion parameter.

Let us now describe the contributions to the asymptotic expansion in the limits $t/s \to 0$ and $s/t \to 0$ for Example 8.12 in the language of alpha parameters. The functions (2.43) and (2.44) for the double box are

$$
\begin{aligned}
D_{8.12} &= (\alpha_1 + \alpha_2 + \alpha_7)(\alpha_3 + \alpha_4 + \alpha_5) \\
&\quad + \alpha_6(\alpha_1 + \alpha_2 + \alpha_3 + \alpha_4 + \alpha_5 + \alpha_7) \,, \qquad (9.16) \\
A_{8.12} &= [\alpha_1\alpha_2(\alpha_3 + \alpha_4 + \alpha_5) + \alpha_3\alpha_4(\alpha_1 + \alpha_2 + \alpha_7) \\
&\quad + \alpha_6(\alpha_1 + \alpha_3)(\alpha_2 + \alpha_4)]s + \alpha_5\alpha_6\alpha_7 t \,. \qquad (9.17)
\end{aligned}
$$

The (h–h) contributions are reproduced by considering all the alpha parameters to be of the same order. Let us again use dimensionless alpha parameters by putting $s = -1$ in the first limit and $t = -1$ in the second limit. In the limit $t/s \to 0$, we reproduce the contributions of the collinear momentum space regions considered previously as follows:

$$
\begin{aligned}
&(1c\text{–}1c) \to \{\alpha_{1,3,5,6,7} \sim t^0 \,,\ \alpha_{2,4} \sim -t\} \,, \\
&(2c\text{–}2c) \to \{\alpha_{2,4,5,6,7} \sim t^0 \,,\ \alpha_{1,3} \sim -t\} \,. \qquad (9.18)
\end{aligned}
$$

In the limit $s/t \to 0$, we have

$$
\begin{aligned}
&(1c\text{–}1c) \to \{\alpha_{1,2,3,4,6,7} \sim s^0 \,,\ \alpha_7 \sim -s\} \,, \\
&(2c\text{–}2c) \to \{\alpha_{1,2,3,4,5,6} \sim s^0 \,,\ \alpha_5 \sim -s\} \,, \\
&(1c\text{–}2c) \to \{\alpha_{1,2,3,4,5,7} \sim s^0 \,,\ \alpha_6 \sim -s\} \,. \qquad (9.19)
\end{aligned}
$$

We see that, in this example, the description of the collinear contributions in the language of alpha parameters is certainly simpler than in momentum space. In practice, an appropriate choice of either momentum space or the alpha representation language for expansion by regions can considerably simplify calculations. When the languages are of the same complexity, one can use them both in order to obtain an extra natural check based on the independence of the results obtained in the two approaches.

9.3 Mathematical Formulation of Expansion by Regions

Let us realize that we are dealing with a heuristic strategy. The very word 'region' is understood in the 'physical' sense. In fact, it indicates relations between components of the loop momenta, expressed in terms of the given masses and kinematical invariants. This is clearly not the mathematical sense, where a region is determined by inequalities. We do not even bother about the decomposition of unity, i.e. the fact that our initial integral over the whole space of the loop momenta is divided into a sum of integrals over all possible regions, which, presumably, have zero measure in the intersection of any pair, with their union being the whole integration space.

Although this pragmatic point of view has proven to be quite successful, let us change our orientation and formulate, at last, the strategy of expansion by regions in a mathematical language. Let us use, for this purpose, the variant of the strategy based on alpha parameters, which is, mathematically, certainly preferable.

Let Γ be a graph and F_Γ the corresponding Feynman integral. Let us suppose, for simplicity, that it is a scalar integral and that it depends on two parameters, for example a kinematical invariant q^2 and a mass m^2. Let us consider the limit $m^2 \ll q^2$. The diagram can have more than two external legs, so that this can be either an off-shell large-momentum limit or some limit typical of Minkowski space, in particular, some version of the Sudakov limit. Generalizations to other cases are, in most cases, straightforward.

Let us use the alpha representation (2.42). The function A (2.44), with the restrictions on the external momenta, specified above, depends only on one invariant, q^2. Let us include the mass term of the exponent in the function A so that now we have

$$F_\Gamma(q^2, m^2; d) = C_\Gamma \int_0^\infty \mathrm{d}\alpha_1 \ldots \int_0^\infty \mathrm{d}\alpha_L \, D^{-d/2} \mathrm{e}^{\mathrm{i}\bar{A}/D} \,, \tag{9.20}$$

where C_Γ is the factor given by (2.42) and

$$\bar{A}(q^2, m^2; \alpha_1, \ldots, \alpha_L) = A(q^2; \alpha_1, \ldots, \alpha_L) - D(\alpha_1, \ldots, \alpha_L) \sum_l m_l^2 \alpha_l \,. \tag{9.21}$$

Let $\nu = (n_1, n_2, \ldots, n_L)$ be a family of non-negative integers corresponding to L lines of the given graph, and let us define $\alpha_l'(\kappa) = \kappa^{n_l} \alpha_l$. The functions (2.43) and (9.21) of $\alpha_l'(\kappa)$, with the mass rescaled as $m \to \kappa m$, take the form

$$D(\alpha_1'(\kappa), \ldots) = \sum_{j=n(\nu,D)} \kappa^j D_j(\alpha_1, \ldots) \,, \tag{9.22a}$$

$$\bar{A}(q^2, \kappa^2 m^2; \alpha_1'(\kappa), \ldots) = \sum_{j=n(\nu,A)} \kappa^j \bar{A}_j(q^2, m^2; \alpha_1, \ldots) \,, \tag{9.22b}$$

where the finite sums run from some minimal numbers $n(\nu, D)$ and $n(\nu, A)$. The initial functions D and $\bar{A}$ of the variables α_l take the form

$$D = D_{n(\nu,D)} + D' \,, \tag{9.23a}$$

$$\bar{A} = \bar{A}_{n(\nu,A)} + \bar{A}' \,, \tag{9.23b}$$

where the first terms on the right-hand side correspond to the terms in (9.22a) and (9.22b) with the minimal powers of κ.

Let us associate with the given multi-index ν the following formal infinite series:

$$F_\Gamma^\nu(q^2, m^2; d) = C_\Gamma \int_0^\infty \mathrm{d}\alpha_1 \ldots \int_0^\infty \mathrm{d}\alpha_L \, \mathcal{T}_\kappa \Bigg[\left(D_{n(\nu,D)} + \kappa D'\right)^{-d/2} \times \exp\left(\mathrm{i}\kappa^{n(\nu,A)-n(\nu,D)} \frac{\bar{A}_{n(\nu,A)} + \kappa \bar{A}'}{D_{n(\nu,D)} + \kappa D'} \right) \Bigg] \Bigg|_{\kappa=1} \,. \tag{9.24}$$

Note that the number $n(\nu, A) - n(\nu, D)$ is always non-negative. In our previous language, (9.24) is nothing but the contribution of the region of alpha parameters where α_l is of order m^{n_l} for $l = 1, \ldots, L$. The experience obtained in the previous chapters tells us that the following statement looks quite plausible.

Conjecture. The asymptotic expansion of the Feynman integral F_Γ in the limit described above is given by

$$F_\Gamma \sim \sum_\nu F_\Gamma^\nu \,, \tag{9.25}$$

where the sum runs over all multi-indices consisting of non-negative integers.

Comments.

1. Experience tells us that the numbers n_l are indeed integers. For limits typical of Euclidean space and for threshold limits with one heavy mass in the threshold (see Chap. 6), the relevant multi-indices consist of $n_l = 2$ or $n_l = 0$.

2. The prescription of expansion by regions that scaleless integrals are set to zero is assumed in the definition of the contribution F_Γ^ν. For example, if the number $n(\nu, A) - n(\nu, D)$ is positive, the resulting contribution is a scaleless integral because the exponent is expanded and the factor $D_{n(\nu,D)}$ is a homogeneous function of the alpha parameters. It is therefore sufficient to consider only contributions with $n(\nu, A) = n(\nu, D)$.

3. The leading order of any given contribution F_Γ^ν,

$$C_\Gamma \int_0^\infty \mathrm{d}\alpha_1 \ldots \int_0^\infty \mathrm{d}\alpha_L \, D_{n(\nu,D)}^{-d/2} \exp\left(\mathrm{i}\frac{\bar{A}_{n(\nu,A)}}{D_{n(\nu,D)}}\right) \,, \tag{9.26}$$

can be obtained without calculation, just by power counting: its dependence on the small expansion parameter m of the problem is of the form $m^{\delta(\nu,D)}$, where $\delta(\nu, D) = \sum n_l - dn(\nu, D)/2$ (up to logarithms, which could arise in limits of the Sudakov type – see Chap. 8.)

4. The description of expansion by regions in the language of this section can be easily applied to almost all of the limits and the corresponding regions considered in Chaps. 4–8. The only exception is the contribution of the potential region in the threshold expansion, which is tricky to write in terms of alpha parameters.

5. The language of this section can easily be formulated for Feynman integrals written in terms of momentum space, with qualifications concerning some problems with the mathematical definition of dimensionally regularized integrals over loop momenta in the case of typical Minkowskian momentum configurations. The potential regions can then be treated on the same footing as all the other regions that we have considered.

6. One may hope that the strategy of expansion by regions can be applied not only to Feynman integrals but also to much more general integrals of rational functions. A reasonable problem would be to develop the strategy of expansion by regions for the phase space integrals that arise in the evaluation of real radiation processes.

9.4 Mathematical Status of Expansion by Regions

From the purely pragmatic point of view, there is no need to prove the strategy of expansion by regions, because it has worked successfully in various situations. For practically oriented people, the simple arguments used in the toy example and the one-loop example in Chap. 3 could be quite enough to convince them that a similar mechanism connected with vanishing massless vacuum Feynman integrals works for any limit and any diagram. Let us realize, however, that this is just a one-loop example[2] in a limit typical of Euclidean space, where we have a general proof for any diagram (see Appendix B.2).

From the mathematical point of view, it is not guaranteed that expansion by regions works in all situations, so that it is necessary either to prove that it does or to invent a counterexample. This indeed appears to be a very good mathematical problem. It hardly possible to generalize the direct arguments used in Chap. 3 to prove the validity of the strategy of expansion by regions for a typically pseudo-Euclidean limit. It looks reasonable to try to generalize the proof for the off-shell large-momentum limit presented in Appendix B.2, based on the alpha representation, and to start from the formulation of the strategy in the form of the conjecture formulated in the previous section. One may hope that the remainder of the expansion can be defined in a way similar to the case of limits typical of Euclidean space. In fact, the language of Sect. 9.3 already provides a natural definition, given by (9.24), of the subtraction operator corresponding to a given multi-index (i.e. 'region') $\nu = (n_1, n_2, \ldots, n_L)$.

Then one would need to carry out an analysis of convergence by means of the introduction of appropriate sectors and sector variables, which would provide a factorization of the corresponding integrand. The sectors described in Appendix B.1 are insufficient for typical Minkowskian situations – this can be seen even in one-loop examples. It should be also stressed that, for each on-shell or threshold limit, this problem of resolution of singularities has to be solved separately, with specific sectors and sector variables. One may hope that the prescription to set all the scaleless integrals to zero[3] may arise automatically, similarly to the off-shell case.

Among the limits typical of pseudo-Euclidean space, the threshold limits with one heavy mass in the threshold considered in Chap. 6 are certainly

[2] Direct arguments of this kind for the diagram in Example 4.2 can be found in [133]. This is, however, again an example of the off-shell large-momentum limit.

[3] Let us emphasize that this prescription, which has at present an ad hoc status for typical Minkowskian limits, does not depend on the use of dimensional regularization. For example, some integrals generated by potential contributions in the threshold expansion are not regularized at all – see the examples in Sect. 7.1. Nevertheless, the rule that one sets such scaleless integrals to zero has been experimentally checked, through numerous examples, as part of the whole list of prescriptions for expansion by regions.

simpler, so that it looks reasonable to justify the strategy of expansion by regions for these limits first. At least the combinatorics of the expansion is similar to that of the off-shell limits, and the new subtraction operators can be defined, through (9.24), by a slight modification of the operators (B.27).

9.5 The Last Advice

9.5.1 When There Are More than Two Scales

In fact we have already considered examples of limits with more than two scales: the threshold expansion with two non-zero masses in a threshold when there are three scales, m, mv and mv^2, in Chap. 7, and the example of Fig. 6.4c, which is a graph contributing to the muon anomalous magnetic moment in QED – see Sect. 6.4. In a general situation with several groups of masses and kinematical invariants which differ in their orders of magnitude, one can proceed sequentially by first decomposing the given set of parameters into two groups and applying the corresponding prescriptions associated with two scales, and then one considers the diagrams on the right-hand side, decomposes the parameters further into two groups, and so on. For example, one can proceed by distinguishing first the group with the smallest parameters or, alternatively, by considering all the parameters apart from the group with the largest parameters to be small. Presumably, the results have to be independent of this order.

To illustrate this procedure, let us consider an example of a typical two-loop vertex diagram contributing to the decay of the Z boson into a b quark and antiquark – see Fig. 9.2. The b quarks are taken as massless so that $p_1^2 = p_2^2 = 0$. Although the external momentum q has to be taken on shell, $q^2 = m_Z^2$, let us consider it as a small quantity keeping in mind that, using the small momentum expansion and evaluating a sufficiently large number of terms in the corresponding expansion, we can arrive, with the help of Padé approximants, at precise results for large values of q^2 – see Sect. 9.1.1. So, we arrive at a problem with three scales, $q^2 \ll m_W^2 \ll m_t^2$. One can start from two different (large-mass) limits with two scales:

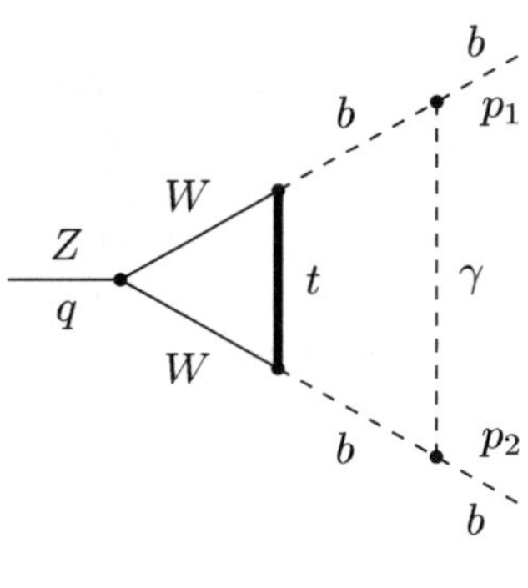

Fig. 9.2. A typical two-loop vertex diagram contributing to the process $Z \to b\bar{b}$

(a) $q^2 \ll m_W^2, m_t^2$ and
(b) $q^2, m_W^2 \ll m_t^2$.

These limits are typical of Euclidean space, so that one can apply the corresponding prescriptions of Chap. 5 in the language of subgraphs. In Limit (a), two subgraphs contribute: the graph itself and the triangle subgraph. The first of them gives the most non-trivial piece, which can be expressed in terms of two-loop bubbles (with numerators) with three lines, with masses $(m_t, m_W, 0)$.

In Limit (b), there are four contributions: the graph itself, the triangle subgraph, the box subgraph and the subgraph consisting of the heavy central line. As was demonstrated in [112] for this and similar diagrams contributing to the process $Z \to b\bar{b}$ (where calculations have been performed with the help of the large-mass expansion and Padé approximants[4] in [134, 110]), both strategies provide very precise results in the physically important region of the parameters involved. A disadvantage of Limit (a) is the presence of the above-mentioned two-loop bubbles with two different non-zero masses. These diagrams can be expressed in terms of polylogarithms. One can, however, continue to expand ad nauseam the diagrams that appear in the first step. In Limit (a), each two-loop bubble is expanded in the ratio of the masses m_W/m_t by the previous technique for the large-mass limit. Here there are two subgraphs that generate non-zero contributions: the bubble itself and the one-loop subgraph consisting of the lines with mass m_t and the zero mass. On the other hand, one can start from Limit (b) (where the largest scale, m_t, is decoupled) and expand every resulting one-loop diagram that depends on the two scales q^2 and m_W^2 in the small momentum limit $q^2 \ll m_W^2$.

One can, however, proceed as in Sect. 6.4 and apply the strategy of regions with three scales of loop momenta: hard ($\sim m_t$), soft ($\sim m_W$) and supersoft ($\sim q$).

9.5.2 Expansion Versus Analytic Evaluation

There are some curious examples where it is easier to evaluate Feynman integrals analytically than to expand them up to a desired order – see Examples 8.9 and 8.12. In such cases, one can use the strategy of expansion by regions for crucial checks.

Sometimes the problem is such that it is necessary to evaluate just specific contributions to the expansion. It can also happen that known analytic results are inappropriate for picking up individual contributions. For example, to calculate the matching coefficients (7.59) of the vector current in QCD and NRQCD one needs only the hard contributions, which could hardly be derived from a complete result, even if it existed. In fact, in any matching calculation, only the hard contributions are relevant. Another example of this kind is the

[4] In [134] one can also find an example of a nested application of asymptotic expansions in a situation with three different scales.

calculation of the reggeon trajectory (see Sect. 8.9) where one needs just the collinear contributions of the scattering diagrams in the Regge limit (e.g. (8.88)), rather than complete results (e.g. (8.89)).

We have seen in Examples 7.3, 7.3a, 8.1 and 8.5 that the method based on the use of the Mellin–Barnes representation (see Sect. 2.5.7) is a successful competitive method with respect to our expansion methods, both for analytical evaluation and for expansion. Let us stress, however, that the MB technique must be applied individually to a given diagram, where non-trivial analytic work has to be done, in particular to find an appropriate way to separate terms in the functions that enter parametric integrals. The technique is hardly applicable to a general diagram, whereas we can write down a result for the terms of the expansion by straightforwardly applying our general rules, without analytic work.

9.5.3 What to Do when Studying a New Limit

I have not tried to cover all possible limits that can arise in elementary-particle physics and have considered in this book the limits that appear to be most typical. The statuses of these limits are different: for the off-shell large-momentum and large-mass limits, as well as for any other typical Euclidean limit, the strategy of expansion by subgraphs is a well-established procedure; for the threshold expansion with two non-zero masses, expansion by regions has played a crucial role, with a lot of physical applications; for one of the threshold limits with one heavy mass, expansion by regions has also been successfully applied; some new results have been obtained by the strategy of expansion by regions for the Sudakov limit; while the strategies described above have not (yet?) provided new results for the limit connected with HQET and the Regge limit. However, one of the goals of the book was to investigate all these limits on the same footing, by means of the two strategies which pretend to be universal, in the hope that expansion by regions will be successfully applied in other situations and, in particular, used to systematically construct effective theories starting from diagrammatic experience.

Suppose now that we have to expand Feynman diagrams in some other limit typical of Minkowski space. The first step is to discover all relevant regions that can generate non-zero contributions in the given limit. It can turn out that these regions are similar to the hard, soft, ultrasoft, potential and collinear regions connected with the limits considered in this book. Performing this step successfully is a matter of experience and (both physical and mathematical) intuition. The physical flavour of the problem is the correspondence of the classes of the regions to certain operators and the subsequent translation of the prescriptions for the diagrammatic expansion into the operator language.

When testing various regions it is necessary to be aware of possible double counting. A combination of searches of the relevant regions both in momentum space and in alpha integrals looks rather reasonable. But how can we decide that we have found all the contributions to the asymptotic expansion? Unfortunately, there is no definite answer to this question. At least one can check the results by comparing them with one-loop examples where explicit analytic results can be obtained. Sometimes comparisons with analytic two-loop results are also available. One can also use checks based on numerical evaluation of the given diagram. Another important partial check is the cancellation of poles in ε, up to a certain order, and analysis of the coefficient of the highest pole, which can be evaluated by an independent method.

Thus, if the situation with the poles is unsatisfactory there are at least two options:

(a) to decide that the strategy of expansion by regions breaks down in the given limit and one has found a counterexample;
(b) to look for missing regions.

But even if the terms of the expansion satisfy the check of poles, success is not guaranteed, because one cannot exclude the existence of a region that enters with simple poles in ε, or even without poles, and is insensitive to this check.

I believe, however, that with the experience obtained by reading this book you will find all the relevant regions and, after the check of poles, as well as the other above mentioned checks, is satisfied, you will stay optimistic.

A Tables

This appendix contains a collection of useful formulae. Almost all of them are used in this book, and some other related formulae are included for completeness.

A.1 Table of Integrals

Each Feynman integral presented here can be evaluated straightforwardly by use of alpha or Feynman parameters. Results are presented for the 'Euclidean' dependence, $-k^2$, of the denominators, which is more natural when the powers of propagators are general complex numbers. As usual, $-k^2$ is understood in the sense of $-k^2 - \mathrm{i}0$, etc. Moreover, denominators with a linear dependence on k are also understood in this sense, e.g. $2p \cdot k \to 2p \cdot k - \mathrm{i}0$, although sometimes this i0 dependence is explicitly indicated to avoid misunderstanding.

$$\int \frac{\mathrm{d}^d k}{(-k^2+m^2)^\lambda} = \mathrm{i}\pi^{d/2} \frac{\Gamma(\lambda+\varepsilon-2)}{\Gamma(\lambda)} \frac{1}{(m^2)^{\lambda+\varepsilon-2}} \; . \tag{A.1}$$

$$\int \mathrm{d}^d k \frac{k^{\alpha_1} \ldots k^{\alpha_{2n}}}{(-k^2+m^2)^\lambda} = \mathrm{i}\pi^{d/2} \frac{\Gamma(\lambda-n+\varepsilon-2)}{2^n \Gamma(\lambda)} \frac{(-1)^n g_{\mathrm{s}}^{\alpha_1 \ldots \alpha_{2n}}}{(m^2)^{\lambda-n+\varepsilon-2}} \; , \tag{A.2}$$

where $g_{\mathrm{s}}^{\alpha_1 \ldots \alpha_{2n}} = g^{\alpha_1 \alpha_2} \ldots g^{\alpha_{2n-1}\alpha_{2n}} + \ldots$ (with $(2n-1)!!$ terms in the sum) is a combination symmetrical with respect to the permutation of any pair of indices. If the number of monomials in the numerator is odd, the corresponding integral is zero.

$$\int \mathrm{d}^d k \frac{(2l\cdot k)^{2n}}{(-k^2+m^2)^\lambda}$$
$$= \mathrm{i}\pi^{d/2} (-1)^n (2n-1)!! \frac{\Gamma(\lambda-n+\varepsilon-2)}{\Gamma(\lambda)} \frac{(l^2)^n}{(m^2)^{\lambda-n+\varepsilon-2}} \; . \tag{A.3}$$

$$\int \frac{\mathrm{d}^d k}{(-k^2+m^2)^{\lambda_1} (-k^2)^{\lambda_2}}$$
$$= \mathrm{i}\pi^{d/2} \frac{\Gamma(\lambda_1+\lambda_2+\varepsilon-2)\Gamma(-\lambda_2-\varepsilon+2)}{\Gamma(\lambda_1)\Gamma(2-\varepsilon)} \frac{1}{(m^2)^{\lambda_1+\lambda_2+\varepsilon-2}} \; . \tag{A.4}$$

$$\int \mathrm{d}^d k \frac{k^{\alpha_1} \dots k^{\alpha_{2n}}}{(-k^2+m^2)^{\lambda_1}(-k^2)^{\lambda_2}}$$
$$= \mathrm{i}\pi^{d/2} \frac{(-1)^n}{2^n} g_{\mathrm{s}}^{\alpha_1 \dots \alpha_{2n}} \frac{\Gamma(\lambda_1+\lambda_2-n+\varepsilon-2)\Gamma(n-\lambda_2-\varepsilon+2)}{\Gamma(\lambda_1)\Gamma(n-\varepsilon+2)(m^2)^{\lambda_1+\lambda_2-n+\varepsilon-2}} . \quad (A.5)$$

$$\int \mathrm{d}^d k \frac{(2l\cdot k)^{2n}}{(-k^2+m^2)^{\lambda_1}(-k^2)^{\lambda_2}} = \mathrm{i}\pi^{d/2}(-1)^n(2n-1)!!$$
$$\times \frac{\Gamma(\lambda_1+\lambda_2-n+\varepsilon-2)\Gamma(n-\lambda_2-\varepsilon+2)(l^2)^n}{\Gamma(\lambda_1)\Gamma(n-\varepsilon+2)(m^2)^{\lambda_1+\lambda_2-n+\varepsilon-2}} . \quad (A.6)$$

$$\int \frac{\mathrm{d}^d k}{(-k^2)^{\lambda_1}[-(q-k)^2]^{\lambda_2}}$$
$$= \mathrm{i}\pi^{d/2} \frac{\Gamma(2-\varepsilon-\lambda_1)\Gamma(2-\varepsilon-\lambda_2)}{\Gamma(\lambda_1)\Gamma(\lambda_2)\Gamma(4-\lambda_1-\lambda_2-2\varepsilon)} \frac{\Gamma(\lambda_1+\lambda_2+\varepsilon-2)}{(-q^2)^{\lambda_1+\lambda_2+\varepsilon-2}} . \quad (A.7)$$

Let $k^{(\alpha_1 \dots \alpha_N)} = k^{\alpha_1} \dots k^{\alpha_n} + \dots$ be traceless with respect to any pair of indices. Then

$$\int \mathrm{d}^d k \frac{k^{(\alpha_1 \dots \alpha_n)}}{(-k^2)^{\lambda_1}(-(q-k)^2)^{\lambda_2}} = \mathrm{i}\pi^{d/2} \frac{A_{\mathrm{T}}(\lambda_1,\lambda_2;n) q^{(\alpha_1 \dots \alpha_n)}}{(-q^2)^{\lambda_1+\lambda_2+\varepsilon-2}} , \quad (A.8)$$

where

$$A_{\mathrm{T}}(\lambda_1,\lambda_2;n) = \frac{\Gamma(\lambda_1+\lambda_2+\varepsilon-2)\Gamma(n+2-\varepsilon-\lambda_1)\Gamma(2-\varepsilon-\lambda_2)}{\Gamma(\lambda_1)\Gamma(\lambda_2)\Gamma(4+n-\lambda_1-\lambda_2-2\varepsilon)} . \quad (A.9)$$

For pure monomials, the corresponding formula has one more finite summation:

$$\int \mathrm{d}^d k \frac{k^{\alpha_1} \dots k^{\alpha_n}}{(-k^2)^{\lambda_1}[-(q-k)^2]^{\lambda_2}}$$
$$= \frac{\mathrm{i}\pi^{d/2}}{(-q^2)^{\lambda_1+\lambda_2+\varepsilon-2}} \sum_{r=0}^{[n/2]} A_{\mathrm{NT}}(\lambda_1,\lambda_2;r,n) \frac{1}{2^r} (q^2)^r \{[g]^r [q]^{n-2r}\}^{\alpha_1 \dots \alpha_n} , \quad (A.10)$$

where

$$A_{\mathrm{NT}}(\lambda_1,\lambda_2;r,n)$$
$$= \frac{\Gamma(\lambda_1+\lambda_2+\varepsilon-2-r)\Gamma(n+2-\varepsilon-\lambda_1-r)\Gamma(2-\varepsilon-\lambda_2+r)}{\Gamma(\lambda_1)\Gamma(\lambda_2)\Gamma(4+n-\lambda_1-\lambda_2-2\varepsilon)} , \quad (A.11)$$

and $\{[g]^r[q]^{n-2r}\}^{\alpha_1 \dots \alpha_n}$ is symmetric in its indices and is composed of the metric tensor and the vector q.

$$\int \mathrm{d}^d k \frac{(2l\cdot k)^n}{(-k^2)^{\lambda_1}[-(q-k)^2]^{\lambda_2}} = \frac{\mathrm{i}\pi^{d/2}}{(-q^2)^{\lambda_1+\lambda_2+\varepsilon-2}}$$
$$\times \sum_{r=0}^{[n/2]} A_{\mathrm{NT}}(\lambda_1,\lambda_2;r,n) \frac{n!}{r!(n-2r)!} (q^2)^r (l^2)^r (2q\cdot l)^{n-2r} , \quad (A.12)$$

$$\int \frac{\mathrm{d}^d k}{(-k^2)^{\lambda_1}(-k^2+2p\cdot k)^{\lambda_2}}$$
$$= \mathrm{i}\pi^{d/2} \frac{\Gamma(\lambda_1+\lambda_2+\varepsilon-2)\Gamma(-2\lambda_1-\lambda_2-2\varepsilon+4)}{\Gamma(\lambda_2)\Gamma(-\lambda_1-\lambda_2-2\varepsilon+4)} \frac{1}{(p^2)^{\lambda_1+\lambda_2+\varepsilon-2}} . \tag{A.13}$$

$$\int \mathrm{d}^d k \frac{k^{(\alpha_1\dots\alpha_n)}}{(-k^2)^{\lambda_1}(-k^2+2p\cdot k)^{\lambda_2}} = \mathrm{i}\pi^{d/2} B_{\mathrm{T}}(\lambda_1,\lambda_2;n) \frac{p^{(\alpha_1\dots\alpha_n)}}{(p^2)^{\lambda_1+\lambda_2+\varepsilon-2}} , \tag{A.14}$$

where

$$B_{\mathrm{T}}(\lambda_1,\lambda_2;n) = \frac{\Gamma(\lambda_1+\lambda_2+\varepsilon-2)\Gamma(-2\lambda_1-\lambda_2+n-2\varepsilon+4)}{\Gamma(\lambda_2)\Gamma(-\lambda_1-\lambda_2+n-2\varepsilon+4)} . \tag{A.15}$$

$$\int \mathrm{d}^d k \frac{k^{\alpha_1}\dots k^{\alpha_n}}{(-k^2)^{\lambda_1}(-k^2+2p\cdot k)^{\lambda_2}} = \frac{\mathrm{i}\pi^{d/2}}{(p^2)^{\lambda_1+\lambda_2+\varepsilon-2}}$$
$$\times \sum_{r=0}^{[n/2]} B_{\mathrm{NT}}(\lambda_1,\lambda_2;r,n) \frac{(-1)^r}{2^r} (p^2)^r \{[g]^r [p]^{n-2r}\}^{\alpha_1\dots\alpha_n} , \tag{A.16}$$

where

$$B_{\mathrm{NT}}(\lambda_1,\lambda_2;r,n)$$
$$= \frac{\Gamma(\lambda_1+\lambda_2+\varepsilon-2-r)\Gamma(-2\lambda_1-\lambda_2+n-2\varepsilon+4)}{\Gamma(\lambda_2)\Gamma(-\lambda_1-\lambda_2+n-2\varepsilon+4)} . \tag{A.17}$$

$$\int \mathrm{d}^d k \frac{(2l\cdot k)^n}{(-k^2)^{\lambda_1}(-k^2+2p\cdot k)^{\lambda_2}} = \frac{\mathrm{i}\pi^{d/2}}{(q^2)^{\lambda_1+\lambda_2+\varepsilon-2}}$$
$$\times \sum_{r=0}^{[n/2]} B_{\mathrm{NT}}(\lambda_1,\lambda_2;r,n)(-1)^r \frac{n!}{r!(n-2r)!} (p^2)^r (l^2)^r (2p\cdot l)^{n-2r} . \tag{A.18}$$

Let $p\cdot q = 0$. Then

$$\int \mathrm{d}^d k \frac{(p\cdot k)^{b_1}(q\cdot k)^{b_2}}{(-k^2)^{\lambda_1}[-(l-k)^2]^{\lambda_2}}$$
$$= \frac{\mathrm{i}\pi^{d/2}}{(-l^2)^{\lambda_1+\lambda_2+\varepsilon-2}} \sum_{r=0}^{[(b_1+b_2)/2]} A_{\mathrm{NT}}(\lambda_1,\lambda_2;r,b_1+b_2) \frac{b_1! b_2!}{4^r} (l^2)^r$$
$$\times \sum_{r_1=\max\{0,r-[b_2/2]\}}^{\min\{r,[b_1/2]\}} \frac{(p\cdot l)^{b_1-2r_1}(q\cdot l)^{b_2-2r+2r_1}(p^2)^{r_1}(q^2)^{r-r_1}}{r_1!(r-r_1)!(b_1-2r_1)!(b_2-2r+2r_1)!} , \tag{A.19}$$

and

$$\int \mathrm{d}^d k \frac{(p\cdot k)^{b_1}(q\cdot k)^{b_2}}{(-k^2)^{\lambda_1}(-k^2+2q\cdot k)^{\lambda_2}}$$
$$= \mathrm{i}\pi^{d/2} \frac{(p^2)^{b_1/2}}{(q^2)^{\lambda_1+\lambda_2+\varepsilon-2-b_1/2-b_2}} B_{pq}(\lambda_1,\lambda_2;b_1,b_2)\,, \qquad \text{(A.20)}$$

for even b_1 (and are equal to zero for odd b_1), where

$$B_{pq}(\lambda_1,\lambda_2;b_1,b_2)$$
$$= \sum_{r=b_1/2}^{b_1/2+[b_2/2]} \frac{(-1)^r}{4^r} \frac{b_1!b_2!}{(b_1/2)!(r-b_1/2)!} B_{\mathrm{NT}}(\lambda_1,\lambda_2;r,b_1+b_2)\,. \qquad \text{(A.21)}$$

$$\int \frac{\mathrm{d}^d k}{(-k^2+m^2)^{\lambda_1}(2p\cdot k)^{\lambda_2}}$$
$$= \frac{\mathrm{i}\pi^{d/2}}{(p^2)^{\lambda_2/2}(m^2)^{\lambda_1+\lambda_2/2+\varepsilon-2}} \frac{\Gamma(\lambda_2/2)\Gamma(\lambda_1+\lambda_2/2+\varepsilon-2)}{2\Gamma(\lambda_1)\Gamma(\lambda_2)}\,. \qquad \text{(A.22)}$$

$$\int \mathrm{d}^d k \frac{k^{(\alpha_1,\ldots,\alpha_n)}}{(-k^2+m^2)^{\lambda_1}(2p\cdot k)^{\lambda_2}}$$
$$= \mathrm{i}\pi^{d/2} \frac{\Gamma((\lambda_2+n)/2)}{2\Gamma(\lambda_1)\Gamma(\lambda_2)} \frac{\Gamma(\lambda_1+(\lambda_2-n)/2+\varepsilon-2)}{(m^2)^{\lambda_1+(\lambda_2-n)/2+\varepsilon-2}} \frac{p^{(\alpha_1,\ldots,\alpha_n)}}{(p^2)^{(\lambda_2+n)/2}}\,. \qquad \text{(A.23)}$$

$$\int \frac{\mathrm{d}^d k}{(-k^2+2p\cdot k)^{\lambda_1}(2p\cdot k)^{\lambda_2}}$$
$$= \frac{\mathrm{i}\pi^{d/2}}{(p^2)^{\lambda_1+\lambda_2+\varepsilon-2}} \frac{\Gamma(\lambda_1+\lambda_2+\varepsilon-2)\Gamma(2\lambda_1+\lambda_2+2\varepsilon-4)}{\Gamma(\lambda_1)\Gamma(2\lambda_1+2\lambda_2+2\varepsilon-4)}\,. \qquad \text{(A.24)}$$

$$\int \frac{\mathrm{d}^d k}{(-k^2)^{\lambda_1}(2p\cdot k+y-\mathrm{i}0)^{\lambda_2}}$$
$$= \mathrm{i}\pi^{d/2} \frac{\Gamma(2-\lambda_1-\varepsilon)\Gamma(2\lambda_1+\lambda_2+2\varepsilon-4)}{\Gamma(\lambda_1)\Gamma(\lambda_2)} (p^2)^{\lambda_1+\varepsilon-2} y^{-2\lambda_1-\lambda_2-2\varepsilon+4}\,. \qquad \text{(A.25)}$$

$$\int \mathrm{d}^d k \frac{k^{(\alpha_1,\ldots,\alpha_n)}}{(-k^2)^{\lambda_1}(2p\cdot k+y-\mathrm{i}0)^{\lambda_2}} = \mathrm{i}\pi^{d/2} y^{-2\lambda_1-\lambda_2+n-2\varepsilon+4}$$
$$\times \frac{p^{(\alpha_1,\ldots,\alpha_n)}}{(p^2)^{-\lambda_1+n-\varepsilon+2}} \frac{\Gamma(2-\lambda_1+n-\varepsilon)\Gamma(2\lambda_1+\lambda_2-n+2\varepsilon-4)}{\Gamma(\lambda_1)\Gamma(\lambda_2)}\,. \qquad \text{(A.26)}$$

Let $p\cdot q = 0$. Then

$$\int \frac{\mathrm{d}^d k}{(-k^2)^{\lambda_1}[-(p-k)^2]^{\lambda_2}(-2q\cdot k - \mathrm{i}0)^{\lambda_3}}$$
$$= \mathrm{i}\pi^{d/2} \frac{\Gamma(-\lambda_1 - \lambda_3/2 - \varepsilon + 2)\Gamma(-\lambda_2 - \lambda_3/2 - \varepsilon + 2)}{\Gamma(-\lambda_1 - \lambda_2 - \lambda_3 - 2\varepsilon + 4)}$$
$$\times \frac{\Gamma(\lambda_1 + \lambda_2 + \lambda_3/2 + \varepsilon - 2)\Gamma(\lambda_3/2)}{2\Gamma(\lambda_1)\Gamma(\lambda_2)\Gamma(\lambda_3)(-p^2)^{\lambda_1+\lambda_2+\lambda_3/2+\varepsilon-2}(q^2)^{\lambda_3/2}} \,. \tag{A.27}$$

Let $p_1^2 = p_2^2 = 0,\ q = p_1 - p_2$. Then

$$\int \frac{\mathrm{d}^d k}{(-k^2 + 2p_1\cdot k)^{\lambda_1}(-k^2 + 2p_2\cdot k)^{\lambda_2}(-k^2)^{\lambda_3}}$$
$$= \mathrm{i}\pi^{d/2} \frac{\Gamma(-\lambda_1 - \lambda_3 - \varepsilon + 2)\Gamma(-\lambda_2 - \lambda_3 - \varepsilon + 2)}{\Gamma(\lambda_1)\Gamma(\lambda_2)\Gamma(-\lambda_1 - \lambda_2 - \lambda_3 - 2\varepsilon + 4)}$$
$$\times \frac{\Gamma(\lambda_1 + \lambda_2 + \lambda_3 + \varepsilon - 2)}{(-q^2)^{\lambda_1+\lambda_2+\lambda_3+\varepsilon-2}} \,, \tag{A.28}$$

$$\int \frac{\mathrm{d}^d k}{(-k^2 + 2p_1\cdot k)^{\lambda_1}(-k^2 + 2p_2\cdot k)^{\lambda_2}(2p_2\cdot k)^{\lambda_3}} = \mathrm{i}\pi^{d/2} \frac{\Gamma(-\lambda_1 - \varepsilon + 2)}{\Gamma(\lambda_1)\Gamma(\lambda_2)}$$
$$\times \frac{\Gamma(\lambda_1 + \lambda_2 + \varepsilon - 2)\Gamma(-\lambda_2 - \lambda_3 - \varepsilon + 2)}{\Gamma(-\lambda_1 - \lambda_2 - \lambda_3 - 2\varepsilon + 4)(-q^2)^{\lambda_1+\lambda_2+\lambda_3+\varepsilon-2}} \,, \tag{A.29}$$

$$\int \frac{\mathrm{d}^d k}{(2p_1\cdot k)^{\lambda_1}(-k^2 + 2p_2\cdot k)^{\lambda_2}(-k^2 + m^2)^{\lambda_3}}$$
$$= \mathrm{i}\pi^{d/2} \frac{\Gamma(\lambda_2 - \lambda_1)\Gamma(\lambda_2 + \lambda_3 + \varepsilon - 2)\Gamma(-\lambda_2 - \varepsilon + 2)}{\Gamma(\lambda_2)\Gamma(\lambda_3)\Gamma(-\lambda_1 - \varepsilon + 2)(-q^2)^{\lambda_1}(m^2)^{\lambda_2+\lambda_3+\varepsilon-2}} \,, \tag{A.30}$$

$$\int \frac{\mathrm{d}^d k}{(2p_1\cdot k)^{\lambda_1}(-k^2 + 2p_2\cdot k)^{\lambda_2}(-k^2 + m^2)^{\lambda_3}(Q^2 - 2p_1\cdot k)^{\lambda_4}}$$
$$= \mathrm{i}\pi^{d/2} \frac{\Gamma(\lambda_2 - \lambda_1)\Gamma(\lambda_2 + \lambda_3 + \varepsilon - 2)\Gamma(-\lambda_2 - \lambda_4 - \varepsilon + 2)}{\Gamma(\lambda_2)\Gamma(\lambda_3)\Gamma(-\lambda_1 - \lambda_4 - \varepsilon + 2)}$$
$$\times \frac{1}{(Q^2)^{\lambda_1+\lambda_4}(m^2)^{\lambda_2+\lambda_3+\varepsilon-2}} \,, \tag{A.31}$$

$$\int \frac{\mathrm{d}^d k}{(2p_1\cdot k + m^2)^{\lambda_1}(2p_2\cdot k + m^2)^{\lambda_2}(-k^2)^{\lambda_3}}$$
$$= \mathrm{i}\pi^{d/2} \frac{\Gamma(\lambda_1 + \lambda_3 + \varepsilon - 2)\Gamma(\lambda_2 + \lambda_3 + \varepsilon - 2)\Gamma(-\lambda_3 - \varepsilon + 2)}{\Gamma(\lambda_1)\Gamma(\lambda_2)\Gamma(\lambda_3)(-q^2)^{-\lambda_3-\varepsilon+2}(m^2)^{\lambda_1+\lambda_2+2\lambda_3+2\varepsilon-4}} \,. \tag{A.32}$$

Let $p_1^2 = 0$, $p_2^2 = -m^2$, $q = p_1 - p_2$. Then

$$\int \frac{\mathrm{d}^d k}{(2p_1\cdot k)^{\lambda_1}(-k^2+2p_2\cdot k+m^2)^{\lambda_2}(-k^2)^{\lambda_3}} = \mathrm{i}\pi^{d/2}\frac{\Gamma(\lambda_2+\lambda_3+\varepsilon-2)}{(m^2)^{\lambda_2+\lambda_3+\varepsilon-2}} \times \frac{\Gamma(-\lambda_1-\lambda_3-\varepsilon+2)\Gamma(-\lambda_2-\varepsilon+2)}{\Gamma(\lambda_2)\Gamma(\lambda_3)\Gamma(-\lambda_1-\lambda_2-\lambda_3-2\varepsilon+4)(-q^2)^{\lambda_1}}\,, \tag{A.33}$$

$$\int \frac{\mathrm{d}^d k}{(2p_1\cdot k)^{\lambda_1}(-k^2+2p_2\cdot k-m^2)^{\lambda_2}(-k^2)^{\lambda_3}(-q^2-2p_1\cdot k)^{\lambda_4}} = \mathrm{i}\pi^{d/2}\frac{\Gamma(\lambda_2+\lambda_3+\varepsilon-2)}{(m^2)^{\lambda_2+\lambda_3+\varepsilon-2}} \times \frac{\Gamma(-\lambda_1-\lambda_3-\varepsilon+2)\Gamma(-\lambda_2-\lambda_4-\varepsilon+2)}{\Gamma(\lambda_2)\Gamma(\lambda_3)\Gamma(-\lambda_1-\lambda_2-\lambda_3-\lambda_4-2\varepsilon+4)(-q^2)^{\lambda_1+\lambda_4}}\,. \tag{A.34}$$

Let $P^2 = M^2$, $p^2 = 0$, $(P-p)^2 = 0$. Then

$$\int \frac{\mathrm{d}^d k}{(-k^2+2P\cdot k)^{\lambda_1}(-k^2+2p\cdot k)^{\lambda_2}(-k^2)^{\lambda_3}} = \mathrm{i}\pi^{d/2}\frac{\Gamma(-\lambda_1-\lambda_2-2\lambda_3-2\varepsilon+4)\Gamma(\lambda_1+\lambda_2+\lambda_3+\varepsilon-2)}{\Gamma(\lambda_1)\Gamma(-\lambda_1-\lambda_2-\lambda_3-2\varepsilon+4)} \times \frac{\Gamma(-\lambda_2-\lambda_3-\varepsilon+2)}{\Gamma(-\lambda_3-\varepsilon+2)(M^2)^{\lambda_1+\lambda_2+\lambda_3+\varepsilon-2}}\,. \tag{A.35}$$

Let $p_1^2 = 0$, $p_2^2 = m^2$, $Q^2 = 2p_1\cdot p_2$. Then

$$\int \frac{\mathrm{d}^d k}{(2p_1\cdot k)^{\lambda_1}(-k^2+2p_2\cdot k)^{\lambda_2}(-k^2)^{\lambda_3}(Q^2-2p_1\cdot k)^{\lambda_4}} = \mathrm{i}\pi^{d/2}\frac{\Gamma(\lambda_3-\lambda_4)\Gamma(-\lambda_1-\lambda_2-2\lambda_3-2\varepsilon+4)}{\Gamma(\lambda_2)\Gamma(\lambda_3)\Gamma(-\lambda_1-\lambda_2-\lambda_3-\lambda_4-2\varepsilon+4)} \times \frac{\Gamma(\lambda_2+\lambda_3+\varepsilon-2)}{(Q^2)^{\lambda_1+\lambda_4}(m^2)^{\lambda_2+\lambda_3+\varepsilon-2}}\,, \tag{A.36}$$

$$\int \frac{\mathrm{d}^d k}{(2p_1\cdot k)^{\lambda_1}(-k^2+2p_2\cdot k)^{\lambda_2}(-k^2)^{\lambda_3}} = \frac{\mathrm{i}\pi^{d/2}}{(Q^2)^{\lambda_1}(m^2)^{\lambda_2+\lambda_3+\varepsilon-2}} \times \frac{\Gamma(\lambda_2+\lambda_3+\varepsilon-2)\Gamma(-\lambda_1-\lambda_2-2\lambda_3-2\varepsilon+4)}{\Gamma(\lambda_2)\Gamma(-\lambda_1-\lambda_2-\lambda_3-2\varepsilon+4)}\,. \tag{A.37}$$

The following integrals are related to two-loop diagrams:

$$\int\int \frac{\mathrm{d}^d k\,\mathrm{d}^d l}{(-k^2+m^2)^{\lambda_1}[-(k+l)^2]^{\lambda_2}(-l^2+m^2)^{\lambda_3}} = \left(\mathrm{i}\pi^{d/2}\right)^2 \frac{\Gamma(\lambda_1+\lambda_2+\varepsilon-2)\Gamma(\lambda_2+\lambda_3+\varepsilon-2)\Gamma(2-\varepsilon-\lambda_2)}{\Gamma(\lambda_1)\Gamma(\lambda_3)} \times \frac{\Gamma(\lambda_1+\lambda_2+\lambda_3+2\varepsilon-4)}{\Gamma(\lambda_1+2\lambda_2+\lambda_3+2\varepsilon-4)\Gamma(2-\varepsilon)(m^2)^{\lambda_1+\lambda_2+\lambda_3+2\varepsilon-4}}\,, \tag{A.38}$$

$$\int\int \frac{\mathrm{d}^d k\, \mathrm{d}^d l}{(-k^2)^{\lambda_1}[-(k+l)^2]^{\lambda_2}(m^2-l^2)^{\lambda_3}}$$
$$= \left(\mathrm{i}\pi^{d/2}\right)^2 \frac{\Gamma(\lambda_1+\lambda_2+\lambda_3+2\varepsilon-4)}{(m^2)^{\lambda_1+\lambda_2+\lambda_3+2\varepsilon-4}}$$
$$\times \frac{\Gamma(\lambda_1+\lambda_2+\varepsilon-2)\Gamma(2-\varepsilon-\lambda_1)\Gamma(2-\varepsilon-\lambda_2)}{\Gamma(\lambda_1)\Gamma(\lambda_2)\Gamma(\lambda_3)\Gamma(2-\varepsilon)} \,. \tag{A.39}$$

A.2 Some Useful Formulae

To traceless expressions and back:

$$k^{\alpha_1}\dots k^{\alpha_N} = \frac{1}{N!}\sum_{r=0}^{[N/2]} \frac{1}{2^r(d/2+N-2r)_r}(k^2)^r\{[g]^r[k]^{(N-2r)}\}^{\alpha_1\dots\alpha_N} \,, \tag{A.40a}$$

$$k^{(\alpha_1\dots\alpha_N)} = \frac{1}{N!}\sum_{r=0}^{[N/2]} \frac{1}{2^r(2-N-d/2)_r}(k^2)^r\{[g]^r[k]^{N-2r}\}^{\alpha_1\dots\alpha_N} \,, \tag{A.40b}$$

where $k^{(\alpha_1\dots\alpha_N)}$ and $\{[g]^r[k]^{N-2r}\}^{\alpha_1\dots\alpha_N}$ are defined before (A.8) and after (A.11), respectively, and $(a)_n$ is the Pochhammer symbol (A.53).

Furthermore,

$$(k\cdot p)^N = \sum_{r=0}^{[N/2]} a_{N,r}(k^2)^r(p^2)^r(k\cdot p)^{(N-2r)} \,, \tag{A.41}$$

$$(k\cdot p)^{(N)} = \sum_{r=0}^{[N/2]} b_{N,r}(k^2)^r(p^2)^r(k\cdot p)^{N-2r} \,, \tag{A.42}$$

$$k_{(\alpha_1\dots\alpha_N)}k^{(\alpha_1\dots\alpha_N)} = \frac{(d-2)_N}{2^N\left((d-2)/2\right)_N}(k^2)^N \,, \tag{A.43}$$

where $(k\cdot p)^{(N)} = k_{(\alpha_1\dots\alpha_N)}p^{(\alpha_1\dots\alpha_N)}$ and

$$a_{N,r} = \frac{N!}{4^r r!(N-2r)!(d/2+N-2r)_r} \,, \tag{A.44}$$

$$b_{N,r} = \frac{1}{4^r r!(N-2r)!(2-N-d/2)_r} \,. \tag{A.45}$$

Summation formulae:

$$[(k_1)^m(k_2)^n * g_\mathrm{s}] \equiv k_1^{\alpha_1}\dots k_1^{\alpha_m}k_2^{\alpha_{m+1}}\dots k_2^{\alpha_{m+n}} g_{\mathrm{s},\,\alpha_1\dots\alpha_{m+n}}$$
$$= \sum_{j\ge 0,\; j+\min\{m,n\}\ \mathrm{even}}^{\min\{m,n\}} \frac{m!n!}{2^{(m+n)/2-j}((m-j)/2)!((n-j)/2)!j!}$$
$$\times (k_1^2)^{(m-j)/2}(k_2^2)^{(n-j)/2}(k_1\cdot k_2)^j \,, \tag{A.46}$$

$$
\begin{aligned}
&[(k_1)^m (k_2)^n * \{[g]^r (k_3)^{m+n-2r}\}] \\
&= \sum_{r_1=\max\{0,2r-n\}}^{\min\{2r,m\}} \sum_{j\geq 0,\ j+r_1 \text{ even}}^{\min\{r_1,2r-r_1\}} \frac{1}{(m-r_1)!(n-2r+r_1)!} \\
&\times \frac{m!n!}{2^{r-j}((r_1-j)/2)!(r-(r_1+j)/2)!j!} (k_1^2)^{(r_1-j)/2} (k_2^2)^{r-(r_1+j)/2} \\
&\times (k_1\cdot k_2)^j (k_1\cdot k_3)^{m-r_1} (k_2\cdot k_3)^{n-2r+r_1} .
\end{aligned} \tag{A.47}
$$

In particular,

$$
\begin{aligned}
&[(k_1)^m (k_2)^n * \{[g]^r (k_3)^{N-2r}\}] \\
&= \binom{n}{N-2r} (k_2\cdot k_3)^{N-2r} [(k_1)^m (k_2)^{n-N+2r} * g_{\mathrm{s}}] ,
\end{aligned} \tag{A.48}
$$

where $k_1\cdot k_3 = 0$, and

$$
\begin{aligned}
&[p^{b_1} q^{b_2} * \{[g]^r (l)^{n-2r}\}] \\
&= \frac{b_1!b_2!}{2^r} \sum_{r_1=\max\{0,r-[b_2/2]\}}^{\min\{r,[b_1/2]\}} \frac{(p\cdot l)^{b_1-2r_1} (q\cdot l)^{b_2-2r+2r_1} (p^2)^{r_1} (q^2)^{r-r_1}}{r_1!(r-r_1)!(b_1-2r_1)!(b_2-2r+2r_1)!} ,
\end{aligned} \tag{A.49}
$$

where $p\cdot q = 0$.

$$
\begin{aligned}
&[(k_1)^m (k_2)^n (k_3)^{l-m-n} * g_{\mathrm{s}}] \\
&= \sum_{j_1\geq 0,\ j_1+m \text{ even}} \ \sum_{j_2\geq 0,\ j_2+n \text{ even}} \ \sum_{j_3\geq 0,\ j_3+l-m-n \text{ even}} a(l,m,n,j_1,j_2,j_3) \\
&\times (k_1^2)^{(m-j_1)/2} (k_2^2)^{(n-j_2)/2} (k_3^2)^{(l-m-n-j_3)/2} \\
&\times (k_1\cdot k_2)^{(j_1+j_2-j_3)/2} (k_1\cdot k_3)^{(j_1-j_2+j_3)/2} (k_2\cdot k_3)^{(-j_1+j_2+j_3)/2} , \\
&a(l,m,n,j_1,j_2,j_3) = \frac{2^{(j_1+j_2+j_3-l)/2} m!n!(l-m-n)!}{((m-j_1)/2)!((n-j_2)/2)!((l-m-n-j_3)/2)!} \\
&\times \frac{\theta(j_1+j_2-j_3)\theta(j_1-j_2+j_3)\theta(-j_1+j_2+j_3)}{((j_1+j_2-j_3)/2)!((j_1-j_2+j_3)/2)!((-j_1+j_2+j_3)/2)!} ,
\end{aligned} \tag{A.50}
$$

where $\theta(n) = 1$ for $n \geq 0$ and $\theta(n) = 0$ otherwise.

The (inverse) Fourier transformation in d dimensions:

$$
\frac{1}{(2\pi)^d} \int \mathrm{d}^d q \frac{\mathrm{e}^{-\mathrm{i}x\cdot q}}{(-q^2 - i0)^\lambda} = \frac{\Gamma(d/2-\lambda)}{4^\lambda \pi^{d/2} \Gamma(\lambda)} \frac{1}{(-x^2 + i0)^{d/2-\lambda}} . \tag{A.51}
$$

A.3 Some Special Functions

The Gauss hypergeometric function is defined by the series

$$ {}_2F_1(a,b;c;z) = \sum_{n=0}^{\infty} \frac{(a)_n (b)_n}{(c)_n n!} z^n \,, \tag{A.52} $$

where

$$ (x)_n = \Gamma(x+n)/\Gamma(x) \tag{A.53} $$

is the Pochhammer symbol. This power series has a radius of convergence equal to one. It is analytically continued to the whole complex plane, with a cut, usually chosen as $[1,\infty)$. The analytic continuation to values of z where $|z| > 1$ is given by

$$ \begin{aligned} {}_2F_1(a,b;c;z) = {} & \frac{\Gamma(c)\Gamma(b-a)}{\Gamma(b)\Gamma(c-a)}(-z)^{-a}\, {}_2F_1\left(a, 1-c+a; 1-b+a; \frac{1}{z}\right) \\ & + \frac{\Gamma(c)\Gamma(a-b)}{\Gamma(a)\Gamma(c-b)}(-z)^{-b}\, {}_2F_1\left(b, 1-c+b; 1-a+b; \frac{1}{z}\right) . \end{aligned} \tag{A.54} $$

Another formula for the analytic continuation is

$$ {}_2F_1(a,b;c;z) = (1-z)^{-a}\, {}_2F_1\left(a, c-b; c; \frac{z}{z-1}\right) . \tag{A.55} $$

The polylogarithms [165] and generalized polylogarithms [153, 96] are defined by

$$ \mathrm{Li}_a(z) = \sum_{n=1}^{\infty} \frac{z^n}{n^a} \tag{A.56} $$

$$ = \int_0^1 \frac{\ln^a(1-zt)}{t}\, \mathrm{d}t \tag{A.57} $$

and

$$ S_{a,b}(z) = \frac{(-1)^{a+b-1}}{(a-1)!b!} \int_0^1 \frac{\ln^{a-1} t \ln^b(1-zt)}{t}\, \mathrm{d}t \,, \tag{A.58} $$

where a and b are positive integers.

B Technical Details: Convergence and Asymptotic Behaviour

The difficulty, as in all this work, is to find a notation which is both concise and intelligible to at least two people of whom one may be the author.

(P.T. Matthews and A. Salam [176])

In this appendix, details connected with the analysis of the convergence of Feynman integrals and a proof of the asymptotic behaviour of the remainder in the off-shell large-momentum expansion are presented.

B.1 Analysis of Convergence

We obtain the alpha representation of an analytically and dimensionally regularized Feynman integral corresponding to a graph Γ starting from the alpha representation (2.42) and substituting the powers of propagators a_l by $a_l + \lambda_l$ with general complex numbers λ_l. For simplicity, let us assume the scalar case and that the powers of propagators are equal to one. (If $a_l > 1$, one can represent such a line by a sequence of a_l lines.) In this case the alpha representation takes a simpler form

$$F_\Gamma(\underline{q}, \underline{m}; d, \underline{\lambda})$$

$$= \int_0^\infty \mathrm{d}\underline{\alpha} \prod_l \alpha_l^{\lambda_l} D(\underline{\alpha})^{-d/2} \exp\left(\mathrm{i}A(\underline{q}, \underline{\alpha})/D(\underline{\alpha}) - \mathrm{i}\sum_l m_l^2 \alpha_l \right) , \quad \text{(B.1)}$$

where the functions A and D are given by (2.44) and (2.43), and from now on we omit the coefficient

$$(-1)^L \mathrm{e}^{\mathrm{i}\pi\left(\sum \lambda_l + h(1-d/2)\right)/2} \pi^{hd/2} \prod_l \Gamma(\lambda_l + 1) ,$$

which is irrelevant to the analysis of convergence and the proof of the asymptotic estimate. In this appendix (as in Sect. 9.1.3), families of variables are denoted by underlined letters, i.e. $\underline{q} = (q_1, \ldots, q_n)$, $\underline{m} = (m_1, \ldots, m_L)$, $\underline{\lambda} = (\lambda_1, \ldots, \lambda_L)$, $\underline{\alpha} = (\alpha_1, \ldots, \alpha_L)$, etc., with $\mathrm{d}\underline{\alpha} = \mathrm{d}\alpha_1 \ldots \mathrm{d}\alpha_L$. Let us also assume here and later that the limit of integration refers to all of the integration variables involved.

The alpha parameters have dimension -2 in mass units. By making the change of variables $\alpha_l \to \mu^{-2}\alpha_l$, where μ is a massive parameter, we can transform to dimensionless alpha parameters. For simplicity, let us take $\mu = 1$ in this appendix. To separate the analysis of the UV and IR convergence as much as possible let us decompose the integration from 0 to ∞ over each alpha parameter into two regions: from 0 to 1 and from 1 to ∞. The integral (B.1) is then divided into 2^L pieces, each of which is determined by a decomposition of the set of lines $\mathcal{L}$ of the given graph into two subsets, $\mathcal{L}_\alpha$ and $\mathcal{L}_\beta$, corresponding to the integrations over the UV region (from 0 to 1) and the IR region (from 1 to ∞), respectively. For a given piece generated by a subset $\mathcal{L}_\alpha$, let us change the variables α_l for $l \in \mathcal{L}_\beta$ according to $\alpha_l = 1/\beta_l$. The corresponding integral then takes the form

$$F_\Gamma^{\mathcal{L}_\alpha}(\underline{q}, \underline{m}; d, \underline{\lambda}) = \int_0^1 \mathrm{d}\underline{\alpha}\, \mathrm{d}\underline{\beta} \prod_{l\in\mathcal{L}_\alpha} \alpha_l^{\lambda_l} \prod_{l\in\mathcal{L}_\beta} \beta_l^{-\lambda_l-\varepsilon} D(\underline{\alpha}, \underline{\beta})^{-d/2}$$

$$\times \exp\left(\mathrm{i} A(\underline{q}, \underline{\alpha}, \underline{\beta})/D(\underline{\alpha}, \underline{\beta}) - \mathrm{i} \sum_{l\in\mathcal{L}_\alpha} m_l^2 \alpha_l - \mathrm{i} \sum_{l\in\mathcal{L}_\beta} m_l^2/\beta_l \right) . \tag{B.2}$$

For brevity, the new functions D and A are denoted by the same letters, although they are now of the form

$$D(\underline{\alpha}, \underline{\beta}) = \left(\prod_{l\in\mathcal{L}_\beta} \beta_l \right) D(\underline{\alpha})|_{\alpha_l \to 1/\beta_l, l\in\mathcal{L}_\beta}$$

$$= \sum_{T\in T^1} \left(\prod_{l\in\mathcal{L}_\alpha \setminus T} \alpha_l \right) \left(\prod_{l\in\mathcal{L}_\beta \cap T} \beta_l \right) , \tag{B.3}$$

$$A(\underline{q}, \underline{\alpha}, \underline{\beta}) = \left(\prod_{l\in\mathcal{L}_\beta} \beta_l \right) A(\underline{q}, \underline{\alpha})|_{\alpha_l \to 1/\beta_l, l\in\mathcal{L}_\beta}$$

$$= \sum_{T\in T^2} \left(\prod_{l\in\mathcal{L}_\alpha \setminus T} \alpha_l \right) \left(\prod_{l\in\mathcal{L}_\beta \cap T} \beta_l \right) \left(q^T\right)^2 . \tag{B.4}$$

Remember that $\pm q^T$ is the sum of the external momenta that flow into one of the connectivity components of a 2-tree T.

For a given piece $F_\Gamma^{\mathcal{L}_\alpha}$, let us change the numbering of the lines in such a way that the UV lines (i.e. those with $\alpha_l \le 1$) have smaller numbers. Thus we perform integration in the domain $0 \le \alpha_l \le 1,\ 1 \le l \le \bar{l}$ and $0 \le \beta_l \le 1,\ \bar{l}+1 \le l \le L$, where $\bar{l} = |\mathcal{L}_\alpha|$. If S is a finite set, we denote by $|S|$ the number of its elements.

As we shall see, the analysis of UV and IR convergence is now decoupled. To analyse the UV convergence let us divide the domain of integration over α_l into sectors. In the following, we shall use sectors of two types associated with nest and forests, respectively. The sectors connected with nests of subgraphs [137] (let us call them N-sectors) are defined by

$$\alpha_1 \le \ldots \le \alpha_{\bar{l}} \tag{B.5}$$

and similar domains obtained by permutations. Without loss of generality, let us consider only the sector (B.5). Let us then change the integration variables according to

$$\alpha_l = t_l \dots t_{\bar{l}} \,. \tag{B.6}$$

The new (N-sector) variables t_l are expressed in terms of α_l by

$$t_l = \begin{cases} \alpha_l/\alpha_{l+1} & \text{if } l < \bar{l} \\ \alpha_{\bar{l}} & \text{if } l = \bar{l} \end{cases} \,. \tag{B.7}$$

The corresponding Jacobian equals $\prod t_l^{l-1}$.

The decomposition of the IR integration, over β_l, is performed in a quite similar way. The following are the corresponding analogues of N-sectors and sector variables:

$$\beta_L \geq \dots \geq \beta_{\bar{l}+1} \,, \tag{B.8}$$

$$\beta_l = \tau_{\bar{l}+1} \cdots \tau_l \,, \tag{B.9}$$

$$\tau_l = \begin{cases} \beta_l/\beta_{l-1} & \text{if } l > \bar{l}+1 \\ \beta_{\bar{l}+1} & \text{if } l = \bar{l}+1 \end{cases} \,, \tag{B.10}$$

and the corresponding Jacobian is $\prod \tau_l^{L-l}$.

So, the initial integral is eventually divided into $(L+1)!$ sectors

$$\alpha_{\pi(1)} \leq \dots \leq \alpha_{\pi(\bar{l})} \leq 1 \leq \beta_{\pi(\bar{l}+1)} \leq \beta_{\pi(L)} \,, \tag{B.11}$$

which are labelled by permutations π of the numbers $1, \dots, L$ and the number $\bar{l}$. As we have stated, we consider only the contribution of the identical permutation, i.e. $\pi(l) = l$, $l = 1, \dots, L$.

Although these sectors provide a resolution of the singularities of the integrand, they can turn out to be too rough for analysing convergence. (But, as we shall see, the N-sectors will be still used when we analyse the estimate of the remainder of the asymptotic expansion.) A more sophisticated set of sectors corresponds to the maximal UV and IR forests. A set f of 1PI subgraphs and single lines with non-coincident end points is called a *UV forest* [194, 222, 35] if the following conditions hold: (*i*) for any pair $\gamma, \gamma' \in f$, we have either $\gamma \subset \gamma'$, $\gamma' \subset \gamma$ or $\mathcal{L}(\gamma \cap \gamma') = \emptyset$; (*ii*) if $\gamma_1, \dots, \gamma^n \in f$ and $\mathcal{L}(\gamma^i \cap \gamma^j) = \emptyset$ for any pair from this family, the subgraph $\cup_i \gamma^i$ is *one-vertex reducible* (i.e. can be made disconnected by deleting a vertex).

Let $\mathcal{F}$ be a *maximal* UV forest (i.e. there are no UV forests that include $\mathcal{F}$) of a given graph Γ. An element $\gamma \in \mathcal{F}$ is called *trivial* if it consists of a single line and is not a loop line. Any maximal UV forest has $h(\Gamma)$ non-trivial and $L - h(\Gamma)$ trivial elements. Let us define the mapping $\sigma : \mathcal{F} \to \mathcal{L}$ such that $\sigma(\gamma) \in \mathcal{L}(\gamma)$ and $\sigma(\gamma) \notin \mathcal{L}(\gamma')$ for any $\gamma' \subset \gamma$, $\gamma' \in \mathcal{F}$. Its inverse σ^{-1} uniquely determines the minimal element $\sigma^{-1}(l)$ of the UV forest $\mathcal{F}$ that contains the line l. Let us denote by γ_+ the minimal element of $\mathcal{F}$ that strictly includes the given element γ.

For a given maximal UV forest $\mathcal{F}$, let us define the corresponding sector (F-sector) as

$$\mathcal{D}_{\mathcal{F}} = \left\{\underline{\alpha}|\alpha_l \leq \alpha_{\sigma(\gamma)} \leq 1,\; l \in \gamma \in \mathcal{F}\right\} . \quad (B.12)$$

The intersection of two different F-sectors has zero measure and the union of all the sectors gives the whole integration domain of the UV alpha parameters (i.e. $\alpha_l \leq 1$) (see [194, 222, 35, 208]). For a given F-sector, let us introduce new variables labelled by the elements of $\mathcal{F}$,

$$\alpha_l = \prod_{\gamma\in\mathcal{F}:\, l\in\gamma} t_\gamma \;, \quad (B.13)$$

where the corresponding Jacobian is $\prod_\gamma t_\gamma^{L(\gamma)-1}$. The inverse formula is

$$t_\gamma = \begin{cases} \alpha_{\sigma(\gamma)}/\alpha_{\sigma(\gamma_+)} & \text{if } \gamma \text{ is not maximal} \\ \alpha_{\sigma(\gamma)} & \text{if } \gamma \text{ is maximal} \end{cases} . \quad (B.14)$$

Consider, for example, the two-loop self-energy diagram of Fig. 2.5 and the following maximal UV forest: $\mathcal{F} = \gamma^1 = \{1\},\ \gamma^2 = \{2\},\ \gamma^3 = \{3\},\ \gamma^4 = \{1,2,5\},\ \gamma^5 = \Gamma$. The mapping σ is $\sigma(\gamma^1) = 1,\ \sigma(\gamma^2) = 2,\ \sigma(\gamma^3) = 3,\ \sigma(\gamma^4) = 5,\ \sigma(\gamma^5) = 4$. The sector associated with this maximal UV forest is given by $\mathcal{D}_{\mathcal{F}} = \{\alpha_{1,2} \leq \alpha_5 \leq \alpha_4,\ \alpha_3 \leq \alpha_4\}$ and the sector variables are $t_{\gamma^1} = \alpha_1/\alpha_5,\ t_{\gamma^2} = \alpha_2/\alpha_5,\ t_{\gamma^3} = \alpha_3/\alpha_4,\ t_{\gamma^4} = \alpha_5/\alpha_4,\ t_{\gamma^5} = \alpha_4$.

The IR F-sectors and variables are introduced in a quite analogous way. New variables τ_γ are associated with maximal IR forests composed of IR-irreducible subgraphs – see [208]. (A subgraph γ of Γ is called *IR irreducible* [62, 208] if the reduced graph $\Gamma/\overline{\gamma}$ is one-vertex-irreducible. It is also implied that the completion $\overline{\gamma} \equiv \Gamma\backslash\gamma$ includes all the external vertices in the same connectivity component.) The UV and IR maximal forests $\mathcal{F}_\alpha$ and $\mathcal{F}_\beta$, composed of lines $\mathcal{L}_\alpha$ and $\mathcal{L}_\beta$, respectively, are then combined in pairs to generate 'generalized maximal forests', with corresponding variables $\{t_\gamma, \tau_{\gamma'}\},\ \gamma \in \mathcal{F}_\beta,\ \gamma \in \mathcal{F}_\beta$. As a result, the initial integration domain is divided into F-sectors associated with generalized maximal forests.

In each of the N- or F-sectors, the function (B.3) takes a factorized form in the new variables [194, 222, 35, 208, 246]:

$$D = \left(\prod_{l=1}^{\bar{l}} t_l^{h(\gamma_l)}\right)\left(\prod_{l=\bar{l}+1}^{L} \tau_l^{L-l+1-h(\Gamma/\gamma_{l-1})}\right)[1 + P_{\rm N}(\underline{t},\underline{\tau})] \quad (B.15)$$

$$= \left(\prod_{\gamma\in\mathcal{F}_\alpha} t_\gamma^{h(\gamma)}\right)\left(\prod_{\gamma\in\mathcal{F}_\beta} \tau_\gamma^{L(\gamma)-h(\Gamma/\overline{\gamma})}\right)[1 + P_{\rm F}(\underline{t},\underline{\tau})] \;, \quad (B.16)$$

where $P_{\rm N}$ and $P_{\rm F}$ are non-negative polynomials, γ_l denotes the subgraph consisting of the lines $\{1,\dots,l\}$, and again $\overline{\gamma} = \Gamma\backslash\gamma$. The factorization of the function (B.4) in the N-sector variables is of the form

$$A = \left(\prod_{l=1}^{\bar{l}} t_l^{h(\gamma_l)}\right) \prod_{l=\bar{l}+1}^{L} \tau_l^{L-l+1-h(\Gamma/\gamma_{l-1})} \left(\tau_{\bar{l}+1}\dots\tau_{l_0}\right)^{-1}$$
$$\times \left[\left(q^{T_0}\right)^2 + P_0(\underline{q},\underline{t},\underline{\tau})\right] , \qquad \text{(B.17)}$$

where l_0 denotes the number such that all the external vertices belong to the same connectivity component of the subgraphs γ_l for $l \geq l_0$. In the Euclidean domain, we have $\left(q^{T_0}\right)^2 < 0$ and $P_0(\underline{q},\underline{t},\underline{\tau}) \leq 0$.

These factorization formulae are proven by constructing an appropriate (2-)tree. In particular, in the case of pure α-variables, one uses the formula

$$\prod_{l \in \mathcal{L}_\alpha \setminus T} \alpha_l = \prod_{\gamma \in \mathcal{F}_\alpha} t_\gamma^{h(\gamma)+c(\gamma \cap T)-c(\gamma)} , \qquad \text{(B.18)}$$

where T is a tree or 2-tree and $c(\gamma)$ is the number of connectivity components of γ, so that the factorization reduces to constructing a (2-)tree that provides the minimal value of the non-negative quantity $c(\gamma \cap T) - c(\gamma)$. In particular, the unity term in the square brackets in (B.15) corresponds to the tree which is constructed as follows: one considers the lines $l = 1, 2, \dots$ consecutively and includes the given line in the tree if a loop is not generated. In (B.16), the minimal power of the sector variables is achieved for the tree which is composed of all trivial elements of the given maximal UV forest $\mathcal{F}$.

The 2-tree T_0 that gives $q_{T_0}^2$ in (B.17) is constructed by a similar procedure with the additional requirement that a line is not included when it could connect all the external vertices of the graph. The factorization in the F-sector variables is a little bit more complicated (see [208]); instead of the contribution of the 2-tree T_0, there is a sum of contributions from some family of 2-trees.

These formulae provide a factorization of the integrand of the alpha representation and make manifest the analysis of the UV and IR convergence. The contribution of the N-sector (B.11) takes the form

$$F_\Gamma^{\bar{l}}(\underline{q},\underline{m};d,\underline{\lambda}) = \int_0^1 \mathrm{d}\underline{t}\, \mathrm{d}\underline{\tau} \left(\prod_{l=1}^{\bar{l}} t_l^{\lambda(\gamma_l)+h(\gamma_l)\varepsilon-[\omega(\gamma_l)/2]-1}\right)$$
$$\times \left(\prod_{l=\bar{l}+1}^{L} \tau_l^{\lambda(\gamma_l')-h(\Gamma/\gamma_{l-1})\varepsilon+[(\omega(\Gamma)-\omega(\gamma_l')+1)/2]-1}\right)$$
$$\times \left[1+P_{\mathrm{N}}(\underline{t},\underline{\tau})\right]^{\varepsilon-2} \exp\left(\mathrm{i}\frac{q_{T_0}^2 + P_0(\underline{q},\underline{t},\underline{\tau})}{1+P_{\mathrm{N}}(\underline{t},\underline{\tau})} \left(\tau_{\bar{l}+1}\dots\tau_{l_0}\right)^{-1}\right.$$
$$\left. -\mathrm{i}\sum_{l=1}^{\bar{l}} m_l^2 \alpha_l(\underline{t}) - \mathrm{i}\sum_{l=\bar{l}+1}^{L} m_l^2/\beta_l(\underline{\tau})\right) , \qquad \text{(B.19)}$$

where

$$\lambda(\gamma) = \sum_{l \in \gamma} \lambda_l \; , \tag{B.20}$$

and, in addition to γ_l, we have introduced the notation $\gamma_l' \equiv \Gamma\backslash\gamma_{l-1}$ for the subgraph composed of the lines $\{l, l+1, \ldots, L\}$. The general case $\bar{l} < l_0$ is assumed. The square brackets in the exponents denote the integer parts of numbers, and $h(\gamma)$ and $\omega(\gamma)$, as before, denote the number of loops and the UV degree of divergence, respectively. This factorization is given here for a general graph. In the scalar case, on which we are concentrating, the degrees of divergence are even numbers so that one can avoid the need to take those integer parts.

The structure of the factorized representation in the F-sector variables is similar, where the product of powers of the sector variables now takes the form

$$\left(\prod_{\gamma \in \mathcal{F}_\alpha} t_\gamma^{\lambda(\gamma)+h(\gamma)\varepsilon-[\omega(\gamma)/2]-1} \right) \times \left(\prod_{\gamma \in \mathcal{F}_\beta} \tau_\gamma^{\lambda(\gamma)-h(\Gamma/\overline{\gamma})\varepsilon+[(\omega(\Gamma)-\omega(\gamma)+1)/2]-1} \right) . \tag{B.21}$$

So the factorized N-sector integrals take the same form as the F-sector integrals if we let the UV subgraph γ be any graph of type γ_l and the IR subgraph γ be any graph of type γ_l', no matter whether they are UV/IR irreducible. Therefore, to analyse the UV and IR convergence, the F-sectors are certainly preferable because it suffices to check convergence in a smaller family of integrals.

The analysis of convergence has therefore been reduced to counting powers in products of one-dimensional integrals over the sector variables. Note that (IR) convergence in the variables τ_l is guaranteed if τ_l^{-1} is present in the exponent. This property can be explained by the fact that the one-dimensional integral $\int_0^\infty \mathrm{d}\tau\, \mathrm{e}^{-\mathrm{i}m^2/\tau} \tau^\lambda \phi(\tau)$, with an infinitely differentiable function ϕ and a sufficient decrease at infinity, is well defined even at arbitrary values of $\mathrm{Re}\,\lambda \leq -2$ (where it is, strictly speaking, divergent): this is true both in the sense of the analytic continuation from the domain $\mathrm{Re}\,\lambda > -1$ and in the sense of the limit $\delta \to +0$ with $m^2 \to m^2 - \mathrm{i}\delta$ (with identical resulting prescriptions in both these variants). In particular, such integrals are well defined for the integer values $\lambda = -1, -2, \ldots$

Thus we have IR convergence when either the subgraph γ_l' (or just γ) has at least one non-zero mass or its completion γ_{l-1} (or $\overline{\gamma}$) does not have all the external vertices in the same connectivity component. Therefore it is sufficient to check the IR convergence for the other IR-irreducible subgraphs. The domain of the regularization parameters λ_l and ε where these sector integrals are convergent is determined by the inequalities

$$\mathrm{Re}\,\lambda(\gamma) + h(\gamma)\,\mathrm{Re}\,\varepsilon > [\omega(\gamma)/2] \; , \tag{B.22a}$$

$$\operatorname{Re}\lambda(\gamma) - h(\Gamma/\overline{\gamma})\operatorname{Re}\varepsilon < [(\omega(\Gamma) - \omega(\gamma) + 1)/2] \,, \tag{B.22b}$$

which correspond to UV-irreducible subgraphs and IR-irreducible subgraphs with essential completions $\overline{\gamma}$, respectively.

It turns out that this domain is non-empty for any graph without massless detachable subgraphs. This statement can be proven [222] by observing that the parameters

$$\lambda_l^{(0)} = (2-\varepsilon)\left(1 + \delta - \frac{|T_l^1|}{|T^1|}\right) - 1 \,, \tag{B.23}$$

where T_l^1 is the set of trees containing the line l, satisfy (B.22a) and (B.22b) for sufficiently small $\delta > 0$. (As before, $|\ldots|$ is the number of elements in the corresponding finite set.) Here again the scalar case is assumed. The generalization to a general diagram is straightforward: one adds $a_l/2$ to the right-hand side of (B.23), where a_l is the degree of the polynomial in the numerator of the lth propagator.

It follows from the factorizations (B.19), when they are written for all the sectors, that the Feynman integral can be continued from the above domain of mutual convergence to the whole hypercomplex plane of the variables $(\underline{\lambda}, \varepsilon)$ as a meromorphic function, with series of UV and IR poles. In this step one uses the well-known property of the integrals $\int_0^\infty \mathrm{d}x\, x^\lambda \phi(x)$ that they are analytic functions of the parameter λ with simple poles at $\lambda = -1, -2, \ldots$ (In distributional language, this is the analytic property of the distribution x_+^λ – see [124].)

It is also clear that, in the case where there is no non-empty mutual-convergence domain, the contribution from any sector can be made convergent by choosing the absolute values of the real parts of the UV/IR analytic-regularization parameters to be sufficiently large (positive and negative for $l \le \bar{l}$ and $l > \bar{l}$, respectively). The analytic regularization can then be switched off, by analytic continuation, and one obtains [61] a dimensionally regularized Feynman integral as the sum of its sector contributions, which were defined in their own initial analyticity domains using the auxiliary analytic regularization.

B.2 Proof of the Asymptotic Estimate

Let us first formulate a statement[1] about the asymptotic behaviour of the remainder (4.28) of the off-shell large-momentum expansion (4.33). To get rid of inessential details, we consider a general scalar, convergent Feynman integral. We comment on various (mostly straightforward) generalizations at the end of this appendix.

[1] This formulation differs slightly from that in [207, 208], where the asymptotic behaviour at large momenta ($\equiv$ short distances) was understood in the distributional sense. The main points of the proof are almost the same in both formulations.

Theorem. Let $F_\Gamma(q,\underline{p},\underline{m})$ be a UV- and IR-convergent Feynman integral corresponding to a 1PI graph Γ and let $\mathcal{R}^{\bar{a}}F_\Gamma$ be the remainder in the large-momentum limit, where q is large with respect to the masses $\underline{m}$ and the rest of the external momenta $\underline{p}$, given by the forest formula (4.28), with subtraction degrees $a(\gamma)=\omega(\gamma)+\bar{a}$ defined for all AI subgraphs. Let q and $\underline{p}$ be Euclidean, i.e. $q^2<0$ and $\left(\sum_i p_i\right)^2<0$ for any subset of the small external momenta p_i. Then

$$\left|\mathcal{R}^{\bar{a}}F_\Gamma(q/\varrho,\underline{p},\underline{m})\right| < C\varrho^{\bar{a}+1+\delta} \tag{B.24}$$

for any $\delta>0$ and a constant C that depends on $(q,\underline{p},\underline{m})$.

Before proving this theorem, let us first introduce an explicit representation for the subtraction operators acting on the integrand in the alpha representation. The operator $\mathcal{M}_\Gamma$ corresponding to the whole graph Γ expands the integrand in the masses $\underline{m}$ and small external momenta $\underline{p}$. Using the homogeneity of the functions A and D, we can write

$$\begin{aligned}
\mathcal{M}^a_\Gamma F_\Gamma(q,\underline{p},\underline{m};d) &= \int_0^\infty \mathrm{d}\underline{\alpha}\, D(\underline{\alpha})^{-d/2}\, \mathcal{T}^a_{\underline{p},\underline{m}} \exp\left[\mathrm{i}W(q,\underline{p},\underline{m},\underline{\alpha})\right] \\
&= \int_0^\infty \mathrm{d}\underline{\alpha}\, D(\underline{\alpha})^{-d/2}\, \mathcal{T}^a_\kappa \exp\left[\mathrm{i}W(q/\kappa,\underline{p},\underline{m},\kappa^2\underline{\alpha})\right]\Big|_{\kappa=1} \\
&= \int_0^\infty \mathrm{d}\underline{\alpha}\, \mathcal{T}^a_\kappa \kappa^{hd} D(\kappa^2\underline{\alpha})^{-d/2} \exp\left[\mathrm{i}W(q/\kappa,\underline{p},\underline{m},\kappa^2\underline{\alpha})\right]\Big|_{\kappa=1} ,
\end{aligned} \tag{B.25}$$

where

$$W(q,\underline{p},\underline{m},\underline{\alpha}) = A(q,\underline{p},\underline{\alpha})/D(\underline{\alpha}) - \sum m_l^2\alpha_l\ . \tag{B.26}$$

Therefore the action of the subtraction operator for an AI subgraph γ on the whole Feynman integral is given by

$$\begin{aligned}
&\mathcal{M}^a(\gamma)F_\Gamma(q,\underline{p},\underline{m};d) \\
&= \int_0^\infty \mathrm{d}\underline{\alpha}\, \mathcal{T}^a_\kappa \kappa^{h(\gamma)d}\, D(\underline{\alpha}(\kappa))^{-d/2} \exp\left\{\mathrm{i}W(q/\kappa,\underline{p},\underline{m},\underline{\alpha}(\kappa))\right\}\Big|_{\kappa=1} ,
\end{aligned} \tag{B.27}$$

where

$$\alpha_l(\kappa) = \begin{cases} \kappa^2\alpha_l & \text{if } l\in\gamma \\ \alpha_l & \text{if } l\notin\gamma \end{cases} , \tag{B.28}$$

with $\alpha_l(1)=\alpha_l$.

Thus the remainder takes the form

$$\begin{aligned}
\mathcal{R}^{\bar{a}}F_\Gamma(q,\underline{p},\underline{m};d) &= \int_0^\infty \mathrm{d}\underline{\alpha} \sum_N \prod_{\gamma\in N} \left(-\mathcal{O}_{\kappa_\gamma}\mathcal{T}^{a(\gamma)}_{\kappa_\gamma}\right) \kappa_\gamma^{h(\gamma)d} \\
&\quad \times D(\underline{\alpha}(\underline{\kappa}))^{-d/2} \exp\left[\mathrm{i}W\left(\prod_{\gamma\in N}\kappa_\gamma^{-1}q,\underline{p},\underline{m},\underline{\alpha}(\underline{\kappa})\right)\right] ,
\end{aligned} \tag{B.29}$$

where

$$\alpha_l(\underline{\kappa}) = \left(\prod_{\gamma \ni l} \kappa_\gamma^2 \right) \alpha_l \,, \tag{B.30}$$

and the operator $\mathcal{O}$ acts such that $\mathcal{O}_\kappa f(\kappa) = f(1)$. As was pointed out in Chap. 4, the sum runs over nests of AI subgraphs, rather than over arbitrary forests.

Although the given Feynman integral is UV and IR finite we introduce dimensional regularization in order to make it possible to generalize the arguments presented here to the case with divergences.

We shall prove our theorem in three steps:

1. The decomposition of the integration domain of the alpha parameters into N-sectors and rearrangement of terms contributing to the operation $\mathcal{R}$ according to an equivalence relation.
2. The factorization of the integrand in the sector variables, which follows from the factorization in the auxiliary sector variables associated with the given equivalence class.
3. Power counting in the sector variables.

1. *Rearrangement in the forest formula.*

We need the following additional notation. Let us call the two external vertices with the large external momentum q q-vertices. A line of a subgraph is called a *cut line* if the number of its connectivity components increases when this line is deleted. Any one-particle-reducible subgraph can be decomposed into cut lines and 1PI parts, which we call 1PI components. Similarly, any subgraph with a path between q-vertices can be naturally decomposed into its AI components, which consist of its maximal AI subgraph and other pieces that are either cut lines or 1PI subgraphs.

Using (B.1) with analytic regularization switched off ($\lambda_l = 0$), we decompose the alpha integral into N-sector integrals. Without loss of generality, we consider only the contribution $\mathcal{R}F_\Gamma^{\bar{l}}$ of the sector (B.11), with a fixed $\bar{l}$. To rearrange the contributions of forests to the operation $\mathcal{R}$, let us introduce an appropriate equivalence relation. We call two forests f_1 and f_2 equivalent if $\overline{f_1} = \overline{f_2}$ where the operation $f \to \overline{f}$ puts a maximal UV forest $\mathcal{F} \equiv \overline{f}$ in correspondence with a given forest f according to the following construction procedure.

Let $\gamma \in f$. First, let us include in f the element Γ. Consider then the nest $\{\gamma \cap \gamma_l \cup \gamma_- \,, \; l = 1, \ldots, L\}$, where γ_- denotes the union of the elements of the forest f which are inside γ. Let us enumerate the distinct elements of this nest in the natural order: $\gamma^1 \subset \gamma^2 \subset \ldots$. The difference between the number of lines in γ^i and γ^{i-1} is therefore equal to one. If the q-vertices are not connected in γ^i, we include in f a subgraph which is either a cut line or the 1PI component of γ^i containing the line $\mathcal{L}(\gamma^i \backslash \gamma^{i-1})$. If the q-vertices are connected by a path in γ^i, we include in f the AI component of γ^i containing the line $\mathcal{L}(\gamma^i \backslash \gamma^{i-1})$.

As a result, the set of all the UV forests f of Γ is decomposed into classes with respect to the above equivalence relation. Each equivalence class is labelled by a maximal UV forest. Therefore the pre-subtraction operation that determines the remainder is represented as

$$\mathcal{R} = \sum_{\mathcal{F}} \mathcal{R}_{\mathcal{F}} \,, \tag{B.31}$$

$$\mathcal{R}_{\mathcal{F}} = \sum_{f:\, \overline{f}=\mathcal{F}} \prod_{\gamma \in f} \left(-\mathcal{M}_\gamma^{a(\gamma)} \right) \,, \tag{B.32}$$

where the first sum runs over all maximal UV forests of Γ (for which $\overline{\mathcal{F}} = \mathcal{F}$).

Now $\mathcal{R}F_\Gamma^{\bar{l}}$, given by the integral over the sector (B.11), is represented as a sum over maximal UV forests, according to (B.31), and we consider the contribution $\mathcal{R}_{\mathcal{F}}F_\Gamma^{\bar{l}}$ of a given $\mathcal{F}$. Let us introduce sector variables by means of (B.7) and (B.10). To derive a factorization of the integrand in these variables, we shall use factorization in the auxiliary F-sector variables t'_γ given by (B.14) that correspond to the given maximal UV forest $\mathcal{F}$; these variables are indicated by a prime. Note that the AI elements of $\mathcal{F}$ can be one-particle-reducible and that the maximal UV forests contributing to (B.31) are in one-to-one correspondence with the maximal UV forests with 1PI non-trivial elements for the graph $\hat{\Gamma}$ obtained from Γ by identifying the two q-vertices. To illustrate the difference between the maximal UV forests of Γ and $\hat{\Gamma}$, consider the same example of Fig. 2.5. Now $\gamma^3 = \{3\}$ cannot be present in $\mathcal{F}$ together with $\gamma^4 = \{1,2,5\}$, because their union is an AI subgraph since it includes both q-vertices. The following is an example of a maximal UV forest of the graph $\hat{\Gamma}$: $\mathcal{F} = \gamma^1 = \{1\}$, $\gamma^2 = \{2\}$, $\gamma^4 = \{1,2,5\}$, $\gamma^3 = \{1,2,3,5\}$, $\gamma^5 = \Gamma$.

Using (B.14), (B.6) and (B.9), we obtain the following expression for the auxiliary F-sector variables in terms of the initial N-sector variables:

$$t'_\gamma = \begin{cases} t_{\sigma(\gamma)} \cdots t_{\sigma(\gamma_+)-1} & \text{if } \sigma(\gamma) < \sigma(\gamma_+) \le \bar{l} \\ 1/\left(t_{\sigma(\gamma_+)} \cdots t_{\sigma(\gamma)-1}\right) & \text{if } \sigma(\gamma_+) < \sigma(\gamma) \le \bar{l} \\ t_{\sigma(\gamma)} \cdots t_{\bar{l}} \tau_{\bar{l}+1} \cdots \tau_{\sigma(\gamma_+)} & \text{if } \sigma(\gamma) \le \bar{l} < \sigma(\gamma_+) \\ 1/\left(t_{\sigma(\gamma_+)} \cdots t_{\bar{l}} \tau_{\bar{l}+1} \cdots \tau_{\sigma(\gamma)}\right) & \text{if } \sigma(\gamma_+) \le \bar{l} < \sigma(\gamma) \\ \tau_{\sigma(\gamma)+1} \cdots \tau_{\sigma(\gamma_+)} & \text{if } \bar{l} < \sigma(\gamma) < \sigma(\gamma_+) \\ 1/\left(\tau_{\sigma(\gamma_+)+1} \cdots \tau_{\sigma(\gamma)}\right) & \text{if } \bar{l} < \sigma(\gamma_+) < \sigma(\gamma) \end{cases} \tag{B.33}$$

if γ is not a maximal element of a given maximal UV forest $\mathcal{F}$, and

$$t'_\gamma = \begin{cases} t_{\sigma(\gamma)} \cdots t_{\bar{l}} & \text{if } \sigma(\gamma) \le \bar{l} \\ 1/\left(\tau_{\bar{l}+1} \cdots \tau_{\sigma(\gamma)}\right) & \text{if } \sigma(\gamma) > \bar{l} \end{cases} \tag{B.34}$$

for a maximal γ. Since the given graph is supposed to be 1PI, there is only one maximal element in every maximal UV forest, namely Γ.

Any maximal UV forest $\mathcal{F}$ can be decomposed as $\mathcal{F}^+ \cup \mathcal{F}^-$, where $\gamma \in \mathcal{F}^+$, if either $\sigma(\gamma) < \sigma(\gamma_+)$ or $\gamma = \Gamma$. We observe that, if $\mathcal{F} = \overline{f}$, the maximal UV forest $\mathcal{F}$ can be constructed by adding elements of $\mathcal{F}^+$ to f. Therefore an element of an equivalence class of UV forests is characterized by the set

$\mathcal{F}^-$ and a subset of the set $\mathcal{F}^+$, so that the contribution of the equivalence class is represented as

$$\mathcal{R}_{\mathcal{F}} = \prod_{\gamma\in\mathcal{F}^+} \left(1 - \mathcal{M}_\gamma^{a(\gamma)}\right) \prod_{\gamma\in\mathcal{F}^-} \left(-\mathcal{M}_\gamma^{a(\gamma)}\right) . \tag{B.35}$$

Note, however, that the operators $\mathcal{M}_\gamma$ are non-zero only for the AI elements which form a nest $\mathcal{N} \subseteq \mathcal{F}$. For all other γ we formally put $\mathcal{M}_\gamma = 0$. Observe also that $\mathcal{F}^-$ is composed of AI elements only.

2. *Factorization of the integrand.*

Using the notation defined above, we have

$$D(\underline{t},\underline{\tau}) \equiv D(\underline{\alpha}(\underline{t}),\beta(\underline{\tau})) = \prod_{l=\bar{l}+1}^{L} \tau_l^{L-l+1} D(\underline{t}') , \tag{B.36}$$

where we can use the factorization (B.16) for $D(\underline{t}')$.

We need a generalization of the factorization (B.17) to the case where we have large and small momenta and where the function A is decomposed as $A_q + A_p$, with A_q given by a sum over 2-trees such that the q-vertices are in different components of each 2-tree. In fact A_q equals the function D for the graph $\hat{\Gamma}$ obtained from Γ by identifying the q-vertices, so that we can use the factorization (B.36) and take into account the connection between the numbers of independent loops $h(\gamma)$ in subgraphs of Γ and $\hat{\Gamma}$. We obtain

$$A_q/D = \left(\prod_{\gamma\in\mathcal{F}:\gamma^1\subseteq\gamma} t'_\gamma\right) \left[q^2 + P_1(q,\underline{p},\underline{t}')\right] \left[1 + P_{\mathrm{F}}(\underline{t}')\right]^{-1} , \tag{B.37}$$

where γ^1 is the minimal AI subgraph in $\mathcal{F}$, and P_1 is a polynomial in q and $\underline{p}$ with coefficients which are non-negative polynomials in $\underline{t}'$. In the Euclidean domain, $P_1 \leq 0$.

The function A_p is given by the sum over the rest of the 2-trees, i.e. those which do not have all the external vertices in the same connectivity component. Its factorization is obtained by minimizing the powers of the variables t'_γ in the same 2-tree T_A as in the case of the factorization (B.17) in the N-sector variables:

$$A_p/D = \left(\prod_{\gamma\in\mathcal{N}:\gamma^q\subseteq\gamma} t'_\gamma\right) \left[\left(q^{T_A}\right)^2 + P_2(\underline{p},\underline{t}')\right] \left[1 + P_{\mathrm{F}}(\underline{t}')\right]^{-1} , \tag{B.38}$$

where γ^q is the minimal subgraph in $\mathcal{F}$ that includes all the external vertices in the same connectivity component, and P_2 has the same properties as P_1. Note that q^{T_A} is independent of the large momentum q. We have $\gamma^1 \subset \gamma^q$ and, generally, these elements are distinct.

According to (B.33) and (B.34), the auxiliary variables $\{t'_\gamma,\ \gamma \in \mathcal{F}^-\}$ are proportional to negative powers of the initial N-sector variables t_l and/or τ_l, and $\{t'_\gamma,\ \gamma \in \mathcal{F}^+\}$ are proportional to positive powers of t_l and τ_l, the only exception being $\gamma = \Gamma$ when $\sigma(\Gamma) > \bar{l}$.

The factorization in the auxiliary variables enables us to take into account the action of the subtraction operators corresponding to the elements of the given maximal forest $\mathcal{F}$. (Only operators corresponding to the nest $\mathcal{N} \subset \mathcal{F}$ are present.) It turns out that the integrand depends in a rather specific way on the auxiliary parameters κ_γ. Since the functions involved in the integrand depend on sums over trees and 2-trees of products

$$\Pi_T(\underline{\alpha}', \underline{\kappa}) = \prod_{l \notin T} \alpha_l \prod_{\gamma \in \mathcal{F}} \kappa_\gamma^{2L(\gamma \setminus T)} \equiv \prod_{\gamma \in \mathcal{F}} \left(\kappa_\gamma^2 t'_\gamma\right)^{L(\gamma \setminus T)} , \tag{B.39}$$

the parameter κ_γ can enter (up to overall factors) only in the combination $\kappa_\gamma^2 t'_\gamma$.

Observe now that, after the action of an operator $\mathcal{M}_\gamma$ for $\gamma \in \mathcal{F}^-$, the corresponding parameter κ_γ is set to zero. Although κ_Γ is not set to zero even in the case where t'_Γ involves negative powers of the IR variables τ_l, the factorizations (B.36), (B.37) and (B.38) with respect to the auxiliary variables provide a factorization in the initial N-sector variables because the variable t'_Γ enters the factorizations (B.36), (B.37) and (B.38) trivially, with the polynomials P_F and $P_{1,2}$ being independent of t'_Γ.

We use the following formula for the remainder of a one-dimensional Taylor expansion:

$$\left(1 - \mathcal{T}_\kappa^a\right) g(\kappa)\big|_{\kappa=1} = \frac{1}{a!} \int_0^1 \mathrm{d}\kappa\, (1-\kappa)^a g^{(a+1)}(\kappa) . \tag{B.40}$$

Let us now consider the contribution of a maximal UV forest $\mathcal{F}$ to the integral over the sector (B.11) with the identical permutation π and a given number $\bar{l}$. Using the arguments presented above, we obtain the following representation for this contribution:

$$\begin{aligned}
\mathcal{R}_\mathcal{F} F_{\bar{l}}^{\bar{a}}(q, \underline{p}, \underline{m}; \varepsilon) = & \int_0^1 \left(\prod_{\gamma \in \mathcal{N}+} \mathrm{d}\kappa_\gamma\, (1-\kappa_\gamma)^{a(\gamma)} \right) \\
& \times \int_0^1 \mathrm{d}\underline{t}\, \mathrm{d}\underline{\tau} \left(\prod_{l=1}^{\bar{l}} t_l^{l-1} \right) \left(\prod_{\bar{l}+1}^{L} \tau_l^{-L+l-2} \right) \left(\prod_{\gamma \in \mathcal{F}} (t'_\gamma)^{-dh(\gamma)/2} \right) \\
& \times \left(\prod_{\gamma \in \mathcal{N}+} \left(\frac{\partial}{\partial \kappa_\gamma} \right)^{a(\gamma)+1} \right) \prod_{\gamma \in \mathcal{N}-} \left(-\mathcal{O}_{\kappa_\gamma} \mathcal{T}_{\kappa_\gamma}^{a(\gamma)} \right) \\
& \times \left[1 + P_F(\underline{\kappa}, \underline{t}')\right]^{-d/2} \exp\left[\mathrm{i}W \left(q \prod_{\gamma \in \mathcal{N}} \kappa_\gamma^{-1}, \underline{p}, \underline{m}, \underline{\kappa}, \underline{t}' \right) \right] ,
\end{aligned} \tag{B.41}$$

where the last two factors depend on $\kappa_\gamma,\ \gamma \in \mathcal{N}$ through the combinations $\kappa_\gamma^2 t'_\gamma$.

The products of the auxiliary sector variables that are factored out in the functions A_q and A_p can be rewritten in terms of the initial variables as

$$\prod_{\gamma\in\mathcal{F}:\gamma'\subseteq\gamma} t'_\gamma = \begin{cases} t_{\sigma(\gamma')}\dots t_{\bar{l}} & \text{if } \sigma(\gamma')\leq \bar{l} \\ \left(\tau_{\bar{l}+1}\dots\tau_{\sigma(\gamma')}\right)^{-1} & \text{if } \sigma(\gamma')>\bar{l} \end{cases} . \tag{B.42}$$

After differentiation with respect to the parameters κ_γ at zero values of κ_γ, $\gamma\in\mathcal{N}^-$, the function $\overline{W}$ that remains in the exponent takes a form similar to the previous function (B.26), with factorizations given by (B.37), (B.38):

$$\overline{W} = [1+P_{\mathrm{F}}(\underline{t}')]^{-1}\left\{\left(\tau_{\bar{l}+1}\dots\tau_{l_1}\right)^{-1}\left[q^2+\overline{P}_1(\underline{p},\underline{t},\underline{\tau})\right]\right.$$
$$\left.+\left(\tau_{\bar{l}+1}\dots\tau_{l_2}\right)^{-1}\left[\left(q^{T_A}\right)^2+\overline{P}_2(\underline{p},\underline{t},\underline{\tau})\right]\right\}-\sum m_l^2\alpha_l(\underline{t},\underline{\tau}) . \tag{B.43}$$

Modified polynomials $\overline{P}_{1,2}$, which also depend on κ_γ, $\gamma\in\mathcal{N}^+$, are obtained from $P_{1,2}$ and have similar properties. Here $l_1=\sigma(\gamma^1)$, $l_2=\sigma(\gamma^q)$. The element γ_1 is the first in the nest $\mathcal{N}$, and γ_q is the minimal element in $\mathcal{N}$ with all the external vertices in the same connectivity component. The mass term is factorized trivially, with the product $\left(\tau_{\bar{l}+1}\dots\tau_{l_m}\right)^{-1}$, where l_m is the maximal number of the massive line.

3. *Power counting.*

Thus we obtain the following expression for the sector integral:

$$\mathcal{R}_{\mathcal{F}}F_{\bar{l}}^{\bar{a}}(q,\underline{p},\underline{m};\varepsilon) = \int_0^1 \prod_{\gamma\in\mathcal{N}^+}\mathrm{d}\kappa_\gamma\,(1-\kappa_\gamma)^{a(\gamma)}$$
$$\times\int_0^1 \mathrm{d}\underline{t}\,\mathrm{d}\underline{\tau}\left(\prod_{l=1}^{\bar{l}} t_l^{N_l-1}\right)\left(\prod_{l=\bar{l}+1}^{L}\tau_l^{\tilde{N}_l-1}\right)G\mathrm{e}^{\mathrm{i}\overline{W}} , \tag{B.44}$$

where G is an infinitely differentiable function and $\overline{W}$ is as described above.

The exponent of the parametric integral is not relevant to the UV analysis of convergence. Thus it is necessary to count powers of all the integration variables t_l, $l=1,\dots,\bar{l}$. We take into account the fact that differentiation with respect to κ_γ for $\gamma\in\mathcal{N}^\pm\equiv\mathcal{F}^\pm\cap\mathcal{N}$ increases (decreases) powers of t_l (and of τ_l). If we ignore contributions proportional to ε, we obtain the following estimate:

$$N_l \geq l + \sum_{\gamma\in\mathcal{N}:\,\sigma(\gamma)\leq l<\sigma(\gamma_+)}\{-2h(\gamma)+[a(\gamma)/2]+1\}$$
$$-\sum_{\gamma\in\mathcal{N}:\,\sigma(\gamma_+)\leq l<\sigma(\gamma)}\{-2h(\gamma)+[a(\gamma)/2]\}$$
$$+\sum_{\gamma\notin\mathcal{N}:\,\sigma(\gamma)\leq l<\sigma(\gamma_+)}\{-2h(\gamma)\} , \tag{B.45}$$

where summation over $\gamma\in\mathcal{F}$ is understood.

Now we use the following (cumbersome but simple) summation formulae:

$$\sum_{\gamma:\sigma(\gamma)\leq l<\sigma(\gamma_+)} N_\gamma = \sum_{\gamma:\sigma(\gamma)\leq l} N_\gamma - \sum_{\gamma:\sigma(\gamma),\sigma(\gamma_+)\leq l} N_\gamma \,,$$

$$\sum_{\gamma:\sigma(\gamma_+)\leq l<\sigma(\gamma)} N_\gamma = \sum_{\gamma:\sigma(\gamma_+)\leq l} N_\gamma - \sum_{\gamma:\sigma(\gamma),\sigma(\gamma_+)\leq l} N_\gamma \,,$$

$$\sum_{\gamma:\sigma(\gamma_+)\leq l} N_\gamma = \sum_{\gamma:\sigma(\gamma)\leq l} \sum_{\gamma':\gamma'_+=\gamma} N_{\gamma'} \,. \tag{B.46}$$

These formulae give

$$\sum_{\gamma:\sigma(\gamma)\leq l<\sigma(\gamma_+)} N_\gamma - \sum_{\gamma:\sigma(\gamma_+)\leq l<\sigma(\gamma)} N_\gamma = \sum_{\gamma:\sigma(\gamma)\leq l} \left(N_\gamma - \sum_{\gamma':\gamma'_+=\gamma} N_{\gamma'} \right) \,. \tag{B.47}$$

To count IR powers, we shall shortly need the analogues of (B.46), which give

$$\sum_{\gamma:\sigma(\gamma)<l\leq\sigma(\gamma_+)} N_\gamma - \sum_{\gamma:\sigma(\gamma_+)<l\leq\sigma(\gamma)} N_\gamma = - \sum_{\gamma:\sigma(\gamma)\geq l} \left(N_\gamma - \sum_{\gamma':\gamma'_+=\gamma} N_{\gamma'} \right) \,. \tag{B.48}$$

In particular, we have

$$\sum_{\gamma:\sigma(\gamma)\leq l<\sigma(\gamma_+)} L(\gamma) - \sum_{\gamma:\sigma(\gamma_+)\leq l<\sigma(\gamma)} L(\gamma) = l \,, \tag{B.49a}$$

$$\sum_{\gamma:\sigma(\gamma)<l\leq\sigma(\gamma_+)} L(\gamma) - \sum_{\gamma:\sigma(\gamma_+)<l\leq\sigma(\gamma)} L(\gamma) = -(L-l+1) \,. \tag{B.49b}$$

We represent l in (B.45) using (B.49a), substitute $a(\gamma)$ by $\omega(\gamma)+\bar{a}$ with the same $\bar{a}$ for all $\gamma \in \mathcal{N}$, use the UV finiteness of Γ, i.e. the conditions $\omega(\gamma) < 0$ for any 1PI subgraph, and arrive at

$$N_l \geq \sum_{\gamma:\sigma(\gamma)\leq l<\sigma(\gamma_+)} 1 \,. \tag{B.50}$$

The last sum is always positive. (Suppose that this is not true. Then $\sigma(\Gamma) > l$. Let γ_i be the maximal elements of $\mathcal{F}\backslash\Gamma$. According to the assumption, $\sigma(\gamma^i) > l$. Repeating the above arguments, we conclude that $\sigma(\gamma) > l$ for all elements of $\mathcal{F}$, but any maximal UV forest consists of exactly L elements.) Therefore $N_l > 0$ for any l and this means a UV convergence.

Note that, using the same summation formulae, the part of the power of t_l proportional to ε can be written as

$$\varepsilon \sum_{\gamma:\,\sigma(\gamma)\le l} \left(h(\gamma) - \sum_{\gamma':\,\gamma'_+=\gamma} h(\gamma') \right) .$$

To analyse the IR convergence, let us remember the exponent (B.43), which involves τ_l^{-1} for $l \le l_1 \equiv \sigma(\gamma_1)$ multiplied by the quantity $q^2 + \ldots$, which does not tend to zero when $\underline{p}$ and $\underline{m}$ vanish. These powers provide (IR) convergence in the corresponding variable τ_l – see the arguments in the previous section. Convergence in the variables τ_l for $l > \max\{l^q, l^m\}$ is analysed as in the case where the operation $\mathcal{R}$ is absent. This convergence is guaranteed by the IR convergence conditions $\omega(\Gamma) - \omega(\gamma) > 0$ applied to the subgraphs γ involving all the external momenta in the same connectivity component.

In the case $l = \bar{l}+1, \ldots, l_1$, the presence of the second term in the exponent (B.43) provides the IR convergence. We need, however, the necessary asymptotic estimate in the limit under consideration.

To do this, let us treat the last subtraction operator, $\mathcal{M}_\Gamma$, as a Taylor expansion in $\underline{p}$ and $\underline{m}$, and let us pull out the factor

$$1 - \mathcal{M}_\Gamma = 1 - \mathcal{T}^{\omega(\Gamma)+\bar{a}}_{\underline{p},\underline{m}} .$$

Using the homogeneity of the Feynman integral with respect to all momenta and masses,

$$F_\Gamma(q/\varrho, \underline{p}, \underline{m}; \varepsilon) = \rho^{-\omega(\Gamma)+2h(\Gamma)\varepsilon} F_\Gamma(q, \varrho\underline{p}, \varrho\underline{m}) \tag{B.51}$$

we see that a sufficiently fast decrease as $\varrho \to 0$ can be achieved when the behaviour of $\mathcal{R}'_{\mathcal{F}} F^{\bar{a}}_{\bar{l}}(q, \varrho\underline{p}, \varrho\underline{m}; \varepsilon)$ with respect to ϱ is smooth enough. (Here $\mathcal{R}'$ is the operation $\mathcal{R}$ without a subtraction operator for the whole graph.) According to a proposition that we shall present shortly, this means that we need sufficiently large values for the powers $\tilde{N}_l$ for $l = l_1+1, \ldots, \max\{l_q, l_m\}$.

Using the factorization presented above, we have

$$\begin{aligned} \tilde{N}_l \ge & -(L-l+1) + [a(\Gamma)/2] \\ & + \sum_{\gamma\in\mathcal{N}:\,\sigma(\gamma)<l\le\sigma(\gamma_+)} \{-2h(\gamma) + [a(\gamma)/2] + 1\} \\ & - \sum_{\gamma\in\mathcal{N}:\,\sigma(\gamma_+)<l\le\sigma(\gamma)} \{-2h(\gamma) + [a(\gamma)/2]\} \\ & + \sum_{\gamma\notin\mathcal{N}:\,\sigma(\gamma)<l\le\sigma(\gamma_+)} \{-2h(\gamma)\} , \end{aligned} \tag{B.52}$$

where the part proportional to ε is omitted. The term $[a(\Gamma)/2]$ has appeared because of the absence of the contribution from $\mathcal{M}_\Gamma$ in $\mathcal{R}'$.

We represent $-(L-l+1)$ in (B.52) using (B.49b), apply the conditions $\omega(\gamma) < 0$ for any 1PI subgraph, and arrive at the following estimate:

$$\tilde{N}_l \geq [a(\Gamma)/2] + \left(\sum_{\gamma:\, \sigma(\gamma) < l \leq \sigma(\gamma_+)} 1 - 1 \right)$$
$$-[\bar{a}/2] \sum_{\gamma:\, \sigma(\gamma) \geq l} \left(1 - \sum_{\gamma':\, \gamma'_+ = \gamma} 1 \right) . \tag{B.53}$$

In the last sum over the elements γ of the nest $\mathcal{N}$, the term $\sum_{\gamma':\, \gamma'_+ = \gamma} 1$ equals zero only if the subgraph γ is the first element of the nest $\mathcal{N}$. Thus the contribution of the last line can be negative only if γ is such an element. But this cannot happen for the numbers l under consideration. The first parentheses in (B.53) give only a non-negative contribution owing to arguments similar to those used in the UV case. We thus obtain the estimate $\tilde{N}_l \geq a(\Gamma)$. Now, by a smooth change of the variables τ_l, we reduce the analysis to the following auxiliary statement [207, 208], which is a generalization of a similar statement in [246].

Proposition. The integral

$$g(\varrho) = \int_0^1 \prod_l \mathrm{d}\tau_l \; \tau_l^{\tilde{N}_l - r_l \varepsilon} \exp\left\{ -\mathrm{i}\varrho / \left(\tau_1 \ldots \tau_L \right) \right\} \phi(\underline{\tau}) , \tag{B.54}$$

where $\phi \in C^\infty$ and $r_1 > r_2 > \ldots$, has the following asymptotic expansion when $\varrho \to 0$:

$$g(\varrho) \sim \sum_{n=\tilde{N}+1}^{\infty} \sum_{j=0}^{r_1} g_{n,j} \varrho^{n - j\varepsilon} + \sum_{n=0}^{\infty} g_n^{(0)} \varrho^n , \tag{B.55}$$

where $\tilde{N} = \min_l \{\tilde{N}_l\}$.

Now, the ε term in the exponent of τ_l is

$$-\varepsilon \sum_{\gamma:\, \sigma(\gamma) \geq l} \left(h(\gamma) - \sum_{\gamma':\, \gamma'_+ = \gamma} h(\gamma') \right) .$$

Observe that in the case $l_1 > \max\{l_q, l_m\}$, the sector contribution considered here turns out to be an analytic function of $\underline{p}, \underline{m}$ so that the conclusion about the desired asymptotic behaviour becomes trivial. In the considered case, where $l_1 \leq \max\{l_q, l_m\}$, we take into account the factor $\rho^{-\omega(\Gamma) + 2h(\Gamma)\varepsilon}$ in (B.51), use the above estimate for the powers $\tilde{N}_l$ for $l = l_1 + 1, \ldots, \max\{l_q, l_m\}$, with the help of the above proposition, and finally arrive at the desired asymptotic estimate (B.24).

Comments.

1. The generalization to the non-scalar case is straightforward.

2. The generalization to Feynman integrals with UV and/or IR divergences is almost straightforward. The main difference is that the integrals over some of the sector variables t_l and τ_l become divergent because of negative powers. The crucial point for the asymptotic estimate is, however, the

analysis of the powers τ_l for $l = l_1 + 1, \ldots, \max\{l_q, l_m\}$, which proves the existence of a sufficiently large number of derivatives with respect to ϱ of $\mathcal{R}' F_\Gamma(q, \varrho \underline{p}, \varrho \underline{m})$ at $\varrho = 0$. The desired estimate for dimensionally regularized Feynman integrals can be obtained in a similar way.

3. A further generalization to renormalized Feynman integrals reduces to obtaining asymptotic estimates for regularized quantities.

4. The generalization to the large-mass limit and the general off-shell limit of momenta and masses is straightforward. The decisive point is the use of another definition of the asymptotically irreducible subgraph.

5. From the above proof of the validity of the asymptotic estimate, one can obtain, as a by-product, a simple proof of the Bogoliubov–Parasiuk theorem [26, 137, 247]. In its generalized version with oversubtractions [246], it states that the BPHZ-renormalized Feynman integral RF_Γ corresponding to a massive graph Γ is finite provided the degrees of oversubtraction $\bar{a}(\gamma) = a(\gamma) - \omega(\gamma)$ satisfy the relations $\sum_i \bar{a}(\gamma_i) \leq \bar{a}(\gamma)$ for any family $\{\gamma_i\}$ of disjoint subgraphs $\gamma_i \subset \gamma$. The BPHZ R-operation is given by the forest formula (2.87) with subtractions in all 1PI subgraphs, subtraction degrees $a(\gamma)$ and a subtraction operator M_γ that expands diagrams into Taylor series of order $a(\gamma)$. In particular, one can chose minimal BPHZ subtractions with $a(\gamma) = \omega(\gamma)$.

To prove this theorem one can use the same strategy as above (see also a similar strategy in [246]), consisting of the above three steps. In the parametric representation, the operator M_γ has a form similar to (B.25) but now it expands integrands in all the external momenta and does not expand in the masses. One does not need IR power counting because the presence of non-zero masses provides it. The UV power counting is almost the same as in the above proof.

References

A young man from the far north-east of the Soviet Union who wants to become a writer comes to Moscow. He tries to pass an exam in Russian literature and enter the Literary Institute. The professor at the exam suddenly realizes that the student has never read any poems by Pushkin or any novels by Tolstoy.
'Look, you are going to be a writer, and still you have not read Tolstoy!', exclaims the professor.
'But I am going to be a writer, not a reader!'

(A Soviet anecdote from the 1970s)

1. G. Amorós, M. Beneke and M. Neubert, Phys. Lett. B **401** (1997) 81.
2. C. Anastasiou, E.W.N. Glover and C. Oleari, Nucl. Phys. B **575** (2000) 416.
3. C. Anastasiou, T. Gehrmann, C. Oleari, E. Remiddi and J.B. Tausk, Nucl. Phys. B **580** (2000) 577.
4. S.A. Anikin and O.I. Zavialov, Teor. Mat. Fiz. **26** (1976) 162; S.A. Anikin, M.C. Polivanov and O.I. Zavialov, Fortschr. Phys. **27** (1977) 459; S.A. Anikin and O.I. Zavialov, Ann. Phys. **116** (1978) 135.
5. T. Appelquist and J. Carazzone, Phys. Rev. D **11** (1975) 2856.
6. L.V. Avdeev and M.Yu. Kalmykov, Nucl. Phys. B **502** (1997) 419.
7. P.A. Baikov, Phys. Lett. B **385** (1996) 404; Nucl. Instrum. Methods A **389** (1997) 347; P.A. Baikov and M. Steinhauser, Comput. Phys. Commun. **115** (1998) 161.
8. P.A. Baikov, Phys. Lett. B **474** (2000) 385.
9. P.A. Baikov and V.A. Smirnov, Phys. Lett. B **477** (2000) 367.
10. G.A. Baker and P. Graves-Morris, *Padé Approximants*, Encyclopedia of Mathematics and its Applications, Vols. 13 and 14 (Addison-Wesley, Reading, MA, 1981).
11. Ya.Ya. Balitsky and L.N. Lipatov, Sov. J. Nucl. Phys. **28** (1978) 822.
12. R. Barbieri, R. Gatto, R. Körgerler and Z. Kunszt, Phys. Lett. B **57** (1975) 455.
13. W.A. Bardeen, A.J. Buras, D.W. Duke and T. Muta, Phys. Rev. D **18** (1978) 3998.
14. W. Beenakker and A. Denner, Nucl. Phys. B **338** (1990) 349.
15. M. Beneke and A. Signer, Phys. Lett. B **471** (1999) 233.
16. M. Beneke, A. Signer and V.A. Smirnov, Phys. Rev. Lett. **80** (1998) 2535.
17. M. Beneke, A. Signer and V.A. Smirnov, Phys. Lett. B **454** (1999) 137.
18. M. Beneke and V.A. Smirnov, Nucl. Phys. B **522** (1998) 321.
19. F.A. Berends, A.I. Davydychev, V.A. Smirnov and J.B. Tausk, Nucl. Phys. B **439** (1995) 536; F.A. Berends, A.I. Davydychev and V.A. Smirnov, Nucl. Phys. B **478** (1996) 59.
20. W. Bernreuther and W. Wetzel, Nucl. Phys. B **197** (1982) 228; W. Bernreuther, Ann. Phys. **151** (1983) 127; Z. Phys. C **20** (1983) 331.
21. J.J. van der Bij and M. Veltman, Nucl. Phys. B **231** (1984) 205.
22. A. Billoire, Phys. Lett. B **92** (1980) 343.
23. T. Binoth and G. Heinrich, Nucl. Phys. B **585** (2000) 741.

24. K.S. Bjoerkevoll, P. Osland and G. Faeldt, Nucl. Phys. B **386** (1992) 303.
25. G.T. Bodwin, E. Braaten and G.P. Lepage, Phys. Rev. D **51** (1995) 1125; Phys. Rev. D **55** (1997) 5853.
26. N.N. Bogoliubov and O.S. Parasiuk, Acta Math. **97** (1957) 227.
27. N.N. Bogoliubov and D.V. Shirkov, *Introduction to Theory of Quantized Fields*, 3rd edition (Wiley, New York, 1983).
28. C.G. Bollini and J.J. Giambiagi, Nuovo Cim. B **12** (1972) 20.
29. E.E. Boos and A.I. Davydychev, Vestnik MGU **28** (1987) 8.
30. E.E. Boos and A.I. Davydychev, Teor. Mat. Fiz. **89** (1991) 56.
31. J. Botts and G. Sterman, Nucl. Phys. B **325** (1989) 62.
32. P. Boucaud et al., Phys. Rev. D **63** (2001) 114003.
33. E. Braaten and Y.Q. Chen, Phys. Rev. D **57** (1998) 4236.
34. E. Braaten, S. Narison and A. Pich, Nucl. Phys. B **373** (1992) 581.
35. P. Breitenlohner and D. Maison, Commun. Math. Phys. **52** (1977) 11, 39, 55.
36. D.J. Broadhurst, Z. Phys. C **47** (1990) 115.
37. D.J. Broadhurst, Z. Phys. C **54** (1992) 599.
38. D.J. Broadhurst, J. Fleischer and O.V. Tarasov, Z. Phys. C **60** (1993) 287.
39. D.J. Broadhurst and S.C. Generalis, Open University report OUT-4102-22 (1988) (unpublished).
40. D.J. Broadhurst, N. Gray and K. Schilcher, Z. Phys. C **52** (1991) 111.
41. D.J. Broadhurst and A.G. Grozin, Phys. Lett. B **267** (1991) 105.
42. W.E. Caswell and A.D. Kennedy, Phys. Rev. D **25** (1982) 392.
43. W.E. Caswell and G.P. Lepage, Phys. Lett. B **167** (1986) 437.
44. W. Celmaster and R.J. Gonsalves, Phys. Rev. D **20** (1979) 1420.
45. H. Cheng and T.T. Wu, *Expanding Protons: Scattering at High Energies* (MIT Press, Cambridge, MA, 1987).
46. K.G. Chetyrkin, Phys. Lett. B **126** (1983) 371.
47. K.G. Chetyrkin, Teor. Mat. Fiz. **75** (1988) 26; Teor. Mat. Fiz. **76** (1988) 207; Max-Planck-Institut preprint MPI-PAE/PTh 13/91 (Munich, 1991).
48. K.G. Chetyrkin, Phys. Lett. B **307** (1993) 169.
49. K.G. Chetyrkin, Phys. Lett. B **404** (1997) 161.
50. K.G. Chetyrkin, S.G. Gorishny and F.V. Tkachov, Phys. Lett. B **119** (1982) 407.
51. K.G. Chetyrkin, R. Harlander, J.H. Kühn and M. Steinhauser, Nucl. Phys. B **503** (1997) 339.
52. K.G. Chetyrkin, R. Harlander, T. Seidensticker and M. Steinhauser, Phys. Rev. D **60** (1999) 114015.
53. K.G. Chetyrkin, A.L. Kataev and F.V. Tkachov, Nucl. Phys. B **174** (1980) 345.
54. K.G. Chetyrkin, B.A. Kniehl and M. Steinhauser, Phys. Rev. Lett. **79** (1997) 2184.
55. K.G. Chetyrkin, B.A. Kniehl and M. Steinhauser, Nucl. Phys. B **510** (1998) 61.
56. K.G. Chetyrkin and J.H. Kühn, Nucl. Phys. B **432** (1994) 337.
57. K.G. Chetyrkin and J.H. Kühn, Phys. Lett. B **406** (1997) 102.
58. K.G. Chetyrkin, J.H. Kühn and A. Kwiatkowski, Phys. Rep. **277** (1997) 189.
59. K.G. Chetyrkin and A.A. Pivovarov, Nuovo Cim. A **100** (1988) 899.
60. K.G. Chetyrkin and T. Seidensticker, Phys. Lett. B **495** (2000) 74.
61. K.G. Chetyrkin and V.A. Smirnov, Teor. Mat. Fiz. **56** (1983) 206.

62. K.G. Chetyrkin and V.A. Smirnov, Phys. Lett. B **144** (1984) 419.
63. K.G. Chetyrkin and V.A. Smirnov, Teor. Mat. Fiz. **64** (1985) 370.
64. K.G. Chetyrkin and V.P. Spiridonov, Sov. J. Nucl. Phys. **47** (1988) 522.
65. K.G. Chetyrkin, V.P. Spiridonov and S.G. Gorishny, Phys. Lett. B **160** (1985) 149.
66. K.G. Chetyrkin and M. Steinhauser, Phys. Rev. Lett. **83** (1999) 4001.
67. K.G. Chetyrkin and M. Steinhauser, Nucl. Phys. B **573** (2000) 617.
68. K.G. Chetyrkin and F.V. Tkachov, Nucl. Phys. B **192** (1981) 159.
69. S. Ciulli and J. Fischer, Nucl. Phys. **24** (1961) 465.
70. P. Colangelo and A. Khodjamirian, hep-ph/0010175.
71. J.C. Collins, Nucl. Phys. B **80** (1974) 341.
72. J.C. Collins, Nucl. Phys. B **92** (1975) 477.
73. J.C. Collins, Phys. Rev. D **22** (1980) 1478.
74. J.C. Collins, *Renormalization* (Cambridge University Press, Cambridge, 1984).
75. J.C. Collins, in *Perturbative QCD*, ed. A.H. Mueller (World Scientific, Singapore, 1989) p. 573.
76. P.D.B. Collins, *An Introduction to Regge Theory and High Energy Physics* (Cambridge University Press, Cambridge, 1977).
77. J.M. Cornwall and G. Tiktopoulos, Phys. Rev. Lett. **35** (1975) 338; Phys. Rev. D **13** (1976) 3370.
78. A. Czarnecki, Phys. Rev. Lett. **76** (1996) 4124.
79. A. Czarnecki and A.G. Grozin, Phys. Lett. B **405** (1997) 142.
80. A. Czarnecki and J.H. Kühn, Phys. Rev. Lett. **77** (1996) 3955.
81. A. Czarnecki and W.J. Marciano, Phys. Rev. D **64** (2001) 013014.
82. A. Czarnecki and K. Melnikov, Phys. Rev. Lett. **78** (1997) 3630.
83. A. Czarnecki and K. Melnikov, Phys. Rev. D **56** (1997) 1638.
84. A. Czarnecki and K. Melnikov, Phys. Rev. Lett. **80** (1998) 2531.
85. A. Czarnecki and K. Melnikov, Phys. Rev. Lett. **87** (2001) 013001.
86. A. Czarnecki and M. Skrzypek, Phys. Lett. B **449** (1999) 354.
87. A. Czarnecki and V.A. Smirnov, Phys. Lett. B **394** (1997) 211.
88. A.I. Davydychev, Phys. Lett. B **263** (1991) 107.
89. A.I. Davydychev, J. Math. Phys. **33** (1992) 358.
90. A.I. Davydychev and A.G. Grozin, Phys. Rev. D **59** (1999) 054023.
91. A.I. Davydychev and P. Osland, Phys. Rev. D **59** (1998) 014006.
92. A.I. Davydychev, P. Osland and O.V. Tarasov, Phys. Rev. D **54** (1996) 4087.
93. A.I. Davydychev and V.A. Smirnov, Nucl. Phys. B **554** (1999) 391.
94. A.I. Davydychev, V.A. Smirnov and J.B. Tausk, Nucl. Phys. B **410** (1993) 325.
95. A.I. Davydychev and J.B. Tausk, Nucl. Phys. B **397** (1993) 123.
96. A. Devoto and D.W. Duke, Riv. Nuovo Cim. **7**, No. 6 (1984) 1.
97. A. Djouadi, Nuovo Cim. A **100** (1988) 357.
98. R.J. Eden, P.V. Landshoff, D.I. Olive and J.C. Polkinghorne, *The Analytic S-matrix* (Cambridge University Press, Cambridge, 1966).
99. E. Eichten and B. Hill, Phys. Lett. B **243** (1990) 427.
100. E. Eichten and R. Jackiw, Phys. Rev. D **4** (1971) 439.
101. S. Eidelman, F. Jegerlehner, A.L. Kataev and O.L. Veretin, Phys. Lett. B **454** (1999) 369.
102. H.H. Elend, Phys. Lett. **20** (1966) 682; (erratum) Phys. Lett. **21** (1966) 720.

103. A. Erdélyi (ed.), *Higher Transcendental Functions*, Vols. 1 and 2 (McGraw-Hill, New York, 1954).
104. V.S. Fadin and V.A. Khoze, Pis'ma Zh. Eksp. Teor. Fiz. **46** (1987) 417; Yad. Fiz. **48** (1988) 487.
105. V.S. Fadin, E.A. Kuraev and L.N. Lipatov, Phys. Lett. B **60** (1975) 50; E.A. Kuraev, L.N. Lipatov and V.S. Fadin, Sov. Phys. JETP **44** (1976) 45; **45** (1977) 199.
106. V.S. Fadin, L.N. Lipatov, A.D. Martin and M. Melles, Phys. Rev. D **61** (2000) 094002.
107. A.F. Falk, B. Grinstein, and M.E. Luke, Nucl. Phys. B **357** (1991) 185.
108. W. Fischler, Nucl. Phys. B **129** (1977) 157.
109. J. Fleischer et al., Eur. Phys. J. C **2** (1998) 747.
110. J. Fleischer, F. Jegerlehner, M. Tentyukov and O. Veretin, Phys. Lett. B **459** (1999) 625.
111. J. Fleischer and M.Yu. Kalmykov, Phys. Lett. B **470** (1999) 168.
112. J. Fleischer, M.Yu. Kalmykov and O.L. Veretin, Phys. Lett. B **427** (1998) 141.
113. J. Fleischer, V.A. Smirnov and O.V. Tarasov, Z. Phys. C **74** (1997) 379.
114. J. Fleischer and O.V. Tarasov, Comp. Phys. Commun. **71** (1992) 193.
115. J. Fleischer and O.V. Tarasov, Z. Phys. C **64** (1994) 413.
116. J. Fleischer and O.V. Tarasov, in *Computer Algebra in Science and Engineering*, eds. J. Fleischer, J. Grabmeier, F.W. Hehl and W. Küchlin (World Scientific, Singapore, 1995).
117. D.Z. Freedman, K. Johnson and J.I. Latorre, Nucl. Phys. B **371** (1992) 353.
118. J. Frenkel, Phys. Lett. B **65** (1976) 383.
119. J. Frenkel and R. Meuldermans, Phys. Lett. B **65** (1976) 64.
120. J. Frenkel and J.C. Taylor, Nucl. Phys. B **116** (1976) 185.
121. J. Gasser, V.E. Lyubovitskij, A. Rusetsky and A. Gall, Phys. Rev. D **64** (2001) 016008.
122. T. Gehrmann and E. Remiddi, Nucl. Phys. B **580** (2000) 485.
123. T. Gehrmann and E. Remiddi, Nucl. Phys. B **601** (2001) 248; Nucl. Phys. B **601** (2001) 287.
124. I.M. Gel'fand and G.E. Shilov, *Generalized Functions*, Vol. 1 (Academic Press, New York, London, 1964).
125. R.J. Gonsalves, Phys. Rev. D **28** (1983) 1542.
126. S.G. Gorishny, Nucl. Phys. B **319** (1989) 633.
127. S.G. Gorishny and S.A. Larin, Nucl. Phys. B **283** (1987) 452.
128. S.G. Gorishny, S.A. Larin and F.V. Tkachov, Phys. Lett. B **124** (1983) 217.
129. N. Gray, D.J. Broadhurst, W. Grafe and K. Schilcher, Z. Phys. C **48** (1990) 673.
130. H.W. Grießhammer, Phys. Rev. D **58** (1998) 094027; Nucl. Phys. B **579** (2000) 313.
131. A.G. Grozin, J. High Energy Phys. **03** (2000) 013.
132. A.G. Grozin, *Heavy Quark Effective Theory. An Introduction* (Springer, Berlin, Heidelberg, to be published in 2002).
133. R. Harlander, Acta Phys. Polon. B **30** (1999) 3443.
134. R. Harlander, T. Seidensticker and M. Steinhauser, Phys. Lett. B **426** (1998) 125.
135. G.H. Hardy, *Divergent Series* (Clarendon Press, Oxford, 1963).

136. W. Heisenberg and H. Euler, Z. Phys. **98** (1936) 714.
137. K. Hepp, Commun. Math. Phys. **2** (1966) 301.
138. A.H. Hoang, Phys. Rev. D **56** (1997) 7276.
139. A.H. Hoang, Phys. Rev. D **59** (1999) 014039.
140. A.H. Hoang, Phys. Rev. D **61** (2000) 034005.
141. A. Hoang et al., Eur. Phys. J. Direct C **3** (2000) 1.
142. A.H. Hoang and T. Teubner, Phys. Rev. D **58** (1998) 114023; Phys. Rev. D **60** (1999) 114027.
143. G. 't Hooft, Nucl. Phys. B **61** (1973) 455.
144. G. 't Hooft and M. Veltman, Nucl. Phys. B **44** (1972) 189.
145. G. 't Hooft and M. Veltman, Nucl. Phys. B **153** (1979) 365.
146. G. 't Hooft and M. Veltman, Nucl. Phys. B **160** (1979) 151.
147. K. Huang, *Quarks, Leptons and Gauge Fields* (World Scientific, Singapore, 1982).
148. F.J. Indurain, *Quantum Chromodynamics* (Springer, Berlin, Heidelberg, 1983).
149. K. Itzykson and J.B. Zuber, *Quantum Field Theory* (McGraw-Hill, New York, 1985).
150. R. Jackiw, Ann. Phys. **48** (1968) 292; Ann. Phys. **51** (1969) 575.
151. R. Karplus and A. Klein, Phys. Rev. **87** (1952) 848.
152. B.A. Kniehl and A.A. Penin, Nucl. Phys. B **563** (1999) 200; Nucl. Phys. B **577** (2000) 197; Phys. Rev. Lett. **85** (2000) 1210, (erratum) 3065; 5094.
153. K.S. Kölbig, J.A. Mignaco and E. Remiddi, BIT **10** (1970) 38; K.S. Kölbig, Math. Comp. **39** (1982) 647.
154. G. Korchemsky, Phys. Lett. B **217** (1989) 330.
155. G. Korchemsky, Phys. Lett. B **220** (1989) 629.
156. G. Kramer and B. Lampe, J. Math. Phys. **28** (1987) 945.
157. D. Kreimer, Phys. Lett. B **273** (1991) 277.
158. J.H. Kühn, S. Moch, A.A. Penin and V.A. Smirnov, hep-ph 0106298
159. J.H. Kühn, A.A. Penin and V.A. Smirnov, Eur. Phys. J. C **17** (2000) 97.
160. S. Laporta, Nuovo Cim. A **106** (1993) 675.
161. S.A. Larin, F.V. Tkachov and J.A.M. Vermaseren, Preprint NIKHEF-H/91-18 (1991).
162. S.A. Larin, T. van Ritbergen and J.A.M. Vermaseren, Nucl. Phys. B **438** (1995) 278.
163. G. Leibbrandt, Rev. Mod. Phys. **47** (1975) 849.
164. G.P. Lepage et al., Phys. Rev. D **46** (1992) 4052.
165. L. Lewin, *Polylogarithms and Associated Functions* (North-Holland, Amsterdam, 1981).
166. S. Libby and G. Sterman, Phys. Rev. D **18** (1978) 3252.
167. L.N. Lipatov, Sov. J. Nucl. Phys. **23** (1976) 642.
168. L.N. Lipatov, Phys. Rep. **286** (1997) 131.
169. J.H. Lowenstein and W. Zimmermann, Nucl. Phys. B **86** (1975) 77.
170. J.H. Lowenstein, Commun. Math. Phys. **47** (1976) 53.
171. M.E. Luke and A.V. Manohar, Phys. Lett. B **286** (1992) 348.
172. M.E. Luke, A.V. Manohar and I.Z. Rothstein, Phys. Rev. D **61** (2000) 074025.
173. L. Magnea and G. Sterman, Phys. Rev. D **42** (1990) 4222.
174. T. Mannel, Rep. Prog. Phys. **60** (1997) 1113.
175. A.V. Manohar and M.B. Wise, *Heavy Quark Physics* (Cambridge University Press, Cambridge, 2000).

176. P.T. Matthews and A. Salam, Rev. Mod. Phys. **23** (1951) 311.
177. K. Melnikov and A. Yelkhovsky, Nucl. Phys. B **528** (1998) 59.
178. K. Melnikov and A. Yelkhovsky, Phys. Rev. D **59** (1999) 114009.
179. A.H. Mueller, Phys. Rep. **73** (1981) 35.
180. T. Nagano, A. Ota and Y. Sumino, Phys. Rev. D **60** (1999) 114014.
181. N. Nakanishi, *Graph Theory and Feynman Integrals* (Gordon and Breach, New York, 1971).
182. W.L. van Neerven, Nucl. Phys. B **268** (1986) 453.
183. M. Neubert, Phys. Rep. **245** (1994) 259.
184. V.A. Novikov, M.A. Shifman, A.I. Vainstein and V.I. Zakharov, Fortschr. Phys. **32** (1984) 585.
185. F.W.J. Olver, *Asymptotics and Special Functions* (Academic Press, New York, 1974).
186. B.A. Ovrut and H.J. Schnitzer, Phys. Lett. B **100** (1981) 403.
187. W. Pauli and F. Villars, Rev. Mod. Phys. **21** (1949) 434.
188. A.A. Penin and A.A. Pivovarov, Phys. Lett. B **435** (1998) 413; Nucl. Phys. B **549** (1999) 217.
189. A.A. Penin and A.A. Pivovarov, Nucl. Phys. B **550** (1999) 375.
190. M.E. Peskin and D.V. Schroeder, *An Introduction to Quantum Field Theory* (Perseus, Reading, MA, 1995).
191. M.E. Peskin and M.J. Strassler, Phys. Rev. D **43** (1991) 1500.
192. M. Peter, Phys. Rev. Lett. **78** (1997) 602; Nucl. Phys. B **501** (1997) 471.
193. A. Pineda and J. Soto, Nucl. Phys. Proc. Suppl. **64** (1998) 428; Phys. Rev. D **59** (1999) 016005.
194. K. Pohlmeyer, J. Math. Phys. **23** (1982) 2511.
195. A.P. Prudnikov, Yu.A. Brychkov and O.I. Marichev, *Integrals and Series*, Vols. 1–3 (Gordon and Breach, New York, 1986–1990).
196. T. van Ritbergen and R.G. Stuart, Phys. Rev. Lett. **82** (1999) 488; Phys. Lett. B **437** (1998) 201.
197. T. van Ritbergen, J.A.M. Vermaseren and S.A. Larin, Phys. Lett. B **400** (1997) 379.
198. J.L. Rosner, Ann. Phys. **44** (1967) 11.
199. A.D. Sakharov, Zh. Eksp. Teor. Fiz. **18** (1948) 631.
200. R. Scharf and J.B. Tausk, Nucl. Phys. B **412** (1994) 523.
201. Y. Schröder, Phys. Lett. B **447** (1999) 321.
202. J. Schwinger, Phys. Rev. **82** (1951) 664.
203. A. Sen, Phys. Rev. D **24** (1981) 3281.
204. A. Sen, Phys. Rev. D **28** (1983) 860.
205. M.A. Shifman, A.I. Vainstein and V.I. Zakharov, Nucl. Phys. B **147** (1979) 385, 448, 519.
206. V. Smilga, Nucl. Phys. B **161** (1979) 449.
207. V.A. Smirnov, Commun. Math. Phys. **134** (1990) 109.
208. V.A. Smirnov, *Renormalization and Asymptotic Expansions* (Birkhäuser, Basel, 1991).
209. V.A. Smirnov, Mod. Phys. Lett. A **10** (1995) 1485; Talk given at 5th International Conference on Physics Beyond the Standard Model, Balholm, Norway, 29 Apr. – 4 May 1997 (hep-ph/9708423).
210. V.A. Smirnov, Phys. Lett. B **394** (1997) 205.
211. V.A. Smirnov, Phys. Lett. B **404** (1997) 101.

212. V.A. Smirnov, Int. J. Mod. Phys. A **12** (1997) 4241.
213. V.A. Smirnov, Phys. Lett. B **460** (1999) 397.
214. V.A. Smirnov, Phys. Lett. B **465** (1999) 226.
215. V.A. Smirnov, Phys. Lett. B **491** (2000) 130; Phys. Lett. B **500** (2001) 330.
216. V.A. Smirnov and E.R. Rakhmetov, Teor. Mat. Fiz. **120** (1999) 64.
217. V.A. Smirnov and O.L. Veretin, Nucl. Phys. B **566** (2000) 469.
218. A. Sommerfeld, *Atombau und Spektrallinien* (Vieweg, Braunschweig, 1960).
219. E.R. Speer, J. Math. Phys. **9** (1968) 1404.
220. E.R. Speer, Commun. Math. Phys. **23** (1971) 23; Commun. Math. Phys. **25** (1972) 336.
221. E.R. Speer, in *Renormalization Theory*, eds. G. Velo and A.S. Wightman (Reidel, Dodrecht, 1976) p. 25.
222. E.R. Speer, Ann. Inst. H. Poincaré **26** (1977) 87.
223. V.P. Spiridonov, preprint INR P-0378 (Moscow, 1984) (unpublished).
224. M. Steinhauser, Ph.D. thesis, Karlsruhe University (Shaker, Aachen, 1996).
225. G. Sterman, Phys. Rev. D **17** (1978) 2773.
226. V.V. Sudakov, Zh. Eksp. Teor. Fiz. **30** (1956) 87.
227. L.R. Surguladze and F.V. Tkachov, Nucl. Phys. B **331** (1990) 35.
228. O.V. Tarasov, Nucl. Phys. B **480** (1996) 397; Phys. Rev. D **54** (1996) 6479.
229. O.V. Tarasov, Nucl. Phys. (Proc. Suppl.) **89** (2000) 237.
230. J.B. Tausk, Phys. Lett. B **469** (1999) 225.
231. B.A. Thacker and G.P. Lepage, Phys. Rev. D **43** (1991) 196.
232. F.V. Tkachov, Phys. Lett. B **124** (1983) 212.
233. N.I. Ussyukina and A.I. Davydychev, Phys. Lett. B **298** (1993) 363; Phys. Lett. B **332** (1994) 159.
234. A.N. Vassiliev, Teor. Mat. Fiz. **81** (1989) 336.
235. A.N. Vassiliev, Yu.M. Pis'mak and Yu.R. Khonkonen, Teor. Mat. Fiz. **47** (1981) 291.
236. J.A.M. Vermaseren, S.A. Larin and T. van Ritbergen, Phys. Lett. B **405** (1997) 327.
237. S. Weinberg, Phys. Rev. **118** (1960) 838.
238. S. Weinberg, Phys. Lett. B **91** (1980) 51.
239. W. Wetzel, Nucl. Phys. B **196** (1982) 259.
240. A.S. Wightman, in *Renormalization Theory*, eds. G. Velo and A.S. Wightman (Reidel, Dodrecht, 1976) p. 1.
241. K.G. Wilson, Phys. Rev. **179** (1969) 1499.
242. K.G. Wilson, Phys. Rev. D **7** (1973) 2911.
243. S. Wolfram, *The Mathematica Book*, 4th edition (Wolfram Media and Cambridge University Press, Cambridge, 1999).
244. O. Yakovlev, Phys. Lett. B **457** (1999) 170.
245. O.I. Zavialov and B.M. Stepanov, Yad. Fiz. **1** (1965) 922.
246. O.I. Zavialov, *Renormalized Quantum Field Theory* (Kluwer Academic, Dodrecht, 1990).
247. W. Zimmermann, Commun. Math. Phys. **15** (1969) 208.
248. W. Zimmernann, Ann. Phys. **77** (1973) 536, 570.

List of Symbols

A – function in the alpha representation
A_μ – electromagnetic vector potential
A_μ^a – gluon field
$\bar{a}$ – number of oversubtractions
a_l – power of propagator
$\mathbf{B}$ – magnetic field
C_i – coefficient function
$C_A = N$ – the quadratic Casimir operator of the adjoint representation of the SU(N) group
$C_F = (N^2-1)/(2N)$ – the quadratic Casimir operator of the fundamental representation of the SU(N) group
c^a – ghost field
D – function in the alpha representation
$D_\mu = \partial_\mu - igA_\mu^a t^a$ – covariant derivative
$D(x)$ – propagator in coordinate space
$\tilde{D}(p)$ – propagator
$d = 4 - 2\varepsilon$ – parameter of dimensional regularization
E – energy
$\mathbf{E}$ – electric field
e – electron charge
F_Γ – Feynman integral
$F_{1,2}$ – form factors
${}_2F_1(a, b; c; z)$ – the Gauss hypergeometric function
$F_{\mu\nu}$ – electromagnetic field tensor
f – forest
$G(\lambda_1, \lambda_2)$ – function in one-loop massless integration formula
$G_C(\mathbf{r}, \mathbf{r}'; E)$ – Coulomb Green function
G_Γ – Feynman amplitude
$G_{\mu\nu}^a$ – gluon field strength tensor
g – coupling constant
$g_{\mu\nu}$ – metric tensor
h – number of loops
$J_i(x)$ – composite operator
k – loop momentum
L – number of lines
$\mathcal{L}$ – Lagrangian
$\mathrm{Li}_a(z)$ – polylogarithm
l – loop momentum
M – mass
$\mathcal{M}$, M – subtraction operator
m – mass
N – number of colours
n_f – number of flavours
$O_i(x)$ – composite operator
p – external or internal momentum
Q – external momentum
$Q^2 = -q^2$ – Euclidean external momentum squared
q – external momentum
$q(x)$ – quark field
R – R-operation
$\mathcal{R}$ – operation that determines a remainder in an asymptotic expansion
$R(s)$ – total cross section for hadron production in e^+e^- annihilation
S – S-matrix
$S_{a,b}(z)$ – generalized polylogarithm
s – Mandelstam variable
T – tree, 2-tree, pseudotree
$\mathcal{T}$ – operator of Taylor expansion
$T_F = 1/2$ – the index of the fundamental representation the SU(N) group
t – Mandelstam variable
t^a – SU(3) generator
t_l – sector variable
u – Mandelstam variable
u_l – auxiliary parameter

V – number of vertices
v – velocity
$\mathcal{W}$ – operator
x – coordinate
$y = m^2 - q^2/4$ – threshold variable
Z_i – counterterm
Z_{ik} – renormalization matrix of composite operators
Z_l – polynomial in propagator

$\alpha = e^2/(4\pi)$ – fine-structure constant
$\alpha_s = g^2/(4\pi)$ – strong coupling
α_l – alpha parameter
$\mathrm{B}(x)$ – beta function (second Euler integral)
$\beta_l = 1/\alpha_l$ – inverse alpha parameter
$\beta = \sqrt{1 - 4m^2/s}$ – velocity
$\beta(g)$ – beta function
Γ – graph
$\Gamma(x)$ – gamma function (first Euler integral)
γ – (sub)graph
$\gamma(\alpha)$ – function in evolution equation
$\gamma_{\mathrm{E}} = 0.577216\ldots$ – Euler constant
γ_μ – gamma matrices
$\Delta(\gamma)$ – counterterm operation
$\delta(x)$ – delta function
$\varepsilon = (4-d)/2$ – parameter of dimensional regularization
$\zeta(x)$ – Riemann zeta function
$\zeta(\alpha)$ – function in evolution equation
Λ – cut-off
λ_l – parameter of analytic regularization
μ – mass parameter within dimensional regularization
ξ – gauge parameter
ξ_i – Feynman parameter
$\xi(\alpha)$ – function in evolution equation
Π_i – projector
$\Pi(q^2)$ – polarization of vacuum
σ – mapping
σ_i – Pauli matrices
τ_l – sector variable
$\phi(x)$ – scalar field
$\chi(x)$ – quark field
$\psi(x)$ – quark field
ω – degree of divergence

Index

Springer Tracts in Modern Physics

160 **Physics with Tau Leptons**
By A. Stahl 2000. 236 figs. VIII, 315 pages

161 **Semiclassical Theory of Mesoscopic Quantum Systems**
By K. Richter 2000. 50 figs. IX, 221 pages

162 **Electroweak Precision Tests at LEP**
By W. Hollik and G. Duckeck 2000. 60 figs. VIII, 161 pages

163 **Symmetries in Intermediate and High Energy Physics**
Ed. by A. Faessler, T.S. Kosmas, and G.K. Leontaris 2000. 96 figs. XVI, 316 pages

164 **Pattern Formation in Granular Materials**
By G.H. Ristow 2000. 83 figs. XIII, 161 pages

165 **Path Integral Quantization and Stochastic Quantization**
By M. Masujima 2000. 0 figs. XII, 282 pages

166 **Probing the Quantum Vacuum**
Pertubative Effective Action Approach in Quantum Electrodynamics and its Application
By W. Dittrich and H. Gies 2000. 16 figs. XI, 241 pages

167 **Photoelectric Properties and Applications of Low-Mobility Semiconductors**
By R. Könenkamp 2000. 57 figs. VIII, 100 pages

168 **Deep Inelastic Positron-Proton Scattering in the High-Momentum-Transfer Regime of HERA**
By U.F. Katz 2000. 96 figs. VIII, 237 pages

169 **Semiconductor Cavity Quantum Electrodynamics**
By Y. Yamamoto, T. Tassone, H. Cao 2000. 67 figs. VIII, 154 pages

170 **d–d Excitations in Transition-Metal Oxides**
A Spin-Polarized Electron Energy-Loss Spectroscopy (SPEELS) Study
By B. Fromme 2001. 53 figs. XII, 143 pages

171 **High-T_c Superconductors for Magnet and Energy Technology**
By B. R. Lehndorff 2001. 139 figs. XII, 209 pages

172 **Dissipative Quantum Chaos and Decoherence**
By D. Braun 2001. 22 figs. XI, 132 pages

173 **Quantum Information**
An Introduction to Basic Theoretical Concepts and Experiments
By G. Alber, T. Beth, M. Horodecki, P. Horodecki, R. Horodecki, M. Rötteler, H. Weinfurter, R. Werner, and A. Zeilinger 2001. 60 figs. XI, 216 pages

174 **Superconductor/Semiconductor Junctions**
By Thomas Schäpers 2001. 91 figs. IX, 145 pages

175 **Ion-Induced Electron Emission from Crystalline Solids**
By Hiroshi Kudo 2002. 85 figs. IX, 161 pages

176 **Infrared Spectroscopy of Molecular Clusters**
An Introduction to Intermolecular Forces
By Martina Havenith 2002. 33 figs. VIII, 120 pages

177 **Applied Asymptotic Expansions in Momenta and Masses**
By Vladimir A. Smirnov 2002. 52 figs. IX, 263 pages

Printing: Mercedes-Druck, Berlin
Binding: Stürtz AG, Würzburg